AF876068

# Springer Protocols Handbooks

For further volumes:
http://www.springer.com/series/8623

Springer Protocols Handbooks collects a diverse range of step-by-step laboratory methods and protocols from across the life and biomedical sciences. Each protocol is provided in the Springer Protocol format: readily-reproducible in a step-by-step fashion. Each protocol opens with an introductory overview, a list of the materials and reagents needed to complete the experiment, and is followed by a detailed procedure supported by a helpful notes section offering tips and tricks of the trade as well as troubleshooting advice. With a focus on large comprehensive protocol collections and an international authorship, Springer Protocols Handbooks are a valuable addition to the laboratory.

# Accelerated Breeding of Cereal Crops

Edited by

**Andriy Bilichak**

*Agriculture and Agri-Food Canada, Morden, MB, Canada*

**John D. Laurie**

*Agriculture and Agri-Food Canada, Lethbridge, AB, Canada*

Humana Press

*Editors*
Andriy Bilichak
Agriculture and Agri-Food Canada
Morden, MB, Canada

John D. Laurie
Agriculture and Agri-Food Canada
Lethbridge, AB, Canada

ISSN 1949-2448 ISSN 1949-2456 (electronic)
Springer Protocols Handbooks
ISBN 978-1-0716-1525-6 ISBN 978-1-0716-1526-3 (eBook)
https://doi.org/10.1007/978-1-0716-1526-3

© Springer Science+Business Media, LLC, part of Springer Nature 2022
This work is subject to copyright. All rights are reserved by the Publisher, whether the whole or part of the material is concerned, specifically the rights of translation, reprinting, reuse of illustrations, recitation, broadcasting, reproduction on microfilms or in any other physical way, and transmission or information storage and retrieval, electronic adaptation, computer software, or by similar or dissimilar methodology now known or hereafter developed.
The use of general descriptive names, registered names, trademarks, service marks, etc. in this publication does not imply, even in the absence of a specific statement, that such names are exempt from the relevant protective laws and regulations and therefore free for general use.
The publisher, the authors, and the editors are safe to assume that the advice and information in this book are believed to be true and accurate at the date of publication. Neither the publisher nor the authors or the editors give a warranty, expressed or implied, with respect to the material contained herein or for any errors or omissions that may have been made. The publisher remains neutral with regard to jurisdictional claims in published maps and institutional affiliations.

This Humana imprint is published by the registered company Springer Science+Business Media, LLC, part of Springer Nature.
The registered company address is: 1 New York Plaza, New York, NY 10004, U.S.A.

# Preface

Growing world population and increased threats associated with a warming climate place high demands on plant breeders to rapidly develop resilient crops with stable yields across different environments. To meet global food demand, growers need to double crop production by 2050, a seemingly unrealistic goal with the current yield increase. Maize, rice, and wheat provide 60% of the energy intake for people around the globe. Although crops like rice and maize have seen significant development of genomic resources and understanding of genetic causes for agronomically important traits, hexaploid wheat lags in this regard. The large genome of wheat has hampered progress, but its polyploid nature offers great untapped potential for the creation of functional variation. With advances in omics technologies and the advent of gene editing, biotechnology in cereal crops is accelerating at an incredible pace.

Leading scientists with expertise ranging from cereal biotechnology and genomics to conventional breeding provide in this book either up-to-date methods used in respective programs or an overview of their research fields. In a stepwise manner, the reader is introduced to well-established and emerging techniques that can potentially streamline the discovery of genes responsible for agriculturally important traits in crops. We start with the recent progress in genomics of selected crops from the *Poaceae* family, highlighting approaches undertaken to create the first pangenome for bread wheat and what can be done to streamline mining for genetic variation. Methods that utilize this variation are increasingly crucial for breeders. The reader is introduced to techniques widely used in conventional breeding, such as marker-assisted selection and mapping of quantitative trait loci, as well as recent advancements of genome-wide association studies. Next, we show how artificial intelligence can help with breeding by using machine learning for genomic selection. Technologies that speed-up the breeding process, such as the production of doubled haploid plants through microspores, are of great interest to breeders. In addition to the variation that occurs naturally, we look at induced mutagenesis through the use of chemicals, radiation, and transposable elements. Rapid high-throughput detection of natural and induced variations can be accomplished using PCR and Targeting Induced Local Lesions in Genomes (TILLING). Transcription factors play an essential role in regulating gene expression; therefore, the precise mapping of their position using methods such as DNA affinity purification sequencing (DAP-seq) is vital for functional genomics studies. With rapid advancements in gene editing, this technology is now making it into the breeders' hands. Included methods describe approaches to construct CRISPR/Cas expression cassettes for monocots using modular assembly systems and genome editing in wheat. A manifestation of genomic sequence at the phenotypic level is performed through proteins and their posttranslational modifications. Therefore, quantitative proteomics in cereals is gaining momentum for a comprehensive understanding of crop genetics. Finally, phenotyping is a crucial component of plant breeding, integrating crop development, growth, and yield. High-throughput phenomics offers a powerful tool for breeders to select traits that are either challenging to see or not detectable by the eye.

Overall, this book provides a collection of methods that, we believe, will be of interest to plant breeders, researchers, postdoctoral fellows, and students working in functional genomics for the development of the next generation of crop plants. Finally, we would like to gratefully acknowledge the authors for their valuable contribution to this book, the associate editor for Springer protocols, Monica Suchy, and Springer Nature for the invitation to contribute.

*Morden, MB, Canada* *Andriy Bilichak*
*Lethbridge, AB, Canada* *John D. Laurie*

## Contents

## Contributors


AMANDA ASKINS • *Department of Biology and Geology, University of South Carolina Aiken, Aiken, SC, USA*

MARGARET BALCERZAK • *Ottawa Research and Development Centre, Ottawa, ON, Canada*

KYLE BODELL • *Lethbridge Research and Development Centre, Agriculture and Agri-Food Canada, Lethbridge, AB, Canada; The Department of Biological Sciences, University of Lethbridge, Lethbridge, AB, Canada*

LIANG CHEN • *State Key Laboratory of Crop Stress Biology for Arid Areas, College of Agronomy, Northwest A&F University, Yangling, Shaanxi, China*

RAMAN DHARIWAL • *Agriculture and Agri-Food Canada, Lethbridge Research and Development Centre, Lethbridge, AB, Canada*

NATHAN GOOD • *Agriculture Victoria, Grains Innovation Park, Horsham, VIC, Australia*

JOEL HAGUE • *Department of Cell and Molecular Biology, University of Rhode Island, West Kingston, RI, USA*

FLORIAN HAHN • *Plant Sciences Department, Rothamsted Research, Harpenden, UK*

C. NATHAN HANCOCK • *Department of Biology and Geology, University of South Carolina Aiken, Aiken, SC, USA*

YIN-GANG HU • *State Key Laboratory of Crop Stress Biology for Arid Areas, College of Agronomy, Northwest A&F University, Yangling, Shaanxi, China; Institute of Water Saving Agriculture in Arid Regions of China, Northwest A&F University, Yangling, Shaanxi, China*

SHAO-SHAN CAROL HUANG • *Center for Genomics and Systems Biology, Department of Biology, New York University, New York, NY, USA*

FENGYING JIANG • *Lethbridge Research and Development Centre, Agriculture and Agri-Food Canada, Lethbridge, AB, Canada*

SATEESH KAGALE • *National Research Council Canada, Saskatoon, SK, Canada*

SURYA KANT • *Agriculture Victoria, Grains Innovation Park, Horsham, VIC, Australia; Centre for Agricultural Innovation, School of Agriculture and Food, Faculty of Veterinary and Agricultural Sciences, The University of Melbourne, Melbourne, VIC, Australia*

ALBERT P. KAUSCH • *Department of Cell and Molecular Biology, University of Rhode Island, West Kingston, RI, USA*

IBRAHIM KHODABOCUS • *Department of Biological Sciences, University of Alberta, Edmonton, AB, Canada*

SANTOSH KUMAR • *Brandon Research and Development Centre, Agriculture and Agri-Food Canada, Brandon, MB, Canada*

ANDRÉ LAROCHE • *Agriculture and Agri-Food Canada, Lethbridge Research and Development Centre, Lethbridge, AB, Canada*

JOHN D. LAURIE • *Lethbridge Research and Development Centre, Agriculture and Agri-Food Canada, Lethbridge, AB, Canada*

MIAOMIAO LI • *Center for Genomics and Systems Biology, Department of Biology, New York University, New York, NY, USA*

QIAOMU LI • *Department of Biological Sciences, University of Alberta, Edmonton, AB, Canada*

DEVANG MEHTA • *Department of Biological Sciences, University of Alberta, Edmonton, AB, Canada*
SARAH M. MENDOZA • *Department of Biology and Geology, University of South Carolina Aiken, Aiken, SC, USA*
MATTHEW J. MILNER • *NIAB, Cambridge, UK*
JHADESWAR MURMU • *Ottawa Research and Development Centre, Ottawa, ON, Canada*
THEIA JENSEN • *Lethbridge Research and Development Centre, Agriculture and Agri-Food Canada, Lethbridge, AB, Canada; The Department of Biological Sciences, University of Lethbridge, Lethbridge, AB, Canada*
VLADIMIR NEKRASOV • *Plant Sciences Department, Rothamsted Research, Harpenden, UK*
KIMBERLY NELSON • *Department of Cell and Molecular Biology, University of Rhode Island, West Kingston, RI, USA*
KIRBY T. NILSEN • *Brandon Research and Development Centre, Agriculture and Agri-Food Canada, Brandon, MB, Canada*
SUDHAKAR PANDURANGAN • *Brandon Research and Development Centre, Agriculture and Agri-Food Canada, Brandon, MB, Canada*
ANGELA N. PLEMMONS • *Department of Biology and Geology, University of South Carolina Aiken, Aiken, SC, USA*
CURTIS J. POZNIAK • *Crop Development Centre and Department of Plant Sciences, College of Agriculture and Bioresources, University of Saskatchewan, Saskatoon, SK, Canada*
RAJA RAGUPATHY • *Agriculture and Agri-Food Canada, Lethbridge Research and Development Centre, Lethbridge, AB, Canada*
HARPINDER SINGH RANDHAWA • *Agriculture and Agri-Food Canada, Lethbridge Research and Development Centre, Lethbridge, AB, Canada*
MARCUS A. SAMUEL • *Department of Biological Sciences, University of Calgary, Calgary, AB, Canada*
LAURA SANJURJO LOURES • *Plant Sciences Department, Rothamsted Research, Harpenden, UK*
JOHANN SCHERNTHANER • *Ottawa Research and Development Centre, Ottawa, ON, Canada*
RAJU SOOLANAYAKANAHALLY • *Agriculture and Agri-Food Canada, Saskatoon, SK, Canada*
BRAULIO J. SOTO-CERDA • *Agriaquaculture Nutritional Genomic Center (CGNA), Temuco, Chile*
RAJAGOPAL SUBRAMANIAM • *Ottawa Research and Development Centre, Ottawa, ON, Canada*
EMILY THODAY-KENNEDY • *Agriculture Victoria, Grains Innovation Park, Horsham, VIC, Australia*
R. GLEN UHRIG • *Department of Biological Sciences, University of Alberta, Edmonton, AB, Canada*
NEHA VAID • *Department of Biological Sciences, University of Calgary, Calgary, AB, Canada*
AKSHAYA VASUDEVAN • *Agriculture and Agri-Food Canada, Ottawa Research and Development Centre, Ottawa, ON, Canada; Department of Biology, University of Ottawa, Ottawa, ON, Canada*
SEAN WALKOWIAK • *Grain Research Laboratory, Canadian Grain Commission, Winnipeg, MB, Canada*
EMMA J. WALLINGTON • *NIAB, Cambridge, UK*
XIANGFENG WANG • *National Maize Improvement Center, College of Agronomy and Biotechnology, China Agricultural University, Beijing, China*

CLARE WORKMAN • *Brandon Research and Development Centre, Agriculture and Agri-Food Canada, Brandon, MB, Canada*
YUETONG XU • *National Maize Improvement Center, College of Agronomy and Biotechnology, China Agricultural University, Beijing, China*
ALEX YONCHAK • *Department of Cell and Molecular Biology, University of Rhode Island, West Kingston, RI, USA*

# Chapter 1

# Recent Advances in Sequencing of Cereal Genomes

**Sean Walkowiak, Curtis J. Pozniak, and Kirby T. Nilsen**


## Abstract

Cereal crops are members of the Poaceae family that are grown primarily for their seed, many of which are staple foods around the globe. Together, maize, rice, wheat, barley, sorghum, millet, oat, rye, and triticale make up more than half of the agricultural production globally. Since diverging from a common ancestor 50–70 Mya, the genomes of cereal crops have been subject to evolutionary forces giving rise to changes in genome size, composition, and complexity. Over the past two decades, efforts to sequence the genomes of the main cereal crops have resulted in highly contiguous, chromosome-scale genome assemblies. Rice has the smallest genome size (420 megabases (Mb)), which led to it being the first assembled cereal genome. However, the immense genome size and complexity for some cereal crops, such as wheat (17 gigabases (Gb)) and oat (12 Gb), have hindered genome sequencing progress. Recently, the reduction in sequencing costs coupled with new technological advances including ultra-long-read sequencing and advanced genome assembly algorithms has now made it possible to achieve chromosome-scale assemblies in all cereals. As a result, we have now entered a new era of genomics for cereal crops. Researchers now face the daunting task of unravelling the function of all genes and non-genic regions of the genomes in cereals, particularly in crops like wheat and oat where little is known about the genetic basis for most economically important traits. Recent advances in high-throughput phenotyping, metabolomics, proteomics, and CRISPR/CAS9 technologies will facilitate this process and pave the way for exciting discoveries in functional genomics of cereals.


**Key words** Sequencing, Genomics, Cereals, Pan-genome, Functional genomics

## 1 Introduction

Cereals are monocotyledonous plants of the family Poaceae, many of which are staple foods grown across the world. Maize, rice, wheat, barley, sorghum, millet, oat, rye, and triticale are collectively grown across nearly a billion hectares globally and contribute significantly to human and livestock caloric and protein intake. Together, cereals account for more than half of global agricultural production (http://www.fao.org/faostat/). As the human population continues to rise, our production needs will continue to increase, despite challenges posed by climate change, increases in pest and disease pressure, and agro-ecosystem damage. To

Andriy Bilichak and John D. Laurie (eds.), *Accelerated Breeding of Cereal Crops*, Springer Protocols Handbooks,
https://doi.org/10.1007/978-1-0716-1526-3_1, © Springer Science+Business Media, LLC, part of Springer Nature 2022

overcome these challenges and feed our current and future populations, breeders must continue to develop improved crop varieties. By incorporating new genes and recombining alleles into new combinations, breeders are able to improve yields and enhance resistance to abiotic and biotic stress. In the past, breeding has relied on selection based on phenotypic expression of economically important traits. Now, breeding strategies aim to improve traits based on the incorporation of genetic and genomic information into selection pipelines, which have largely been facilitated by the advent of new sequencing technologies that have been increasing in throughput and lowering in cost [1]. We are currently entering a new age of genomics for cereal crops that will undoubtedly have a profound impact on shaping the future of agriculture.

The Poaceae family is comprised of ~10,000 individual species that have been evolving separately since diverging from a common ancestor approximately 50–70 million years ago [2]. Over this time period, substantial changes to genome size, composition, and complexity have occurred between species. The most striking difference is the variation in genome size that exists between the cultivated cereals ranging from 420 Mb in rice, to 16 Gb in wheat. Differences in genome composition include varying number of chromosomes, ploidy levels, and repeat content such as transposable elements (TEs), while complexity is shaped by genomic rearrangements, introgression, and translocation. Despite these differences, the relatively close evolutionary history within the cereals makes them an interesting model for comparative genomic analysis due to the high level of synteny between genomes [3, 4].

In this chapter, we discuss challenges associated with the sequencing of cereal genomes and recent methods that are being used to overcome those challenges. We present the evolution of sequencing technologies over the past two decades and discuss these improvements in the context of developing complete, whole-genome assemblies. We highlight some of the major advances that have been made in the principal cereal crops of global economic importance, and briefly summarize the genomic progress made in each crop, with emphasis on the completion of chromosome-scale reference assemblies. Finally, we touch on the applications of genomic resources towards solving complex challenges in cereal crop production.

## 2 Genome Sequencing in Cereal Grasses: Methods and Challenges

### 2.1 Sequencing

In general, genomics pipelines typically involve sequencing, assembly, annotation, and databasing [5]. The data is then applied to specific research questions or breeding needs. Since research and breeding are adaptive processes, they create new biological samples,

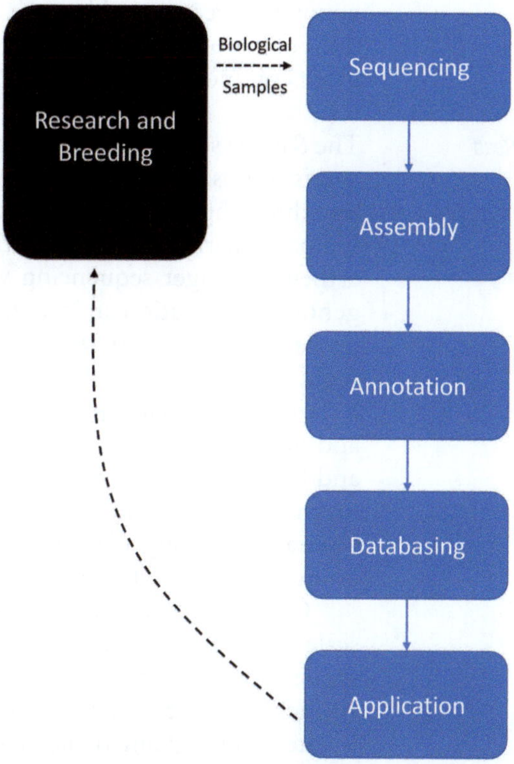


**Fig. 1** Generic genomics pipeline

data, and questions that can be further addressed with genomics. A general overview of the process is presented in Fig. 1.

Despite the high-throughput and low cost of modern sequencing technologies, the size and complexity of many cereal genomes create several technical challenges in genomics pipelines. As genome size increases, so too does the amount of sequencing data that is required, and increasing complexity of genomes necessitates sequencing data both that is accurate and that can span long distances. Processing and digital reconstruction of the ordered nucleotide sequence from this large volume of data requires significant computational resources and technical skills, and the tools available are limited since most of the software used in genomics are not designed for exceedingly large and complex genomes.

Given the costs and technical challenges, historically genomics efforts have focused around a single representative genotype for each species, or reference genome. Recently, efforts have focused on improving the quality of current reference genomes using modern technologies such as long reads and in vivo chromosome conformation technologies (i.e. Hi-C) to fill gaps and correct misassemblies [5]. As the sequencing and assembly technologies and tools improve, there is growing interest in generating multiple genome assemblies, or pan-genomes, which are able to capture the genomic diversity within each crop. This is widening the applications of genomics in cereals, making its use in research and breeding more common.

#### *2.1.1 Short-Read Sequencing*

The Sanger sequencing method was a 'gold standard' in the 1980s and is still used today for many targeted applications; however, the low throughput makes it difficult to apply to genomes of any size, but particularly for large repetitive genomes such as wheat [6]. Nevertheless, Sanger sequencing was instrumental in providing early genome information and was used for sequencing of amplicons and genes. In the mid-2000s, significant advances in high-throughput sequencing technologies were made, giving rise to a number of different sequencing platforms, with 454, SOLID, Ion Torrent, and Solexa platforms being the most used. The high throughput and low cost of these technologies changed the field of genomics, making genome sequencing more accessible than ever before. The Solexa sequencing platform was later acquired by Illumina, and eventually became the predominant sequencing platform.

Currently, there are several different benchtop and production-scale Illumina sequencers that generate millions to billions of sequencing reads and up to 6000 Gb of nucleotide data per run. In each case, the sequencing method is similar. The first step is to isolate healthy plant tissues (typically leaves from seedlings). The second step is to isolate DNA using a commercially available DNA isolation kit or common laboratory method (i.e. CTAB); note nuclei preparation [7] is recommended to avoid the sequencing of chloroplast and mitochondrial DNA. The next step is the library preparation. This step will vary depending on the study and the number of samples being sequenced in a single run (i.e. multiplexing). While library preparation is often performed by the sequencing provider as an added fee, they can be performed in a standard molecular laboratory using available library preparation kits; note that the DNA may need to be sheared prior to preparing the library, depending on the kit that is selected. Once the sequencing library is prepared, it can be sequenced using the sequencing flow cells designed for each of the different Illumina sequencing platforms. It is recommended that paired-end sequencing is performed, which results in two reads for each DNA fragment being sequenced, one from each strand. Creating libraries of different DNA fragment sizes is recommended, as the estimated distance between the paired reads can be used in data analysis pipelines. Typically, the read lengths vary from 100 to 300 bp for each paired-end read; therefore, a small fragment size of 450–500 bp would have the paired-end reads overlap, thereby creating a consolidated read that is the entire length of the DNA fragment. While short-read technologies have high throughput, low cost, and read accuracies >99.9%, their lengths can be problematic in downstream applications, particularly in repetitive genomes such as cereals.

#### 2.1.2 Special Applications of Short-Read Sequencing

Illumina short-read technologies have revolutionized genome sequencing; nevertheless, the high ploidy and repetitive nature of many of the cereal genomes can be difficult to parse using standard short-read methods. If desired, alternate library preparation methods may be applied to capture long-range information using short reads, for example, 10X Genomics Chromium and Hi-C (Dovetail and Chicago) methods; 10X Genomics and Hi-C are useful for resolving genome information across larger physical distances using short-read sequencing; however, long-range information can also be obtained by Oxford Nanopore Technologies (ONT) and Pacific Biosciences (PacBio) long-read sequencing or by Bio-Nano optical maps.

For some applications, high-quality whole-genome information is not required. For example, low-coverage genome sequencing and genotype-by-sequencing are able to capture genome-wide information at a reduced level, which can then be applied to genotyping or genome imputation using available reference genome sequences [1]. Pooling DNA from individuals that share common traits can also help identify common sequences or variants in populations using bulked segregant analysis, thereby reducing the amount of sequencing required. Likewise, the majority of the genomes for cereals are repetitive, such as transposable elements (TEs), and sequencing the gene space may reduce sequencing costs and would be better suited to identifying new genes, genetic variants, and alleles. This information can be generated by enriching the DNA for genic regions of the genome using exome capture techniques [8]. Recently, more targeted capture techniques have also been applied for gene families involved in disease resistance, AgRenSeq [9] and MutRenSeq [10], which have been combined with disease phenotyping data to rapidly clone disease resistance genes to major pathogens in wheat. A similar approach called GlutEnSeq [11] was developed to target enrich gluten genes. A disadvantage of gene capture methods is that the assays may not be able to capture all genes; also, for genes that are captured, the sequencing coverage may be limited and not span the entire gene. Another approach to target the gene space is to perform RNA sequencing, which can also facilitate gene annotation of the assembled genomes; however, a disadvantage of RNA sequencing is that it only captures the available mRNA from the tissues that were used. Therefore, for annotation purposes, sampling from multiple representative tissues at various time stages is required.

#### 2.1.3 Long-Read Sequencing and Optical Mapping

Long-read sequencing technologies, such as PacBio and ONT, are gaining momentum and are the latest approach to genome assembly. Long reads can span repetitive and difficult-to-assemble regions, such as telomeres and centromeres, which was not previously possible using short-read approaches. As a result, there have been recent efforts to improve and finish existing assemblies using long reads [12]. These methods require high-molecular weight DNA that is of a greater length than the fragmented or sheared DNA used in Illumina sequencing. While some commercial kits are available for isolation of high-molecular weight DNA, laboratory methods of DNA extraction are also suitable (i.e. CTAB), provided special care is taken to avoid shearing the DNA while handling the sample, using wide-bore pipette tips, for example. For both ONT and PacBio sequencing, sequencing libraries are made using commercially available kits, and the samples are sequenced on their respective sequencing platforms. The PacBio sequencing platforms (Sequel, Sequel II, and Sequel IIe) are all production scale, while ONT has both benchtop (MinIon and GridIon) and production-scale platforms (PromithIon). Both of these PacBio and ONT have the advantage of generating reads from several kb to over a Mb in the case of ONT, which can be highly advantageous in assembly contiguity and detection of structural variants. However, read accuracies are typically much lower than short-read technologies. The HiFi technology from PacBio can significantly improve read accuracy by first circularizing the DNA and then performing iterative sequencing of the circular molecule (circular consensus sequencing), which is reported to generate long-read accuracies >99% [1].

In some cases, long-read information may only be needed to generate high-quality genomic information for a particular region of interest. Recently, targeted sequencing of genomic regions has been achieved in wheat through CRISPR-assisted targeted enrichment sequencing methods, which take advantage of the specificity of the CRISPR guide system to digest the DNA and the boundary of regions of interest, thus making the boundary available for the initiation of long-read sequencing [13]. This approach can be advantageous in cloning genes within quantitative trait loci, as it reduces the sequencing cost by avoiding whole-genome sequencing, and it provides long reads for a region of interests that will allow for accurate de novo assembly of new genes. In the future, targeted sequencing may also be possible by selectively ejecting reads from ONT reactions that correspond to regions that are not of interest; however, to our knowledge, this method has yet to be applied to cereals [14].

Another method of gathering long-range information is BioNano optical mapping [5]. The BioNano system uses hybridization of enzymes to label high-molecular weight DNA. Since the sequence at the hybridization site is known, the distance between hybridization events can be used to guide ordering and to precisely orient sequences as well as detect structural variants. This system requires specialized instruments and sample label kits, though service providers also exist.

### 2.2 Analysis of Genome Sequencing Data

Once sequencing is complete, the next stage is data processing using various bioinformatics approaches, which includes read processing, genome assembly and comparison, annotation, and databasing (Fig. 1). Most of the tools required for these steps are designed to be performed on high-performance computer servers running a Linux operating system, in a command-line environment. While this is the preferred approach to analysis of DNA sequencing data, there are several tools available to those with beginner or intermediate level bioinformatics skills that have graphic user interfaces that are designed to work in Windows operating systems, including CLC Genomics Workbench, Geneious, and Galaxy. However, these software can be limiting and are not amenable to diverse applications that may be needed for customization of data analysis for breeding and research purposes.

#### 2.2.1 Read Processing

Once the sequencing data is generated, it needs to be quality checked and processed to improve read quality. FastQC is a very easy-to-use software and provides a wealth of information on your sequencing output for short reads, including read quality and contaminants. Due to their length, other software are available that provide similar information for long reads, such as NanoStat for ONT reads. This information can be used to help identify and trim low-quality data or contaminants from reads, while still retaining the high-quality information from the reads, or to completely remove low-quality reads all together. There are a variety of tools that are available for read process, including Trimmomatic and PoreChop. If desired, the accuracy of the long reads can also be improved prior to assembly using short-read data; alternatively, read accuracy of long reads can be improved at the assembly stage.

#### 2.2.2 Genome Assembly and Comparison

There are two main approaches to genome assembly, de novo and reference-guided assembly. The de novo approach generates a new genome assembly using only the sequencing information, while the reference-guided approach compares and orders the new sequencing information to an available reference genome.

There are several genome assembly tools available; however, the size, ploidy, and repetitive nature of cereal genomes have been problematic for most traditional genome assemblers. Assemblers that can utilize the information generated from different library preparation methods from short reads (i.e. different insertion sizes, 10X Genomics) and/or long-read information have recently made assembly of highly contiguous cereal grass genomes possible. The DeNovoMagic assembly pipeline, a proprietary assembler from NRGene, has produced some of the best cereal genomes to date [15–18] and provides an option to those who would prefer to have the genome assembled by a service provider. Other open-source assemblers, including W2RAP [19] and TRITEX [20, 21], have also been able to produce highly contiguous genome assemblies of cereal species using short reads. 10X Genomics also has its own de novo assembler, Supernova, which has been used in targeted analysis of small regions. Long-read-only assemblies can also be generated using tools such as HGAP, Flye, Miniasm, MECAT, and Canu. It is also possible to combine both long and short reads using Canu. Similarly, MaSuRCA is a whole-genome assembly software that combines graph-based and overlap-consensus approaches, allowing it to use a mixture of short- and long-read data and has been applied to wheat and its wild progenitors [22]. Once a draft genome is assembled, the contigs can be ordered and oriented into chromosome pseudomolecules and gaps between contigs estimated or filled using genetic maps [23], Hi-C data [6, 15, 16], long reads, optical maps [24], and/or an available reference genome(s).

While de novo assembly is able to provide the most complete information about a genome, including novel sequences that may be genome specific, it requires a significant amount of data and compute resources to achieve. Depending on the application, it may be much more advantageous to perform reference-based genome assembly, which requires less data and is more straightforward computationally; however, a reference genome is required. There are several tools available for aligning sequencing reads to a reference genome, including Novoalign (Novocraft) and BWA for short reads, longranger for 10X Genomics data, MiniMap2 [25] for long reads, and HISAT [26], HISAT2 [27], and STAR [28] for RNA sequencing, just to name a few. One advantage of reference-based genome analyses is that many genomes can be compared to a common sequence, which can facilitate downstream comparative genomic analyses, including genome imputation, variant calling, and the identification of regions or groups of genes and variants that are typically inherited together (i.e. haplotype blocks).

#### *2.2.3 Genome Annotation*

For newly de novo assembled genomes, an important step is to annotate the genome. Genes are often the functional unit underlying trait variation in cereals; as such, gene annotation is critical for identifying new genes and alleles that may be underlying cause of traits. Gene annotation can be performed by projecting gene models from related accessions using tools such as Exonerate and GMAP, but the limitation of this approach is genes that are not already annotated are missed. If genes from related accessions are unavailable, they can be predicted using trained datasets from other plant species, using tools such as FGENESH and Augustus. RNA sequencing from multiple tissues can also be used for annotating genes. Early efforts to sequence genes involved the construction of expressed sequence tag (EST) libraries, which were cDNA libraries that were often cloned into plasmids and sequenced using the

Sanger method [29]. EST databases were instrumental in identifying genes and gene variants that can be used for marker development, which led to the early genetic maps and DNA SNP array technologies. With the advent of high-throughput sequencing, it is now possible to perform sequencing of cDNA libraries or mRNA directly, at a much greater depth and reduced cost. The RNA sequencing reads are mapped onto the genome and are predicted using tools such as StringTie [30] and TransDecoder. If required, additional information on alternative gene structure and splicing variants can be predicted, which can be facilitated by long-read sequencing of single mRNA transcripts through isoform sequencing (IsoSeq). Once genes are annotated, domains and gene functions can be predicted using tools such as InterPro scan, and mapping predicted InterPro domains to gene ontology terms that indicate gene function. In addition to predicting gene sequences, other features such as repeats and mobile elements such as transposons are also important, as they contribute to genome plasticity and evolution. These features can be predicted using tools such as TandemRepeatFinder and LTRharvest.

#### *2.2.4 Genome Databasing*

Once genomes are assembled and annotated, it is important that the data be made available and accessible, so that they can be applied in research and breeding programs. Typically, the raw sequencing read data is deposited in international databases, such as the Sequence Read Archive (SRA) at the National Center for Biotechnology Information (NCBI) or at the European Nucleotide Archive (ENA) at the European Bioinformatics Institute (EBI). In addition to the raw read data, genome assemblies and their annotations can also be deposited in databases at NCBI (GenBank) and EBI (Ensembl). Currently, Ensembl plants (https://plants.ensembl.org/) contain the most up-to-date databases for many of the cereal genomes, including rice, wheat, barley, maize, and sorghum. The database also contains an interactive viewer that allows you to explore the genomes, options to retrieve gene sequences by accession, search for sequences of interest using alignment tools (i.e. BLAST), locations of variants from SNP arrays, variants from sequenced mutagenesis populations, whole-genome alignments between cereal crop genomes, and many other tools that may be useful to breeders and researchers. Additional tools are also available in other cereal specific databases, such as GrainGenes (https://wheat.pw.usda.gov/GG3/), which contains the latest oat assembly; the Triticeae Toolbox (https://triticeaetoolbox.org/), which contains data and marker information for wheat, barley, and oat; and Maize GDB (https://www.maizegdb.org/) (Table 1). Once the genomes and associated analysis files are made available, they can be used for applied research and breeding.

**Table 1**
**Example databases and tools for cereal genomics**

| Database | Cereal crop | Description | URL |
|---|---|---|---|
| Ensembl Plants | Rice, maize, wheat, barley, sorghum | Multi-cereal genome viewer and database | https://plants.ensembl.org/ |
| GrainGenes | Wheat, barley, rye, oat | Marker information for many crops, oat genome viewer | https://wheat.pw.usda.gov/GG3 |
| Triticeae Toolbox | Wheat, barley, oat | Marker information | https://triticeaetoolbox.org/ |
| MaizeGDB | Maize | A diverse database for maize genome visualization, population mining | https://www.maizegdb.org/ |

## 3 Status Update on Genome Sequencing and Assembly for Cereals

The 'big three' cereal crops (maize, rice, and wheat) each have more production area than all other cereal grasses combined. As a result, they have been the focus of extensive genomics research. Given their international importance to food security and agricultural production, we provide a brief history of genomics in each of these crops in addition to other cereal crops such as barley, oat, sorghum, millet, rye, and triticale.

### 3.1 Rice

Rice (*Oryza sativa* L.) production has been increasing steadily over the past two decades, reaching approximately 760 million tonnes from a total harvested area of 165 million ha in 2018 (http://www.fao.org/faostat/). In contrast to wheat, rice has a 35-fold smaller genome size [3]. Rice is composed of two morphologically distinct subspecies (ssp. *indica* and *japonica*) that diverged from a common ancestor around ~0.4 Mya [31]. Rice is a diploid crop ($2x = 2x = 24$) with the smallest genome size of the cultivated cereals (420 Mb). Rice has >100,000 diverse germplasm accessions available (http://www.IRRI.org), and has a high transformation efficiency. Together, these factors made rice a model crop for the cereals and an attractive target for early sequencing efforts. Early draft genome assemblies of rice were initiated by a number of independent research groups, which led to the release of whole-genome shotgun assemblies, in addition to chromosome-scale assemblies for chromosomes 1, 4, and 10 [2, 32–35]. A coordinated global effort led to the establishment of the International Rice Genome Sequencing Consortium (IRGSC) in 1997), which resulted in rice being the first of the cereal crops to be sequenced and assembled to the chromosome scale (Table 2). The approach to generate the original assembly for *Oryza sativa* L. ssp. *japonica* v. Nipponbare was

**Table 2**
**Examples of available chromosome-scale assemblies for cereal species**

| Scientific name | Crop | Ploidy | Assembly size (Mb) | Reference accession | Number of genes | Reference |
|---|---|---|---|---|---|---|
| *Avena atlantica* | na | Diploid $2n = 2x = 14$ | 3690 | CC 7277 | 51,100 | [67] |
| *Avena eriantha* | na | Diploid $2n = 2x = 14$ | 3780 | CN 19328 | 49,105 | [67] |
| *Avena sativa* | Common oat | Hexaploid $2n = 6x = 42$ | 10,982 | OT3098_v1 | | https://wheat.pw.usda.gov/GG3/node/922 |
| *Cenchrus americanus* | Pearl millet | Diploid $2n = 2x = 14$ | 1760 | Tift23D$_2$B$_1$-P1-P5 | 38,579 | [70] |
| *Digitaria exilis* | Fonio millet | Tetraploid $2n = 4x = 36$ | 716 | CM05836 | 59,844 | [69] |
| *Hordeum vulgare* | Barley | Diploid $2n = 2x = 7$ | 4790 | Morex | 39,734 | [61] |
| *Oryza sativa* L. ssp. *indica* | Rice | Diploid $2n = 2x = 24$ | 390 | Shuhui498 | 38,714 | [39] |
| *Oryza sativa* L. ssp. *japonica* | Rice | Diploid $2n = 2x = 24$ | 370 | Nipponbare | 37,544 | [36] |
| *Oryza sativa* L. ssp. *japonica* | Rice | Diploid $2n = 2x = 24$ | 373 | Nipponbare | 37,960 | [37] |
| *Panicum miliaceum* L | Broomcorn millet | Tetraploid $2n = 4x = 36$ | 855 | 00000390 | 55,930 | [102] |
| *Secale cereale* | Rye | Diploid $2n = 2x = 14$ | 6740 | Lo7 | 34,441 | [62] |
| *Sorghum bicolor* | Sorghum | Diploid $2x = 2x = 20$ | 626 | BTx623 | 27,640 | [68] |
| *Triticum aestivum* | Wheat | Hexaploid $2n = 6x = 42$ | 13,800 | Chinese spring | 107,891 | [17] |
| *Triticum turgidum* L. var. *durum* | Durum wheat | Tetraploid $2n = 4x = 28$ | 10,450 | Svevo | 66,559 | [15] |
| *Triticum turgidum* ssp. *dicoccoides* | Emmer wheat | Tetraploid $2n = 4x = 28$ | 10,500 | Zavitan | 65,012 | [16] |

(continued)

**Table 2 (continued)**

| Scientific name | Crop | Ploidy | Assembly size (Mb) | Reference accession | Number of genes | Reference |
|---|---|---|---|---|---|---|
| *Zea maize* | Maize | Diploid $2n = 2x = 20$ | 2104 | B73_RefGen_v4 | 39,324 | [52] |
| *Zea maize* | Maize | Diploid $2n = 2x = 20$ | 2048 | B73_RefGen_v1 | 32,000 | [45] |

constructed by first cloning genome fragments into bacterial artificial chromosomes (BACs), BAC fingerprinting, then sequencing, and assembling each BAC fragment separately; this resulted in an assembly of 370 Mb that contained 37,544 genes [36]. Since that time, work continued to improve the rice genome assembly, which led to the release of an updated and revised and corrected assembly, Os-Nipponbare-Reference-IRGSP-1.0 [37]. With the advent of long-read sequencing technologies, new genomes for other rice accessions continue to be released at an unprecedented rate [38–41]. The rice genome sequences have been instrumental in advancing the understanding of the basis of heterosis, the processes of domestication, the dissection of complex agronomic traits, gene duplication, and genetic diversity [42, 43]. In addition, the genome and gene sequences of rice have been useful to aid genome assembly and gene annotation of other cereals due to the shared gene orthology and genome synteny across members of related grasses [3].

### 3.2 Maize

Maize (*Zea mays*) is a diploid crop ($2n = 2x = 20$) and is currently the world's most important cereal crop both in terms of production and harvested area. Currently, maize is grown on over 100 million ha with a total production of approximately 1 billion tonnes (http://www.fao.org/faostat/). Maize is a staple food, but is also used as animal feed and for industrial purposes, including the production of biofuel. Efforts to sequence the maize genome were initiated in 2002 with the launch of the Maize Genome Sequencing Project [44]. The maize genome is more than five times greater than rice, is ~2.5 Gb in size, and contains between 30,000 and 40,000 genes [45]. Similar to rice, the first maize reference genome (Table 2) was constructed by sequencing a minimum tiling path of overlapping BACs using a Sanger sequencing strategy of the reference line B73 [45]. Since then, numerous other maize reference genomes have now been released for lines such as Mo17 [46, 47], W22 [48], SK [49], PH207 [50], *mexicana* [47], and HZS [51]. Recently, the B73 genome was re-assembled using modern approaches centred around PacBio long reads coupled with BioNano optical maps, resulting in an updated genome with 39,000 genes [52]. This new RefGen v4 assembly boasts a 52-fold increase in contig length over the original assembly, highlighting the benefits of incorporating long reads into the assembly pipeline (Table 2).

### 3.3 Wheat

Sequencing and assembly of the wheat genome was a daunting task due to its immense genome size (~16 Gb), complex repetitive structure >80%, and high ploidy [6]. Hexaploid bread wheat (*Triticum aestivum* L., genome AABBDD, $2n = 6x = 42$) and tetraploid durum wheat (*Triticum turgidum* L. ssp. *durum* (Desf.) Husn., genome AABB, $2n = 4x = 28$) are the most widely grown wheat species. In 2014, the first attempt to assemble the entire wheat genome using short reads resulted in a highly fragmented survey sequence [6]. In 2017, the DeNovoMagic pipeline by NRGene was used to construct the first highly contiguous wheat assembly, which was generated for wild emmer wheat *Triticum turgidum* ssp. *dicoccoides* (genome AABB), an ancestor durum wheat [16]. This assembly approach was later applied to both bread [17] and durum wheat [15] (Table 2). The reference-quality genome assemblies of wild emmer wheat, durum wheat, and bread wheat have been invaluable to the research and breeding communities and have facilitated the characterization of traits from domestication, such as free-threshing grain [16], and in-demand traits used in modern breeding, such as insect resistance [17, 18] and heavy metal contamination in grain [15]. Recently, the 10+ Wheat Genomes project assembled ten reference-quality assemblies and five scaffolded assemblies for bread wheat, which represent diversity from global breeding programs, which were used to assess genome variation used in modern breeding [18].

In addition to the main cultivated wheats, wheat's wild ancestors and other wild grasses are often used to introduce new genes into wheat. The 'A' genome of wheat is derived from a wild grass species *T. urartu*, which was sequenced and assembled by Illumina and PacBio technologies, and supplemented with 10X Genomics synthetic long reads and BioNano optical maps [53]. The 'A' and 'B' genomes were the first two of the wheat genomes to hybridize, resulting in wild emmer wheat, which eventually was domesticated and led to modern tetraploid wheats, including durum wheat. While the 'A' genome is known, the 'B' genome donor has been lost in history, but is likely related to *T. speltoides*. The 'D' genome was then introduced after an additional hybridization even between tetraploid wheat and *Ae. tauschii*, which was sequenced and assembled using Illumina, PacBio, and BioNano technologies [54].

Other undomesticated but related grass species have also been used to introduce new wheat genes into elite wheat cultivars. The 10+ Wheat Genomes project was able to use comparative genomics to identify several regions within modern wheat lines that are likely derived from other grass species, including *Aegilops ventricosa*, *Triticum timopheevii*, and *Thinopyrum ponticum*, which were associated with resistance to a wide array of diseases [18]. Recently the genome of *Thinopyrum elongatum* was sequenced, which facilitated the characterization of a gene involved in *Fusarium* head blight resistance that has been introgressed into modern bread wheat; interestingly, this gene likely originated from a fungus and was horizontally transferred into *Th. elongatum*, prior to introgression into bread wheat [55].

### 3.4 Rye, Barley, Oat, Sorghum, Millet, Triticale, and Other Poaceae

#### 3.4.1 Barley

Barley (*Hordeum vulgare* ssp. *vulgare*) was amongst the earliest crops to be domesticated [56] and is the only cultivated species out of 35 members within the genus *Hordeum* [57]. Barley is an agriculturally and industrially important cereal crop that is primarily used for animal feed, while approximately 20% is used in the malting industry towards the production of alcoholic and non-alcoholic beverages [58]. Barley is an inbreeding diploid species ($2n = 2x = 7$) with a large genome size ~5 Gb and high repetitive sequence content >80% [59]. Because it is diploid and has a close evolutionary relationship to wheat, barley has been considered a model species for hexaploid wheat. Early efforts to assemble the barley genome using map-based ordering was hindered by difficulties in generating high-resolution genetic maps [59]. The International Barley Genome Sequencing Consortium was established in 2006 with the aim to develop the first high-quality reference genome. A physical map was constructed consisting of 4.98 Gb of sequence, composed of 67,000 BAC clones anchored by a high-resolution genetic map [60]. The first chromosome-scale assembly for barley (Table 2) was completed in 2017 for the cultivar Morex [61]. Recently, Morex was reassembled using PacBio long reads; at the same time, the TRITEX pipeline was used to generate chromosome-scale sequence assemblies for 20 varieties [21].

#### 3.4.2 Rye and Triticale

Rye (*Secale cereale* L.) is grown in northern latitudes because of its superior resilience against abiotic and biotic stress, particularly winter hardiness. While rye production is less than rice, maize, and wheat, rye's hardiness makes it a desired crop in many growing regions. Recently, the ~7–8 Gb diploid rye genome (Table 2) was completed using the NRGene DeNovoMagic pipeline [62]. This genome sequence will be an invaluable resource for the improvement of both rye and wheat. Rye is used to introgress genes into wheat to improve resistance to abiotic and biotic stresses; in addition, wheat and rye genomes can also hybridize, resulting in a separate species, *Triticale*.

*Triticale*, which is a hybrid between rye and wheat, is desired for its disease resistance and ability to withstand harsh environmental conditions, which it inherited from rye, while at the same time having improved grain quality, which is inherited from wheat.

While the genomes of rye and wheat have been sequenced, to our knowledge, the genome of triticale is yet to be sequenced and assembled. *Triticale* genomics would be an interesting area of study, as there would be introgressions from both parental species used to construct the hybrid [63]. In addition, *Triticale* can be either tetraploid, hexaploid, or octaploid depending on the ploidy of the wheat species used in the initial cross [64], making it an ideal study subject for the complex behaviours of the genome after hybridization.

#### 3.4.3 Oat

Cultivated oat (*Avena sativa* L. and *A. byzantina* L.) is a globally important crop, sown to approximately ten million ha with a world production of approximately 23 million tonnes (http://www.fao.org/faostat/). In terms of production, oat has the seventh greatest production amongst cereals globally. Cultivated oat is an allohexaploid with three subgenomes (genome AACCDD, $2n = 6x = 42$) and has a genome size of approximately ~12 Gb [65]. The oat genome is exceedingly complex due to high repetitive sequence content, genome rearrangements, and translocations that each pose significant challenges for genome assembly [66]. As a result of these challenges, progress in sequencing the oat genome has lagged behind some of the other cereal crops. In recent years, however, major sequencing efforts have shifted back into oat, as advances in sequencing technology and assembly algorithms have made the assembly of large and complex cereal genomes a reality. In 2019, the first chromosome-scale assemblies for two diploid oat species (*Avena atlantica* and *Avena eriantha*) were released [67]. Both were sequenced using PacBio long reads coupled with scaffolding performed using Hi-C. Each assembly was annotated using transcriptome read mapping and contains approximately 50,000 genes. In 2018, it was reported that the genome of oat cultivar Belinda was sequenced and assembled using the DeNovoMagic pipeline by NRGene. In 2020, the fully assembled reference sequence (Table 2) for oat line OT3098 was made publically available (https://wheat.pw.usda.gov/GG3/node/922). Not only is this a significant achievement for the global oat community, this resource will be of particular relevance to oat breeding and research efforts in Canada where OT3098 originated. Because of the diversity in genomic composition within cultivated oat, oat is an attractive model for future pan-genome analysis, and efforts to sequence and assemble 20 diverse oat accessions are currently underway.

#### 3.4.4 Sorghum, Millet, and Other Members of the Poaceae

While production of millet is less than many of the other cereals, its production is important in developing countries, where it is desired for its short growing season and ability to grow in dry regions with high temperatures. There are several different genus and species of grains that are collectively called millet. *Sorghum bicolor*, or great millet, is often considered a separate cereal. The smaller size of the millet genomes facilitated sequencing efforts of different millet species, including the ~730 Mb sorghum genome (Table 2) that was sequenced by the Sanger method in 2009 [68]. Sequencing of additional millets, including pearl, finger, broomcorn, and fonio millets, was also later sequenced by Illumina [69], and PacBio approaches [70].

Recent genomic progress has also been made for minor cereal crops and other related grasses from Poaceae. While these species do not contribute significantly to global food production, they can be important sources of new genes for crops with higher production levels, or they can occupy niches that other cereal crops do not fill. For example, the genome of *Coix lacryma-jobi*, or Job's tears, was recently sequenced using PacBio and Illumina technologies [71]. This herbal crop species is cultivated in Asia for medicinal purposes.

## 4 Application of Reference Genomes

### *4.1 Transitioning Historic Genomics Research to Available Genome Assemblies*

One of the most immediate applications of a chromosome-scale reference sequence is to 'anchor' the vast amount of historical genetic information that has been generated for a species of interest and unify it on a common physical genetic map. One prominent example of this is the genetic mapping work that was performed prior to the availability of the first genome sequences of wheat. Genetic maps are constructed using populations of lines derived from a cross between two parental lines. These maps are used to determine both the order and genetic distance (*cM*) between markers. However, because genetic distance is a measure of recombination and not an absolute value, its relationship with physical distance (measured in base pairs) is variable. By comparing genetic to physical distance, the recombination landscape that exists along all chromosomes can be evaluated, which in general shows greater amounts of recombination towards the telomeres, and reduced recombination around the centromeric regions [72]. As a result, the genetic distance between markers can vary depending on the recombination rate in the population being studied. By aligning marker sequences from existing genetic maps to the available reference genome, a precise measure of physical distance between markers can be determined, and a standard set of positional coordinates can be established, facilitating the comparison of genomic regions from different studies.

The first genetic maps in wheat were constructed using restriction fragment length polymorphisms (RFLPs); however, this approach suffered from having a limited number of polymorphisms [73]. RFLP mapping is also a laborious process that involves the digestion of genomic DNA with restriction enzymes followed by gel separation using electrophoresis. Eventually, the focus shifted away from RFLP analysis and towards microsatellites, or simple sequence repeats (SSRs), which are tandem repeats that exist throughout the genome. SSRs are useful because they are multi-allelic and have much higher level of polymorphism than RFLPs. SSR screening uses PCR amplification to detect length polymorphism and is therefore much more amenable to throughput and automation. The first SSR linkage map for wheat was constructed by Roder et al. [73] and consisted of 279 loci. By integrating 4 individual genetic maps in wheat, Somers et al. [74] constructed a single genetic consensus map of 1234 microsatellite markers containing the consensus order and recombination distance. SSR technology has now largely been replaced by single nucleotide polymorphism (SNP) assays, which are a much more abundant source of genetic markers in wheat, and are amenable to high-throughput screening platforms such as Illumina iSelect and Affymetrix Axiom SNP array technology and KASP platforms. Arrays are currently available for wheat in a variety of formats that include the 9 K [75], 35 K [76], 90 K [77], 660 K [78], and 820 K [79]. SNP consensus maps are now available for durum [80] and hexaploid wheat [77]. The release of the first chromosome-scale assembly for wheat was used to anchor the physical location of DNA markers to a standardized reference [17]. For the first time, researchers were able to identify physical locations for genetic regions along chromosomes, and the genes contained within. This resource has greatly improved the speed of gene discovery and opens a whole new era for wheat functional genomics research and breeding.

### 4.2 Facilitating Breeding through Marker-Assisted Selection, Genome-Wide Association Studies, and Genomic-Assisted Selection

Molecular markers, such as RFLPs, SSRs, and SNPs, are often associated with traits of interest. In quantitative genetics, phenotypes for in-demand traits, such as yield and disease resistance, can be statistically associated with these molecular markers. In many cases, the cost and throughput of using the molecular marker make it more efficient than phenotyping for the trait itself, allowing breeders to select lines solely on the basis of molecular marker information. The availability of large marker arrays with polymorphisms distributed across all chromosomes has allowed for genome-wide association analysis and the identification of quantitative trait loci or QTL, which are a series of markers that are associated with a trait, with the most associated markers likely being in close proximity to the causal factor. Comparison of QTL from different studies may also help refine genetic intervals if the causal gene is the same, a task which is difficult to do with genetic maps alone since markers are often not shared across genetic maps. In the case of traits that are controlled by many genes that are distributed broadly across the genome, the genome-wide information can also be used to predict trait performances, providing a genomic-estimated breeding value—a concept referred to as genomic selection [81].

One of the challenges of using molecular markers for association analyses is that the associations can be inconsistent in different genetic backgrounds. Two common reasons why markers may be ineffective is that linkage between the trait and the marker has been lost in some individuals, or the marker could be in non-specific region of the genome. Using reference-guided analysis of sequencing data, variants can be identified that are in unique regions that will not be impacted by similar sequences in repetitive or homeologous regions of the genome, which has been a challenge in polyploid species in the past. In addition, genomes allow for new markers to be identified, which can have improved associations with the trait due to increased proximity to the causal factor, or ideally giving rise to variation that is causing the trait. The advent of low-cost sequencing may also allow for genomes to be imputed using low-coverage information, which, when combined with the pedigree information and genomic information on the parents, will predict all genes in the individual and replace the practice of running individual markers.

### 4.3 Gene Cloning and Mutagenesis

Over the past two decades, much of the emphasis has been placed on generating reference sequences for each of the cereal crops. Now that researchers have access to high-quality genomic resources for the major cultivated cereals, the next major challenge becomes how to assign functions to each gene in the genome. Modern genomics is rapidly changing the speed at which important genes can be cloned. In crops such as wheat, the classical approach to clone a gene was an in-depth process that in some cases took a decade or longer [82]. Although the release of wheat genome sequences has facilitated functional gene characterization, the majority of the 100,000 genes are yet to be functionally validated. The availability of a reference genome assembly allows for physical interval of QTL regions to be determined and for gene candidates within those regions to be identified. These genes can then be investigated in greater detail through functional genetic approaches. For example, the availability of reference sequences has resulted in cloning of important genes such as *Yr15* [83], *Sr60* [84], *Sm-1* [18], and *SuSr-D1* [85] in bread wheat and *Cdu-B1* and *SSt1* in durum wheat [15, 13].

Reverse genetics using genome editing technologies, such as CRISPR/Cas9, have the potential to rapidly accelerate genetic research in crop species. This technology is based on two key components, CRISPR (Clustered Regularly Interspersed Palindromic Repeats) and Cas9 (CRISPR-associated endonuclease 9) [86]. Cas9 is an endonuclease capable of creating double-stranded breaks in genomic DNA. When Cas9 is complexed with a single-guide RNA (sgRNA) that contains a nucleotide sequence specific to a target locus in a region of interest in the genome, the result is a highly efficient genome editing with minimal off-target effects. To date, this method has proven effective in creating gene knockouts via targeted deletions; however, the ability to insert larger sequences of DNA has remained a challenge [87]. The success of this system relies on the specificity of the sgRNA, but prior to the availability of genome sequences for wheat, the ability to test the binding specificity of sgRNA was limited. Recently, several tools have been developed based on wheat genome sequence data to help identify sgRNA sequences with the highest target specificity such as E-CRISP [88], CRISPRdirect [89], and WheatCRISPR [90].

Using the available reference genomes and the latest tools in gene editing, CRISPR can now be used to modify traits in advanced lines. This may be particularly valuable for rice, which is highly amenable to transformation; for example, CRISPR was recently used to create a semi-dwarf rice [91]. Recently, a rice genome-wide CRISPR mutant library was generated with mutations in 12,802 genes [92]. Similar resources have also been developed in maize [93].

Forward genetic approaches, such as chemical mutagenesis by ethyl methanesulfonate (EMS) or gamma irradiation, have also been used to generate populations of mutants [8]. These mutant lines can be phenotypically screened for traits of interest and characterized using genomic approaches to identify candidate genes. Approaches to broadly sequence mutant populations have also been employed, allowing for mutant lines to be preselected based on mutations they carry and then screened for phenotypes using targeted assays. While mutagenesis approaches lack the specificity of CRISPR/Cas9 technologies, they may not meet the same regulatory requirements for genetic modification in some areas, which may have beneficial regulatory and/or crop registration implications.

### 4.4 Moving Past the Reference Genome and Towards the Future: Pan-Genomes of Cereal Crops

Initial genome sequencing efforts focused on developing a 'reference-quality' genome assembly that could serve as a 'gold standard' for the crop community. This approach has been effective, since a lot can be accomplished using a single chromosome-scale assembly, especially when accompanied by annotated gene models. However, the reality is that a single organism does not sufficiently capture the full extent of the diversity that exists within a species; thus, there is a need for several representative genomes. The characterization of diversity within a species is a concept that has been coined the pan-genome. Recently, efforts to characterize the pan-genome of several cereal crops have been initiated/completed by the development of chromosome-scale assemblies from multiple representative genotypes [94–96, 18, 21].

To maximize the amount of information from sequencing projects, it is useful to consider a conceptual model that employs a strategic approach involving tiered levels of genomic data (Fig. 2). Each level within the tiered approach retrieves different types of genomic information, ranging from highest (Tier 1) to lowest (Tier 3) level sequence information. Because of the reduced cost

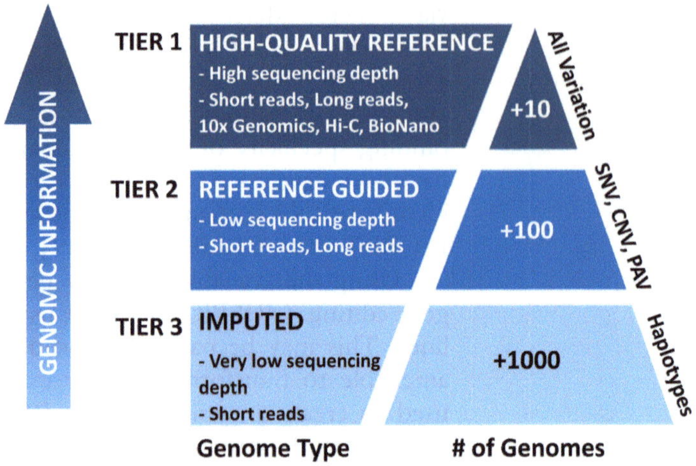


**Fig. 2** Tiered sequencing approach for generating high-quality reference genome assemblies (Tier 1), reference-guided genome assemblies (Tier 2), and imputed genome assemblies (Tier 3)

associated with lower tiers, the number of individual genotypes that can be screened increases. Tier 1 represents the chromosome-scale reference genome, ideally containing the full assemblies from multiple representative genotypes. Tier 2 represents lower-cost whole-genome sequencing (WGS) approaches, often involving sequence alignment to reference genomes, which gives information related to single nucleotide variation (SNV), copy number variation (CNV), or presence-absence variation (PAV). Finally, Tier 3 consists of reduced representation sequencing/genotyping strategies, which gives information on genetic diversity that exists. Tier 3 strategies include genotyping-by-sequencing (GBS), whole-genome skim sequencing, and SNP genotyping arrays.

Using the tiered approach, high-quality genomic information, such as chromosome-scale reference genome assemblies, are only required from a smaller set of individuals. These reference-quality genomes are then the foundation that is used to gather information about more diverse germplasm, or large populations of individuals derived from known parents (i.e. populations generated by biparental crosses and mutagenesis). To give an example of the scale of what can be achieved using Tier 3 genotyping approaches, recently a diversity analysis of 80,000 wheat accessions was surveyed using DArTseq [97]; similarly, thousands of genomes have been genotyped using SNP marker arrays [75–79] and exome capture sequencing methods [98, 99], thereby providing new insights into the genetic diversity that exists within the species. Using genome-wide sequencing information from these approaches in Tier 3, variants can be identified and used to impute the corresponding genome regions from available genome sequences that were assembled at increased level of depth from the other tiers.

## 5 Conclusions and Future Directions

Early genomics research has focused primarily on model organisms, such as *E. coli* for bacteria, *S. cerevisiae* (yeast) for eukaryotes, *C. elegans* and *D. melanogaster* for animals, and *A. thaliana* for plants. *B. distachyon* was also used as a model for grasses, but was the focus of less intense research efforts compared to other genomics models [100]. The pioneering work using these organisms created an important starting point for genetics and genomics research. To this day, much of our understanding of genes and their functions are based on information available from model organisms. Their study has involved detailed investigations of gene expression, subcellular localization, mutagenesis, overexpression, physical and genetic interaction, and analysis of diversity, with many of these details supported by studies using multiple techniques [101]. As a result, the genetic and metabolic networks and pathways in these organisms are well understood.

Genomics of non-model organisms, especially in the case of cereals, which can have large polyploid genomes, was challenged by technical limitations with the available sequencing platforms and the costs associated with sequencing. Without an available reference genome sequence, genomics research was stuck in time, relying on information from genetic linkage maps and gene orthology to model organisms. The challenges of acquiring a reference genome have largely been overcome with the advent of high-throughput sequencing (including long-range short-read technologies such 10X Genomics and Hi-C), long-read sequencing, and optical mapping. As a result, it is now possible to perform detailed genomics and pan-genomics studies in cereal crops. Equipped with genome sequences, the cereal research community has already made significant advances in gene discovery and marker development for applied breeding; however, cereal genomics still lack the functional genomic resources, such as those available for model organisms. With recent advances in high-throughput phenotyping, metabolomics, proteomics, and CRISPR/CAS9 technologies, these information may also be more easily attainable and bring cereal genomics to the next level, functional genomics.

## References


1. Wang N, Yuan Y, Wang H, Yu D, Liu Y, Zhang A, Gowda M, Nair SK, Hao Z, Lu Y, San Vicente F, Prasanna BM, Li X, Zhang X (2020) Applications of genotyping-by-sequencing (GBS) in maize genetics and breeding. Sci Rep 10(1):16308. https://doi.org/10.1038/s41598-020-73321-8
2. Goff SA, Ricke D, Lan T-H, Presting G, Wang R, Dunn M, Glazebrook J, Sessions A, Oeller P, Varma H, Hadley D, Hutchison D, Martin C, Katagiri F, Lange BM, Moughamer T, Xia Y, Budworth P, Zhong J, Miguel T, Paszkowski U, Zhang S, Colbert M, Sun W-L, Chen L, Cooper B, Park S, Wood TC, Mao L, Quail P, Wing R, Dean R, Yu Y, Zharkikh A, Shen R, Sahasrabudhe S, Thomas A, Cannings R, Gutin A, Pruss D, Reid J, Tavtigian S, Mitchell J, Eldredge G, Scholl T, Miller RM, Bhatnagar S, Adey N, Rubano T, Tusneem N,

Robinson R, Feldhaus J, Macalma T, Oliphant A, Briggs S (2002) A draft sequence of the rice genome (*Oryza sativa* L. ssp. *japonica*). Science 296(5565):92. https://doi.org/10.1126/science.1068275

3. Paterson AH, Freeling M, Sasaki T (2005) Grains of knowledge: genomics of model cereals. Genome Res 15(12):1643–1650. https://doi.org/10.1101/gr.3725905
4. Zhao T, Schranz ME (2019) Network-based microsynteny analysis identifies major differences and genomic outliers in mammalian and angiosperm genomes. Proc Natl Acad Sci 116 (6):2165. https://doi.org/10.1073/pnas.1801757116
5. Hu H, Scheben A, Edwards D (2018) Advances in integrating genomics and bioinformatics in the plant breeding pipeline. Agriculture 8:6. https://doi.org/10.3390/agriculture8060075
6. International Wheat Genome Sequencing Consortium (IWGSC) (2014) A chromosome-based draft sequence of the hexaploid bread wheat (*Triticum aestivum*) genome. Science 345(6194):1251788. https://doi.org/10.1126/science.1251788
7. Zhang M, Zhang Y, Scheuring CF, Wu C-C, Dong JJ, Zhang H-B (2012) Preparation of megabase-sized DNA from a variety of organisms using the nuclei method for advanced genomics research. Nat Protoc 7 (3):467–478. https://doi.org/10.1038/nprot.2011.455
8. Krasileva KV, Vasquez-Gross HA, Howell T, Bailey P, Paraiso F, Clissold L, Simmonds J, Ramirez-Gonzalez RH, Wang X, Borrill P, Fosker C, Ayling S, Phillips AL, Uauy C, Dubcovsky J (2017) Uncovering hidden variation in polyploid wheat. PNAS 114(6): E913–E921. https://doi.org/10.1073/pnas.1619268114
9. Arora S, Steuernagel B, Gaurav K, Chandramohan S, Long Y, Matny O, Johnson R, Enk J, Periyannan S, Singh N, Asyraf Md Hatta M, Athiyannan N, Cheema J, Yu G, Kangara N, Ghosh S, Szabo LJ, Poland J, Bariana H, Jones JDG, Bentley AR, Ayliffe M, Olson E, Xu SS, Steffenson BJ, Lagudah E, Wulff BBH (2019) Resistance gene cloning from a wild crop relative by sequence capture and association genetics. Nat Biotechnol 37(2):139–143. https://doi.org/10.1038/s41587-018-0007-9
10. Steuernagel B, Witek K, Jones JDG, Wulff BBH (2017) MutRenSeq: a method for rapid cloning of plant disease resistance genes. Methods Mol Biol 1659:215–229. https://doi.org/10.1007/978-1-4939-7249-4_19
11. Jouanin A, Borm T, Boyd LA, Cockram J, Leigh F, Santos BACM, Visser RGF, Smulders MJM (2019) Development of the GlutEnSeq capture system for sequencing gluten gene families in hexaploid bread wheat with deletions or mutations induced by γ-irradiation or CRISPR/Cas9. J Cereal Sci 88:157–166. https://doi.org/10.1016/j.jcs.2019.04.008
12. Miga KH, Koren S, Rhie A, Vollger MR, Gershman A, Bzikadze A, Brooks S, Howe E, Porubsky D, Logsdon GA, Schneider VA, Potapova T, Wood J, Chow W, Armstrong J, Fredrickson J, Pak E, Tigyi K, Kremitzki M, Markovic C, Maduro V, Dutra A, Bouffard GG, Chang AM, Hansen NF, Wilfert AB, Thibaud-Nissen F, Schmitt AD, Belton J-M, Selvaraj S, Dennis MY, Soto DC, Sahasrabudhe R, Kaya G, Quick J, Loman NJ, Holmes N, Loose M, Surti U, Ra R, Graves Lindsay TA, Fulton R, Hall I, Paten B, Howe K, Timp W, Young A, Mullikin JC, Pevzner PA, Gerton JL, Sullivan BA, Eichler EE, Phillippy AM (2020) Telomere-to-telomere assembly of a complete human X chromosome. Nature 585(7823):79–84. https://doi.org/10.1038/s41586-020-2547-7
13. Nilsen KT, Walkowiak S, Xiang D, Gao P, Quilichini TD, Willick IR, Byrns B, N'Diaye A, Ens J, Wiebe KJPotNAoS (2020) Copy number variation of TdDof controls solid-stemmed architecture in wheat. Proc Natl Acad Sci U S A 117 (46):28708–28718
14. Kovaka S, Fan Y, Ni B, Timp W, Schatz MC (2020) Targeted nanopore sequencing by real-time mapping of raw electrical signal with UNCALLED. Nat Biotechnol. https://doi.org/10.1038/s41587-020-0731-9
15. Maccaferri M, Harris NS, Twardziok SO, Pasam RK, Gundlach H, Spannagl M, Ormanbekova D, Lux T, Prade VM, Milner SG, Himmelbach A, Mascher M, Bagnaresi P, Faccioli P, Cozzi P, Lauria M, Lazzari B, Stella A, Manconi A, Gnocchi M, Moscatelli M, Avni R, Deek J, Biyiklioglu S, Frascaroli E, Corneti S, Salvi S, Sonnante G, Desiderio F, Marè C, Crosatti C, Mica E, Özkan H, Kilian B, De Vita P, Marone D, Joukhadar R, Mazzucotelli E, Nigro D, Gadaleta A, Chao S, Faris JD, Melo ATO, Pumphrey M, Pecchioni N, Milanesi L, Wiebe K, Ens J, MacLachlan RP, Clarke JM, Sharpe AG, Koh CS, Liang KYH, Taylor GJ, Knox R, Budak H, Mastrangelo AM, Xu SS, Stein N, Hale I, Distelfeld A, Hayden MJ,

Tuberosa R, Walkowiak S, Mayer KFX, Ceriotti A, Pozniak CJ, Cattivelli L (2019) Durum wheat genome highlights past domestication signatures and future improvement targets. Nat Genet 51(5):885–895. https://doi.org/10.1038/s41588-019-0381-3

16. Avni R, Nave M, Barad O, Baruch K, Twardziok SO, Gundlach H, Hale I, Mascher M, Spannagl M, Wiebe K, Jordan KW, Golan G, Deek J, Ben-Zvi B, Ben-Zvi G, Himmelbach A, MacLachlan RP, Sharpe AG, Fritz A, Ben-David R, Budak H, Fahima T, Korol A, Faris JD, Hernandez A, Mikel MA, Levy AA, Steffenson B, Maccaferri M, Tuberosa R, Cattivelli L, Faccioli P, Ceriotti A, Kashkush K, Pourkheirandish M, Komatsuda T, Eilam T, Sela H, Sharon A, Ohad N, Chamovitz DA, Mayer KFX, Stein N, Ronen G, Peleg Z, Pozniak CJ, Akhunov ED, Distelfeld A (2017) Wild emmer genome architecture and diversity elucidate wheat evolution and domestication. Science 357(6346):93. https://doi.org/10.1126/science.aan0032
17. The International Wheat Genome Sequencing Consortium (IWGSC), Appels R, Eversole K, Feuillet C, Keller B, Rogers J, Stein N, Pozniak CJ, Choulet F, Distelfeld A, Poland JJS (2018) Shifting the limits in wheat research and breeding using a fully annotated reference genome. Science 361(6403):eaar7191
18. Walkowiak S, Gao L, Monat C, Haberer G, Kassa MT, Brinton J, Ramirez-Gonzalez RH, Kolodziej MC, Delorean E, Thambugala DJN (2020) Multiple wheat genomes reveal global variation in modern breeding. Nature 588:1–7
19. Clavijo BJ, Garcia Accinelli G, Wright J, Heavens D, Barr K, Yanes L, Di-Palma F (2017) W2RAP: a pipeline for high quality, robust assemblies of large complex genomes from short read data. bioRxiv:110999. https://doi.org/10.1101/110999
20. Monat C, Padmarasu S, Lux T, Wicker T, Gundlach H, Himmelbach A, Ens J, Li C, Muehlbauer GJ, Schulman AH, Waugh R, Braumann I, Pozniak C, Scholz U, Mayer KFX, Spannagl M, Stein N, Mascher M (2019) TRITEX: chromosome-scale sequence assembly of Triticeae genomes with open-source tools. Genome Biol 20(1):284. https://doi.org/10.1186/s13059-019-1899-5
21. Jayakodi M, Padmarasu S, Haberer G, Bonthala VS, Gundlach H, Monat C, Lux T, Kamal N, Lang D, Himmelbach A, Ens J, Zhang X-Q, Angessa TT, Zhou G, Tan C, Hill C, Wang P, Schreiber M, Boston LB, Plott C, Jenkins J, Guo Y, Fiebig A, Budak H, Xu D, Zhang J, Wang C, Grimwood J, Schmutz J, Guo G, Zhang G, Mochida K, Hirayama T, Sato K, Chalmers KJ, Langridge P, Waugh R, Pozniak CJ, Scholz U, Mayer KFX, Spannagl M, Li C, Mascher M, Stein N (2020) The barley pan-genome reveals the hidden legacy of mutation breeding. Nature 588 (7837):284–289. https://doi.org/10.1038/s41586-020-2947-8
22. Zimin AV, Puiu D, Luo MC, Zhu T, Koren S, Marçais G, Yorke JA, Dvořák J, Salzberg SL (2017) Hybrid assembly of the large and highly repetitive genome of Aegilops tauschii, a progenitor of bread wheat, with the MaSuRCA mega-reads algorithm. Genome Res 27(5):787–792. https://doi.org/10.1101/gr.213405.116
23. Mascher M, Muehlbauer GJ, Rokhsar DS, Chapman J, Schmutz J, Barry K, Muñoz-Amatriaín M, Close TJ, Wise RP, Schulman AH, Himmelbach A, Mayer KFX, Scholz U, Poland JA, Stein N, Waugh R (2013) Anchoring and ordering NGS contig assemblies by population sequencing (POPSEQ). Plant J 76 (4):718–727. https://doi.org/10.1111/tpj.12319
24. Zhu T, Wang L, Rodriguez JC, Deal KR, Avni R, Distelfeld A, McGuire PE, Dvorak J, Luo M-C (2019) Improved genome sequence of wild emmer wheat Zavitan with the aid of optical maps. G3 (Bethesda) 9(3):619–624. https://doi.org/10.1534/g3.118.200902
25. Li H (2018) Minimap2: pairwise alignment for nucleotide sequences. Bioinformatics 34 (18):3094–3100. https://doi.org/10.1093/bioinformatics/bty191
26. Kim D, Langmead B, Salzberg SL (2015) HISAT: a fast spliced aligner with low memory requirements. Nat Methods 12 (4):357–360. https://doi.org/10.1038/nmeth.3317
27. Kim D, Paggi JM, Park C, Bennett C, Salzberg SL (2019) Graph-based genome alignment and genotyping with HISAT2 and HISAT-genotype. Nat Biotechnol 37 (8):907–915. https://doi.org/10.1038/s41587-019-0201-4
28. Dobin A, Davis CA, Schlesinger F, Drenkow J, Zaleski C, Jha S, Batut P, Chaisson M, Gingeras TR (2012) STAR: ultrafast universal RNA-seq aligner. Bioinformatics 29(1):15–21. https://doi.org/10.1093/bioinformatics/bts635
29. Stein N (2007) Triticeae genomics: advances in sequence analysis of large genome cereal

crops. Chromosom Res 15(1):21–31. https://doi.org/10.1007/s10577-006-1107-9

30. Pertea M, Kim D, Pertea GM, Leek JT, Salzberg SL (2016) Transcript-level expression analysis of RNA-seq experiments with HISAT, StringTie and Ballgown. Nat Protoc 11:1650. https://doi.org/10.1038/nprot.2016.095
31. Ma J, Bennetzen JL (2004) Rapid recent growth and divergence of rice nuclear genomes. Proc Natl Acad Sci U S A 101(34):12404. https://doi.org/10.1073/pnas.0403715101
32. Yu J, Hu S, Wang J, Wong GK-S, Li S, Liu B, Deng Y, Dai L, Zhou Y, Zhang X, Cao M, Liu J, Sun J, Tang J, Chen Y, Huang X, Lin W, Ye C, Tong W, Cong L, Geng J, Han Y, Li L, Li W, Hu G, Huang X, Li W, Li J, Liu Z, Li L, Liu J, Qi Q, Liu J, Li L, Li T, Wang X, Lu H, Wu T, Zhu M, Ni P, Han H, Dong W, Ren X, Feng X, Cui P, Li X, Wang H, Xu X, Zhai W, Xu Z, Zhang J, He S, Zhang J, Xu J, Zhang K, Zheng X, Dong J, Zeng W, Tao L, Ye J, Tan J, Ren X, Chen X, He J, Liu D, Tian W, Tian C, Xia H, Bao Q, Li G, Gao H, Cao T, Wang J, Zhao W, Li P, Chen W, Wang X, Zhang Y, Hu J, Wang J, Liu S, Yang J, Zhang G, Xiong Y, Li Z, Mao L, Zhou C, Zhu Z, Chen R, Hao B, Zheng W, Chen S, Guo W, Li G, Liu S, Tao M, Wang J, Zhu L, Yuan L, Yang H (2002) A draft sequence of the rice genome (*Oryza sativa* L. ssp. *indica*). Science 296(5565):79. https://doi.org/10.1126/science.1068037
33. Sasaki T, Matsumoto T, Yamamoto K, Sakata K, Baba T, Katayose Y, Wu J, Niimura Y, Cheng Z, Nagamura Y, Antonio BA, Kanamori H, Hosokawa S, Masukawa M, Arikawa K, Chiden Y, Hayashi M, Okamoto M, Ando T, Aoki H, Arita K, Hamada M, Harada C, Hijishita S, Honda M, Ichikawa Y, Idonuma A, Iijima M, Ikeda M, Ikeno M, Ito S, Ito T, Ito Y, Ito Y, Iwabuchi A, Kamiya K, Karasawa W, Katagiri S, Kikuta A, Kobayashi N, Kono I, Machita K, Maehara T, Mizuno H, Mizubayashi T, Mukai Y, Nagasaki H, Nakashima M, Nakama Y, Nakamichi Y, Nakamura M, Namiki N, Negishi M, Ohta I, Ono N, Saji S, Sakai K, Shibata M, Shimokawa T, Shomura A, Song J, Takazaki Y, Terasawa K, Tsuji K, Waki K, Yamagata H, Yamane H, Yoshiki S, Yoshihara R, Yukawa K, Zhong H, Iwama H, Endo T, Ito H, Hahn JH, Kim H-I, Eun M-Y, Yano M, Jiang J, Gojobori T (2002) The genome sequence and structure of rice chromosome 1. Nature 420(6913):312–316. https://doi.org/10.1038/nature01184
34. Feng Q, Zhang Y, Hao P, Wang S, Fu G, Huang Y, Li Y, Zhu J, Liu Y, Hu X, Jia P, Zhang Y, Zhao Q, Ying K, Yu S, Tang Y, Weng Q, Zhang L, Lu Y, Mu J, Lu Y, Zhang LS, Yu Z, Fan D, Liu X, Lu T, Li C, Wu Y, Sun T, Lei H, Li T, Hu H, Guan J, Wu M, Zhang R, Zhou B, Chen Z, Chen L, Jin Z, Wang R, Yin H, Cai Z, Ren S, Lv G, Gu W, Zhu G, Tu Y, Jia J, Zhang Y, Chen J, Kang H, Chen X, Shao C, Sun Y, Hu Q, Zhang X, Zhang W, Wang L, Ding C, Sheng H, Gu J, Chen S, Ni L, Zhu F, Chen W, Lan L, Lai Y, Cheng Z, Gu M, Jiang J, Li J, Hong G, Xue Y, Han B (2002) Sequence and analysis of rice chromosome 4. Nature 420(6913):316–320. https://doi.org/10.1038/nature01183
35. Consortium TRCS (2003) In-depth view of structure, activity, and evolution of rice chromosome 10. Science 300(5625):1566. https://doi.org/10.1126/science.1083523
36. Sasaki T, International Rice Genome Sequencing P (2005) The map-based sequence of the rice genome. Nature 436(7052):793–800. https://doi.org/10.1038/nature03895
37. Kawahara Y, de la Bastide M, Hamilton JP, Kanamori H, McCombie WR, Ouyang S, Schwartz DC, Tanaka T, Wu J, Zhou S, Childs KL, Davidson RM, Lin H, Quesada-Ocampo L, Vaillancourt B, Sakai H, Lee SS, Kim J, Numa H, Itoh T, Buell CR, Matsumoto T (2013) Improvement of the Oryza sativa Nipponbare reference genome using next generation sequence and optical map data. Rice 6(1):4. https://doi.org/10.1186/1939-8433-6-4
38. Choi JY, Lye ZN, Groen SC, Dai X, Rughani P, Zaaijer S, Harrington ED, Juul S, Purugganan MD (2020) Nanopore sequencing-based genome assembly and evolutionary genomics of circum-basmati rice. Genome Biol 21(1):21. https://doi.org/10.1186/s13059-020-1938-2
39. Du H, Yu Y, Ma Y, Gao Q, Cao Y, Chen Z, Ma B, Qi M, Li Y, Zhao X, Wang J, Liu K, Qin P, Yang X, Zhu L, Li S, Liang C (2017) Sequencing and de novo assembly of a near complete indica rice genome. Nat Commun 8(1):15324. https://doi.org/10.1038/ncomms15324
40. Tanaka T, Nishijima R, Teramoto S, Kitomi Y, Hayashi T, Uga Y, Kawakatsu T (2020) *De novo* genome assembly of the *indica* rice variety IR64 using linked-read sequencing and nanopore sequencing. G3: Genes|Genomes|

Genetics 10(5):1495. https://doi.org/10.1534/g3.119.400871

41. Brozynska M, Copetti D, Furtado A, Wing RA, Crayn D, Fox G, Ishikawa R, Henry RJ (2017) Sequencing of Australian wild rice genomes reveals ancestral relationships with domesticated rice. Plant Biotechnol J 15 (6):765–774. https://doi.org/10.1111/pbi.12674
42. Song S, Tian D, Zhang Z, Hu S, Yu J (2018) Rice genomics: over the past two decades and into the future. Genomics Proteomics Bioinformatics 16(6):397–404. https://doi.org/10.1016/j.gpb.2019.01.001
43. Yu J, Wang J, Lin W, Li S, Li H, Zhou J, Ni P, Dong W, Hu S, Zeng C, Zhang J, Zhang Y, Li R, Xu Z, Li S, Li X, Zheng H, Cong L, Lin L, Yin J, Geng J, Li G, Shi J, Liu J, Lv H, Li J, Wang J, Deng Y, Ran L, Shi X, Wang X, Wu Q, Li C, Ren X, Wang J, Wang X, Li D, Liu D, Zhang X, Ji Z, Zhao W, Sun Y, Zhang Z, Bao J, Han Y, Dong L, Ji J, Chen P, Wu S, Liu J, Xiao Y, Bu D, Tan J, Yang L, Ye C, Zhang J, Xu J, Zhou Y, Yu Y, Zhang B, Zhuang S, Wei H, Liu B, Lei M, Yu H, Li Y, Xu H, Wei S, He X, Fang L, Zhang Z, Zhang Y, Huang X, Su Z, Tong W, Li J, Tong Z, Li S, Ye J, Wang L, Fang L, Lei T, Chen C, Chen H, Xu Z, Li H, Huang H, Zhang F, Xu H, Li N, Zhao C, Li S, Dong L, Huang Y, Li L, Xi Y, Qi Q, Li W, Zhang B, Hu W, Zhang Y, Tian X, Jiao Y, Liang X, Jin J, Gao L, Zheng W, Hao B, Liu S, Wang W, Yuan L, Cao M, McDermott J, Samudrala R, Wang J, Wong GK-S, Yang H (2005) The genomes of Oryza sativa: a history of duplications. PLoS Biol 3 (2):e38. https://doi.org/10.1371/journal.pbio.0030038
44. Chandler VL, Brendel V (2002) The maize genome sequencing project. Plant Physiol 130(4):1594. https://doi.org/10.1104/pp.015594
45. Schnable PS, Ware D, Fulton RS, Stein JC, Wei F, Pasternak S, Liang C, Zhang J, Fulton L, Graves TA, Minx P, Reily AD, Courtney L, Kruchowski SS, Tomlinson C, Strong C, Delehaunty K, Fronick C, Courtney B, Rock SM, Belter E, Du F, Kim K, Abbott RM, Cotton M, Levy A, Marchetto P, Ochoa K, Jackson SM, Gillam B, Chen W, Yan L, Higginbotham J, Cardenas M, Waligorski J, Applebaum E, Phelps L, Falcone J, Kanchi K, Thane T, Scimone A, Thane N, Henke J, Wang T, Ruppert J, Shah N, Rotter K, Hodges J, Ingenthron E, Cordes M, Kohlberg S, Sgro J, Delgado B, Mead K, Chinwalla A, Leonard S, Crouse K, Collura K, Kudrna D, Currie J, He R, Angelova A, Rajasekar S, Mueller T, Lomeli R, Scara G, Ko A, Delaney K, Wissotski M, Lopez G, Campos D, Braidotti M, Ashley E, Golser W, Kim H, Lee S, Lin J, Dujmic Z, Kim W, Talag J, Zuccolo A, Fan C, Sebastian A, Kramer M, Spiegel L, Nascimento L, Zutavern T, Miller B, Ambroise C, Muller S, Spooner W, Narechania A, Ren L, Wei S, Kumari S, Faga B, Levy MJ, McMahan L, Van Buren P, Vaughn MW, Ying K, Yeh C-T, Emrich SJ, Jia Y, Kalyanaraman A, Hsia A-P, Barbazuk WB, Baucom RS, Brutnell TP, Carpita NC, Chaparro C, Chia J-M, Deragon J-M, Estill JC, Fu Y, Jeddeloh JA, Han Y, Lee H, Li P, Lisch DR, Liu S, Liu Z, Nagel DH, McCann MC, SanMiguel P, Myers AM, Nettleton D, Nguyen J, Penning BW, Ponnala L, Schneider KL, Schwartz DC, Sharma A, Soderlund C, Springer NM, Sun Q, Wang H, Waterman M, Westerman R, Wolfgruber TK, Yang L, Yu Y, Zhang L, Zhou S, Zhu Q, Bennetzen JL, Dawe RK, Jiang J, Jiang N, Presting GG, Wessler SR, Aluru S, Martienssen RA, Clifton SW, McCombie WR, Wing RA, Wilson RK (2009) The B73 maize genome: complexity, diversity, and dynamics. Science 326 (5956):1112. https://doi.org/10.1126/science.1178534
46. Sun S, Zhou Y, Chen J, Shi J, Zhao H, Zhao H, Song W, Zhang M, Cui Y, Dong X, Liu H, Ma X, Jiao Y, Wang B, Wei X, Stein JC, Glaubitz JC, Lu F, Yu G, Liang C, Fengler K, Li B, Rafalski A, Schnable PS, Ware DH, Buckler ES, Lai J (2018) Extensive intraspecific gene order and gene structural variations between Mo17 and other maize genomes. Nat Genet 50(9):1289–1295. https://doi.org/10.1038/s41588-018-0182-0
47. Yang N, Xu X-W, Wang R-R, Peng W-L, Cai L, Song J-M, Li W, Luo X, Niu L, Wang Y, Jin M, Chen L, Luo J, Deng M, Wang L, Pan Q, Liu F, Jackson D, Yang X, Chen L-L, Yan J (2017) Contributions of Zea mays subspecies mexicana haplotypes to modern maize. Nat Commun 8(1):1874. https://doi.org/10.1038/s41467-017-02063-5
48. Springer NM, Anderson SN, Andorf CM, Ahern KR, Bai F, Barad O, Barbazuk WB, Bass HW, Baruch K, Ben-Zvi G, Buckler ES, Bukowski R, Campbell MS, Cannon EKS, Chomet P, Dawe RK, Davenport R, Dooner HK, Du LH, Du C, Easterling KA, Gault C, Guan J-C, Hunter CT, Jander G, Jiao Y, Koch KE, Kol G, Köllner TG, Kudo T, Li Q, Lu F, Mayfield-Jones D, Mei W, McCarty DR, Noshay JM, Portwood JL, Ronen G, Settles

AM, Shem-Tov D, Shi J, Soifer I, Stein JC, Stitzer MC, Suzuki M, Vera DL, Vollbrecht E, Vrebalov JT, Ware D, Wei S, Wimalanathan K, Woodhouse MR, Xiong W, Brutnell TP (2018) The maize W22 genome provides a foundation for functional genomics and transposon biology. Nat Genet 50(9):1282–1288. https://doi.org/10.1038/s41588-018-0158-0

49. Yang N, Liu J, Gao Q, Gui S, Chen L, Yang L, Huang J, Deng T, Luo J, He L, Wang Y, Xu P, Peng Y, Shi Z, Lan L, Ma Z, Yang X, Zhang Q, Bai M, Li S, Li W, Liu L, Jackson D, Yan J (2019) Genome assembly of a tropical maize inbred line provides insights into structural variation and crop improvement. Nat Genet 51(6):1052–1059. https://doi.org/10.1038/s41588-019-0427-6

50. Hirsch CN, Hirsch CD, Brohammer AB, Bowman MJ, Soifer I, Barad O, Shem-Tov D, Baruch K, Lu F, Hernandez AG, Fields CJ, Wright CL, Koehler K, Springer NM, Buckler E, Buell CR, de Leon N, Kaeppler SM, Childs KL, Mikel MA (2016) Draft assembly of elite inbred line PH207 provides insights into genomic and transcriptome diversity in maize. Plant Cell 28(11):2700. https://doi.org/10.1105/tpc.16.00353

51. Li C, Song W, Luo Y, Gao S, Zhang R, Shi Z, Wang X, Wang R, Wang F, Wang J, Zhao Y, Su A, Wang S, Li X, Luo M, Wang S, Zhang Y, Ge J, Tan X, Yuan Y, Bi X, He H, Yan J, Wang Y, Hu S, Zhao J (2019) The Huang-ZaoSi maize genome provides insights into genomic variation and improvement history of maize. Mol Plant 12(3):402–409. https://doi.org/10.1016/j.molp.2019.02.009

52. Jiao Y, Peluso P, Shi J, Liang T, Stitzer MC, Wang B, Campbell MS, Stein JC, Wei X, Chin C-S, Guill K, Regulski M, Kumari S, Olson A, Gent J, Schneider KL, Wolfgruber TK, May MR, Springer NM, Antoniou E, McCombie WR, Presting GG, McMullen M, Ross-Ibarra-J, Dawe RK, Hastie A, Rank DR, Ware D (2017) Improved maize reference genome with single-molecule technologies. Nature 546(7659):524–527. https://doi.org/10.1038/nature22971

53. Ling H-Q, Ma B, Shi X, Liu H, Dong L, Sun H, Cao Y, Gao Q, Zheng S, Li Y, Yu Y, Du H, Qi M, Li Y, Lu H, Yu H, Cui Y, Wang N, Chen C, Wu H, Zhao Y, Zhang J, Li Y, Zhou W, Zhang B, Hu W, van Eijk MJT, Tang J, Witsenboer HMA, Zhao S, Li Z, Zhang A, Wang D, Liang C (2018) Genome sequence of the progenitor of wheat a subgenome Triticum urartu. Nature 557(7705):424–428. https://doi.org/10.1038/s41586-018-0108-0

54. Luo M-C, Gu YQ, Puiu D, Wang H, Twardziok SO, Deal KR, Huo N, Zhu T, Wang L, Wang Y, McGuire PE, Liu S, Long H, Ramasamy RK, Rodriguez JC, Van SL, Yuan L, Wang Z, Xia Z, Xiao L, Anderson OD, Ouyang S, Liang Y, Zimin AV, Pertea G, Qi P, Bennetzen JL, Dai X, Dawson MW, Müller H-G, Kugler K, Rivarola-Duarte L, Spannagl M, Mayer KFX, Lu F-H, Bevan MW, Leroy P, Li P, You FM, Sun Q, Liu Z, Lyons E, Wicker T, Salzberg SL, Devos KM, Dvořák J (2017) Genome sequence of the progenitor of the wheat D genome Aegilops tauschii. Nature 551(7681):498–502. https://doi.org/10.1038/nature24486

55. Wang H, Sun S, Ge W, Zhao L, Hou B, Wang K, Lyu Z, Chen L, Xu S, Guo J, Li M, Su P, Li X, Wang G, Bo C, Fang X, Zhuang W, Cheng X, Wu J, Dong L, Chen W, Li W, Xiao G, Zhao J, Hao Y, Xu Y, Gao Y, Liu W, Liu Y, Yin H, Li J, Li X, Zhao Y, Wang X, Ni F, Ma X, Li A, Xu SS, Bai G, Nevo E, Gao C, Ohm H, Kong L (2020) Horizontal gene transfer of *Fhb7* from fungus underlies *Fusarium* head blight resistance in wheat. Science 368(6493):eaba5435. https://doi.org/10.1126/science.aba5435

56. Purugganan MD, Fuller DQ (2009) The nature of selection during plant domestication. Nature 457(7231):843–848. https://doi.org/10.1038/nature07895

57. Sato K (2020) History and future perspectives of barley genomics. DNA Res 27:4. https://doi.org/10.1093/dnares/dsaa023

58. Blake T, Blake V, Bowman J, Abdel-Haleem H (2011) Barley: production, improvement and uses. In: Ullrich SE (ed) Barley: production, improvement and uses. Wiley-Blackwell, pp 522–531

59. Schulte D, Close TJ, Graner A, Langridge P, Matsumoto T, Muehlbauer G, Sato K, Schulman AH, Waugh R, Wise RP, Stein N (2009) The international barley sequencing consortium--at the threshold of efficient access to the barley genome. Plant Physiol 149(1):142–147. https://doi.org/10.1104/pp.108.128967

60. The International Barley Genome Sequencing Consortium (2012) A physical, genetic and functional sequence assembly of the barley genome. Nature 491(7426):711–716. https://doi.org/10.1038/nature11543

61. Mascher M, Gundlach H, Himmelbach A, Beier S, Twardziok SO, Wicker T,

Radchuk V, Dockter C, Hedley PE, Russell J, Bayer M, Ramsay L, Liu H, Haberer G, Zhang X-Q, Zhang Q, Barrero RA, Li L, Taudien S, Groth M, Felder M, Hastie A, Šimková H, Staňková H, Vrána J, Chan S, Muñoz-Amatriaín M, Ounit R, Wanamaker S, Bolser D, Colmsee C, Schmutzer T, Aliyeva-Schnorr L, Grasso S, Tanskanen J, Chailyan A, Sampath D, Heavens D, Clissold L, Cao S, Chapman B, Dai F, Han Y, Li H, Li X, Lin C, McCooke JK, Tan C, Wang P, Wang S, Yin S, Zhou G, Poland JA, Bellgard MI, Borisjuk L, Houben A, Doležel J, Ayling S, Lonardi S, Kersey P, Langridge P, Muehlbauer GJ, Clark MD, Caccamo M, Schulman AH, Mayer KFX, Platzer M, Close TJ, Scholz U, Hansson M, Zhang G, Braumann I, Spannagl M, Li C, Waugh R, Stein N (2017) A chromosome conformation capture ordered sequence of the barley genome. Nature 544:427. https://doi.org/10.1038/nature22043

62. Rabanus-Wallace MT, Hackauf B, Mascher M, Lux T, Wicker T, Gundlach H, Báez M, Houben A, Mayer KFX, Guo L, Poland J, Pozniak CJ, Walkowiak S, Melonek J, Praz C, Schreiber M, Budak H, Heuberger M, Steuernagel B, Wulff B, Börner A, Byrns B, Čížková J, Fowler DB, Fritz A, Himmelbach A, Kaithakottil G, Keilwagen J, Keller B, Konkin D, Larsen J, Li Q, Myśków B, Padmarasu S, Rawat N, Sesiz U, Sezgi B, Sharpe A, Šimková H, Small I, Swarbreck D, Toegelová H, Tsvetkova N, Voylokov AV, Vrána J, Bauer E, Bolibok-Bragoszewska H, Doležel J, Hall A, Jia J, Korzun V, Laroche A, Ma X-F, Ordon F, Özkan H, Rakoczy-Trojanowska M, Scholz U, Schulman AH, Siekmann D, Stojałowski S, Tiwari V, Spannagl M, Stein N (2019) Chromosome-scale genome assembly provides insights into rye biology, evolution, and agronomic potential. bioRxiv:2019.2012.2011.869693. https://doi.org/10.1101/2019.12.11.869693

63. Evtushenko EV, Lipikhina YA, Stepochkin PI, Vershinin AV (2019) Cytogenetic and molecular characteristics of rye genome in octoploid triticale (× Triticosecale Wittmack). Comp Cytogenet 13(4):423–434

64. Ryöppy PH (1997) Haploidy in triticale. In: Jain SM, Sopory SK, Veilleux RE (eds) In vitro haploid production in higher plants. Current plant science and biotechnology in agriculture, vol 26. Springer, Dordrecht. https://doi.org/10.1007/978-94-017-1862-2_5

65. Bennett MD, Leitch IJ (1995) Nuclear DNA amounts in angiosperms. Ann Bot 76 (2):113–176. https://doi.org/10.1006/anbo.1995.1085

66. Sanz MJ, Jellen EN, Loarce Y, Irigoyen ML, Ferrer E, Fominaya A (2010) A new chromosome nomenclature system for oat (*Avena sativa* L. and *A. byzantina* C. Koch) based on FISH analysis of monosomic lines. Theor Appl Genet 121(8):1541–1552. https://doi.org/10.1007/s00122-010-1409-3

67. Maughan PJ, Lee R, Walstead R, Vickerstaff RJ, Fogarty MC, Brouwer CR, Reid RR, Jay JJ, Bekele WA, Jackson EW, Tinker NA, Langdon T, Schlueter JA, Jellen EN (2019) Genomic insights from the first chromosome-scale assemblies of oat (Avena spp.) diploid species. BMC Biol 17(1):92. https://doi.org/10.1186/s12915-019-0712-y

68. Paterson AH, Bowers JE, Bruggmann R, Dubchak I, Grimwood J, Gundlach H, Haberer G, Hellsten U, Mitros T, Poliakov A, Schmutz J, Spannagl M, Tang H, Wang X, Wicker T, Bharti AK, Chapman J, Feltus FA, Gowik U, Grigoriev IV, Lyons E, Maher CA, Martis M, Narechania A, Otillar RP, Penning BW, Salamov AA, Wang Y, Zhang L, Carpita NC, Freeling M, Gingle AR, Hash CT, Keller B, Klein P, Kresovich S, McCann MC, Ming R, Peterson DG, Mehboobur R, Ware D, Westhoff P, KFX M, Messing J, Rokhsar DS (2009) The Sorghum bicolor genome and the diversification of grasses. Nature 457 (7229):551–556. https://doi.org/10.1038/nature07723

69. Abrouk M, Ahmed HI, Cubry P, Šimoníková D, Cauet S, Pailles Y, Bettgenhaeuser J, Gapa L, Scarcelli N, Couderc M, Zekraoui L, Kathiresan N, Čížková J, Hřibová E, Doležel J, Arribat S, Bergès H, Wieringa JJ, Gueye M, Kane NA, Leclerc C, Causse S, Vancoppenolle S, Billot C, Wicker T, Vigouroux Y, Barnaud A, Krattinger SG (2020) Fonio millet genome unlocks African orphan crop diversity for agriculture in a changing climate. Nat Commun 11(1):4488. https://doi.org/10.1038/s41467-020-18329-4

70. Varshney RK, Shi C, Thudi M, Mariac C, Wallace J, Qi P, Zhang H, Zhao Y, Wang X, Rathore A, Srivastava RK, Chitikineni A, Fan G, Bajaj P, Punnuri S, Gupta SK, Wang H, Jiang Y, Couderc M, Katta MAVSK, Paudel DR, Mungra KD, Chen W, Harris-Shultz KR, Garg V, Desai N, Doddamani D, Kane NA, Conner JA, Ghatak A, Chaturvedi P, Subramaniam S,

Yadav OP, Berthouly-Salazar C, Hamidou F, Wang J, Liang X, Clotault J, Upadhyaya HD, Cubry P, Rhoné B, Gueye MC, Sunkar R, Dupuy C, Sparvoli F, Cheng S, Mahala RS, Singh B, Yadav RS, Lyons E, Datta SK, Hash CT, Devos KM, Buckler E, Bennetzen JL, Paterson AH, Ozias-Akins P, Grando S, Wang J, Mohapatra T, Weckwerth W, Reif JC, Liu X, Vigouroux Y, Xu X (2017) Pearl millet genome sequence provides a resource to improve agronomic traits in arid environments. Nat Biotechnol 35(10):969–976. https://doi.org/10.1038/nbt.3943

71. Kang S-H, Kim B, Choi B-S, Lee HO, Kim N-H, Lee SJ, Kim HS, Shin MJ, Kim H-W, Nam K, Kang KD, Kwon S-J, Oh T-J, Lee S-C, Kim C-K (2020) Genome assembly and annotation of soft-shelled Adlay (Coix lacryma-jobi variety ma-yuen), a cereal and medicinal crop in the Poaceae Family. Front Plant Sci 11:630–630. https://doi.org/10.3389/fpls.2020.00630
72. Gardiner L-J, Wingen LU, Bailey P, Joynson R, Brabbs T, Wright J, Higgins JD, Hall N, Griffiths S, Clavijo BJ, Hall A (2019) Analysis of the recombination landscape of hexaploid bread wheat reveals genes controlling recombination and gene conversion frequency. Genome Biol 20(1):69. https://doi.org/10.1186/s13059-019-1675-6
73. Röder MS, Korzun V, Wendehake K, Plaschke J, Tixier M-H, Leroy P, Ganal MW (1998) A microsatellite map of wheat. Genetics 149(4):2007
74. Somers DJ, Isaac P, Edwards K (2004) A high-density microsatellite consensus map for bread wheat (Triticum aestivum L.). Theor Appl Genet 109(6):1105–1114. https://doi.org/10.1007/s00122-004-1740-7
75. Cavanagh CR, Chao S, Wang S, Huang BE, Stephen S, Kiani S, Forrest K, Saintenac C, Brown-Guedira GL, Akhunova A, See D, Bai G, Pumphrey M, Tomar L, Wong D, Kong S, Reynolds M, da Silva ML, Bockelman H, Talbert L, Anderson JA, Dreisigacker S, Baenziger S, Carter A, Korzun V, Morrell PL, Dubcovsky J, Morell MK, Sorrells ME, Hayden MJ, Akhunov E (2013) Genome-wide comparative diversity uncovers multiple targets of selection for improvement in hexaploid wheat landraces and cultivars. Proc Natl Acad Sci U S A 110 (20):8057–8062. https://doi.org/10.1073/pnas.1217133110
76. Allen AM, Winfield MO, Burridge AJ, Downie RC, Benbow HR, Barker GLA, Wilkinson PA, Coghill J, Waterfall C, Davassi A, Scopes G, Pirani A, Webster T, Brew F, Bloor C, Griffiths S, Bentley AR, Alda M, Jack P, Phillips AL, Edwards KJ (2017) Characterization of a wheat Breeders' Array suitable for high-throughput SNP genotyping of global accessions of hexaploid bread wheat (Triticum aestivum). Plant Biotechnol J 15(3):390–401. https://doi.org/10.1111/pbi.12635
77. Wang SC, Wong DB, Forrest K, Allen A, Chao SM, Huang BE, Maccaferri M, Salvi S, Milner SG, Cattivelli L, Mastrangelo AM, Whan A, Stephen S, Barker G, Wieseke R, Plieske J, Lillemo M, Mather D, Appels R, Dolferus R, Brown-Guedira G, Korol A, Akhunova AR, Feuillet C, Salse J, Morgante M, Pozniak C, Luo MC, Dvorak J, Morell M, Dubcovsky J, Ganal M, Tuberosa R, Lawley C, Mikoulitch I, Cavanagh C, Edwards KJ, Hayden M, Akhunov E, Sequencing IWG (2014) Characterization of polyploid wheat genomic diversity using a high-density 90 000 single nucleotide polymorphism array. Plant Biotechnol J 12(6):787–796. https://doi.org/10.1111/pbi.12183
78. Sun C, Dong Z, Zhao L, Ren Y, Zhang N, Chen F (2020) The wheat 660K SNP array demonstrates great potential for marker-assisted selection in polyploid wheat. Plant Biotechnol J 18(6):1354–1360. https://doi.org/10.1111/pbi.13361
79. Winfield MO, Allen AM, Burridge AJ, Barker GLA, Benbow HR, Wilkinson PA, Coghill J, Waterfall C, Davassi A, Scopes G, Pirani A, Webster T, Brew F, Bloor C, King J, West C, Griffiths S, King I, Bentley AR, Edwards KJ (2016) High-density SNP genotyping array for hexaploid wheat and its secondary and tertiary gene pool. Plant Biotechnol J 14 (5):1195–1206. https://doi.org/10.1111/pbi.12485
80. Maccaferri M, Ricci A, Salvi S, Milner SG, Noli E, Martelli PL, Casadio R, Akhunov E, Scalabrin S, Vendramin V, Ammar K, Blanco A, Desiderio F, Distelfeld A, Dubcovsky J, Fahima T, Faris J, Korol A, Massi A, Mastrangelo AM, Morgante M, Pozniak C, N'Diaye A, Xu S, Tuberosa R (2015) A high-density, SNP-based consensus map of tetraploid wheat as a bridge to integrate durum and bread wheat genomics and breeding. Plant Biotechnol J 13(5):648–663. https://doi.org/10.1111/pbi.12288
81. Robertsen CD, Hjortshøj RL, Janss LL (2019) Genomic selection in cereal breeding.

Agronomy 9:2. https://doi.org/10.3390/agronomy9020095

82. Bettgenhaeuser J, Krattinger SG (2019) Rapid gene cloning in cereals. Theor Appl Genet 132(3):699–711. https://doi.org/10.1007/s00122-018-3210-7
83. Klymiuk V, Yaniv E, Huang L, Raats D, Fatiukha A, Chen S, Feng L, Frenkel Z, Krugman T, Lidzbarsky G, Chang W, Jääskeläinen MJ, Schudoma C, Paulin L, Laine P, Bariana H, Sela H, Saleem K, Sørensen CK, Hovmøller MS, Distelfeld A, Chalhoub B, Dubcovsky J, Korol AB, Schulman AH, Fahima T (2018) Cloning of the wheat Yr15 resistance gene sheds light on the plant tandem kinase-pseudokinase family. Nat Commun 9(1):3735. https://doi.org/10.1038/s41467-018-06138-9
84. Chen S, Rouse MN, Zhang W, Zhang X, Guo Y, Briggs J, Dubcovsky J (2020) Wheat gene Sr60 encodes a protein with two putative kinase domains that confers resistance to stem rust. New Phytol 225(2):948–959. https://doi.org/10.1111/nph.16169
85. Hiebert CW, Moscou MJ, Hewitt T, Steuernagel B, Hernández-Pinzón I, Green P, Pujol V, Zhang P, Rouse MN, Jin Y, McIntosh RA, Upadhyaya N, Zhang J, Bhavani S, Vrána J, Karafiátová M, Huang L, Fetch T, Doležel J, Wulff BBH, Lagudah E, Spielmeyer W (2020) Stem rust resistance in wheat is suppressed by a subunit of the mediator complex. Nat Commun 11(1):1123. https://doi.org/10.1038/s41467-020-14937-2
86. Doudna JA, Charpentier E (2014) The new frontier of genome engineering with CRISPR-Cas9. Science 346(6213):1258096. https://doi.org/10.1126/science.1258096
87. Hou Z, Zhang Y (2019) Inserting DNA with CRISPR. Science 365(6448):25–26. https://doi.org/10.1126/science.aay2056
88. Heigwer F, Kerr G, Boutros M (2014) E-CRISP: fast CRISPR target site identification. Nat Methods 11(2):122–123. https://doi.org/10.1038/nmeth.2812
89. Naito Y, Hino K, Bono H, Ui-Tei K (2015) CRISPRdirect: software for designing CRISPR/Cas guide RNA with reduced off-target sites. Bioinformatics 31(7):1120–1123. https://doi.org/10.1093/bioinformatics/btu743
90. Cram D, Kulkarni M, Buchwaldt M, Rajagopalan N, Bhowmik P, Rozwadowski K, Parkin IAP, Sharpe AG, Kagale S (2019) WheatCRISPR: a web-based guide RNA design tool for CRISPR/Cas9-mediated genome editing in wheat. BMC Plant Biol 19(1):474. https://doi.org/10.1186/s12870-019-2097-z
91. Hu X, Cui Y, Dong G, Feng A, Wang D, Zhao C, Zhang Y, Hu J, Zeng D, Guo L, Qian Q (2019) Using CRISPR-Cas9 to generate semi-dwarf rice lines in elite landraces. Sci Rep 9(1):19096. https://doi.org/10.1038/s41598-019-55757-9
92. Meng X, Yu H, Zhang Y, Zhuang F, Song X, Gao S, Gao C, Li J (2017) Construction of a genome-wide mutant library in rice using CRISPR/Cas9. Mol Plant 10(9):1238–1241. https://doi.org/10.1016/j.molp.2017.06.006
93. Liu H-J, Jian L, Xu J, Zhang Q, Zhang M, Jin M, Peng Y, Yan J, Han B, Liu J, Gao F, Liu X, Huang L, Wei W, Ding Y, Yang X, Li Z, Zhang M, Sun J, Bai M, Song W, Chen H, Xa S, Li W, Lu Y, Liu Y, Zhao J, Qian Y, Jackson D, Fernie AR, Yan J (2020) High-throughput CRISPR/Cas9 mutagenesis streamlines trait gene identification in maize. Plant Cell 32(5):1397–1413. https://doi.org/10.1105/tpc.19.00934
94. Zhao Q, Feng Q, Lu H, Li Y, Wang A, Tian Q, Zhan Q, Lu Y, Zhang L, Huang T, Wang Y, Fan D, Zhao Y, Wang Z, Zhou C, Chen J, Zhu C, Li W, Weng Q, Xu Q, Wang Z-X, Wei X, Han B, Huang X (2018) Pan-genome analysis highlights the extent of genomic variation in cultivated and wild rice. Nat Genet 50(2):278–284. https://doi.org/10.1038/s41588-018-0041-z
95. Zhou Y, Chebotarov D, Kudrna D, Llaca V, Lee S, Rajasekar S, Mohammed N, Al-Bader N, Sobel-Sorenson C, Parakkal P, Arbelaez LJ, Franco N, Alexandrov N, Hamilton NRS, Leung H, Mauleon R, Lorieux M, Zuccolo A, McNally K, Zhang J, Wing RA (2020) A platinum standard pan-genome resource that represents the population structure of Asian rice. Sci Data 7(1):113. https://doi.org/10.1038/s41597-020-0438-2
96. Haberer G, Kamal N, Bauer E, Gundlach H, Fischer I, Seidel MA, Spannagl M, Marcon C, Ruban A, Urbany C, Nemri A, Hochholdinger F, Ouzunova M, Houben A, Schön C-C, Mayer KFX (2020) European maize genomes highlight intraspecies variation in repeat and gene content. Nat Genet 52(9):950–957. https://doi.org/10.1038/s41588-020-0671-9
97. Sansaloni C, Franco J, Santos B, Percival-Alwyn L, Singh S, Petroli C, Campos J, Dreher K, Payne T, Marshall D, Kilian B, Milne I, Raubach S, Shaw P, Stephen G, Carling J, Pierre CS, Burgueño J, Crosa J, Li H, Guzman C, Kehel Z, Amri A, Kilian A,

Wenzl P, Uauy C, Banziger M, Caccamo M, Pixley K (2020) Diversity analysis of 80,000 wheat accessions reveals consequences and opportunities of selection footprints. Nat Commun 11(1):4572. https://doi.org/10.1038/s41467-020-18404-w

98. Pont C, Leroy T, Seidel M, Tondelli A, Duchemin W, Armisen D, Lang D, Bustos-Korts D, Goué N, Balfourier F, Molnár-Láng M, Lage J, Kilian B, Özkan H, Waite D, Dyer S, Letellier T, Alaux M, Russell J, Keller B, van Eeuwijk F, Spannagl M, Mayer KFX, Waugh R, Stein N, Cattivelli L, Haberer G, Charmet G, Salse J, Wheat, Barley Legacy for Breeding Improvement c (2019) Tracing the ancestry of modern bread wheats. Nat Genet 51(5):905–911. https://doi.org/10.1038/s41588-019-0393-z

99. He F, Pasam R, Shi F, Kant S, Keeble-Gagnere G, Kay P, Forrest K, Fritz A, Hucl P, Wiebe K, Knox R, Cuthbert R, Pozniak C, Akhunova A, Morrell PL, Davies JP, Webb SR, Spangenberg G, Hayes B, Daetwyler H, Tibbits J, Hayden M, Akhunov E (2019) Exome sequencing highlights the role of wild-relative introgression in shaping the adaptive landscape of the wheat genome. Nat Genet 51(5):896–904. https://doi.org/10.1038/s41588-019-0382-2

100. Draper J, Mur LAJ, Jenkins G, Ghosh-Biswas GC, Bablak P, Hasterok R, Routledge APM (2001) *Brachypodium distachyon*;. A new model system for functional genomics in grasses. Plant Physiol 127(4):1539. https://doi.org/10.1104/pp.010196

101. Huala E, Dickerman AW, Garcia-Hernandez-M, Weems D, Reiser L, LaFond F, Hanley D, Kiphart D, Zhuang M, Huang W, Mueller LA, Bhattacharyya D, Bhaya D, Sobral BW, Beavis W, Meinke DW, Town CD, Somerville C, Rhee SY (2001) The Arabidopsis Information Resource (TAIR): a comprehensive database and web-based information retrieval, analysis, and visualization system for a model plant. Nucleic Acids Res 29 (1):102–105. https://doi.org/10.1093/nar/29.1.102

102. Zou C, Li L, Miki D, Li D, Tang Q, Xiao L, Rajput S, Deng P, Peng L, Jia W, Huang R, Zhang M, Sun Y, Hu J, Fu X, Schnable PS, Chang Y, Li F, Zhang H, Feng B, Zhu X, Liu R, Schnable JC, Zhu J-K, Zhang H (2019) The genome of broomcorn millet. Nat Commun 10(1):436. https://doi.org/10.1038/s41467-019-08409-5

# Chapter 2

# Mapping Quantitative Trait Loci in Wheat: Historic Perspective, Tools, and Methods for Analysis

**Raman Dhariwal and Harpinder Singh Randhawa**

## Abstract

Most of the economically important traits in crop plants, such as grain size and yield, are quantitative. Methods for studying the genetics of these traits in plants involve a combination of modern molecular techniques and powerful statistical methods. These methods have gradually evolved over the past several decades as the availability of new resources, such as polymerase chain reaction-based markers, increased. However, recent advances in genotyping technologies and genomic tools such as high-throughput single nucleotide polymorphism markers and ease in the development of mapping populations such as doubled haploids have brought not only an ever-increasing interest in QTL mapping studies but also some challenges. Most of the computational tools that revolutionized the field of QTL mapping in the late twentieth and early twenty-first centuries are now proved ineffective. The emergence of open-source bioinformatics tools, like statistical language R, has enabled the handling of large genotyping and phenotyping datasets easy and efficient. In this chapter, we discussed a historic prospective and basics of QTL mapping and described a step-by-step protocol of composite interval mapping with emphasis on the use of open-source platforms.

**Key words** Quantitative traits, Linkage and QTL mapping, Genetic maps, Genome, Circos diagram

## 1 Introduction

Most of the economically important traits in crop plants such as flowering time, drought and heat tolerance, resistance to diseases, and grain size and yield are quantitatively inherited [1, 2]. Quantitative traits have been a focal area of genetic studies for more than a century as they possess stunning phenotypic variation [3] and can be modulated for improvement of commercial crop cultivars. Around a decade after Mendel's experiments on peas [4], Galton was the first to establish initial concepts of the heritability of quantitative traits [5–7]. During the early days of research in this field, biologists used statistical methods mainly based on the means, variances, covariances, correlations, etc. of relatives, without any understanding of the genes involved [3]. Independent rediscovery

Andriy Bilichak and John D. Laurie (eds.), *Accelerated Breeding of Cereal Crops*, Springer Protocols Handbooks,
https://doi.org/10.1007/978-1-0716-1526-3_2, © Springer Science+Business Media, LLC, part of Springer Nature 2022

of Mendel's work, which was later known as the Mendelian theory of particulate inheritance, by Hugo Marie de Vries, Carl Franz Joseph Erich Correns, and Erich Tschermak von Seysenegg in 1900 was groundbreaking for genetic research. This was the time when significant developments in the field of cytology, including the discovery of chromosomes and principles of cell division, were established as a framework suitable for accommodating Mendel's "numbers and ratios" [8].

By the early twentieth century, many biologists working on heredity began to reapply Mendel's theory. In 1902, Theodor Boveri of the University of Würzburg and Walter Sutton of Columbia University found that chromosomes segregate in a way similar to Mendel's factors [9, 10]. They hypothesized that chromosomes carry heritable factors; this theory was popularized as the Boveri-Sutton chromosome theory. In the following years, classical work of Johannsen [11], Nilsson-Ehle [12], and East [13] on quantitative traits helped biologists in understanding inheritance of quantitative traits based on multiple-factor/gene hypothesis and environmental effects [14]. Nilsson-Ehle's famous red- and white-seeded wheat hybridization experiments laid the foundation for the famous multiple-factor hypothesis [12], and East's experiments on size inheritance in *Nicotiana* [13] explained better the continuous variation of quantitative characters [15] and confirmed numerous predictions of this theory [14]. These experiments showed that quantitative trait inheritance results from the segregation of multiple genetic factors modified by environmental effects. Almost at the same time as Johannsen and Nilsson-Ehle, Morgan and colleagues, while working on *Drosophila* white-eyed mutants, established the chromosome theory of inheritance in which Morgan explicitly stated that chromosomes carry heritable factors or genes [16–19]. Immediately, after Morgan's theory came to the horizon, Alfred Henry Sturtevant, an undergraduate student who was working with Morgan, used the proportion of crossing overs as an index of the distance among six factors/genes to linearly map them on X chromosome of fruit fly [20], becoming the first to develop chromosome linkage map of any organism.

In 1919, the British statistician Ronald Fisher presented the quantitative genetics theory [21], which was based on the assumption that the genetic variance in a population was due to a large number of Mendelian factors, each making a small additive contribution to a phenotype [22]. His work provided a solid framework for the study of phenotypic variation in populations that has prevailed to this day [22]. The pioneering genetic studies of Sax [23] and other succeeding scientists [24–27] showed that it was even possible occasionally to map putative quantitative trait loci (QTL) [28], earlier known as polygenes [29], to chromosomes [14]. These studies provided a solid background of theory, principles, and observations for further work [30] but were constrained by the limited number of markers, primarily major genes and morphological markers [23–25], or translocations [26] or even whole chromosomes/monosomics [27]. These methods were not only somewhat laborious but also difficult to apply in populations [3]. It was not until the 1970s when the catchy acronym "QTL" [28] appeared and liberated the subject from the earlier heavy statistical associations with polygenes [3], and the molecular advances really started to integrate into the genetics followed by a new era of the rapid technological advances [22].

Use of recombinant inbred line populations [31, 32] and large number of restriction fragment length polymorphism (RFLP) markers in different species [32, 14], followed by the development and use of relatively cheap and safe polymerase chain reaction (PCR)-based markers, revolutionized the quantitative genetics. The landmark discovery of interval mapping by Lander and Botstein [14], which overcame several disadvantages of traditional QTL analysis, a single marker analysis method based on either t-test or analysis of variance at marker loci, launched an avalanche of large-scale genome-wide QTL mapping studies [33] in different species. Due to recent advances in genotyping technologies, i.e., development of high-throughput and cost-effective markers such as genotyping-by-sequencing (GBS) and array-based single nucleotide polymorphism (SNP) markers, availability of different mapping populations, and computational tools, there has been an ever-increasing interest in QTL analysis of different traits in several species [34].

A number of QTL analysis methods are now available that accommodate the different types of traits and mapping populations as well as the objectives of the studies [34]. These methods can be broadly divided into three broad categories based on the principle and genetic material used for QTL analysis: (i) QTL mapping, QTL discovery using linkage analysis in a biparental population; (ii) association mapping or AM, also called genome-wide association studies or GWAS, QTL discovery using linkage disequilibrium or LD analysis in a set of diverse germplasm; and (iii) joint linkage association mapping or JLAM, QTL discovery using both linkage and LD analyses in several biparental populations simultaneously. Linkage-based QTL mapping methods can be further divided into the following categories and subcategories: (1) single marker analysis or SMA [35, 36] without the use of linkage map, (a) *t*-test method and (b) ANOVA method, and (2) interval mapping, (a) simple interval mapping or SIM [14], (b) composite interval mapping or CIM [37], (c) inclusive composite interval mapping or ICIM [38], (d) multiple trait mapping or MTM [39], (e) multiple interval mapping or MIM [40], and (f) Bayesian interval mapping or BIM [41]. A number of CIM variants such as expression QTL or eQTL analysis [42, 43], protein QTL or pQTL analysis [43, 44], and metabolite or mQTL analysis [44] have also been proposed and used to facilitate the objectives of genetical genomics [42–46], though these methods are largely similar to conventional CIM for any phenotypic trait.

The literature on QTL analysis theory and methods, particularly for crop plants, has also grown tremendously over the past four decades. Readers are advised to consult different review articles published elsewhere for further details about different QTL mapping methods [34, 36, 47–54]. A number of studies involving AM/GWAS and JLAM have also been published in different crop species but are beyond the scope of this chapter; however, a separate chapter on association mapping is presented in this book [55]. For further details about AM/GWAS and JLAM methods, readers are referred to detailed reviews [56–60].

Among all the methods of QTL mapping available so far, composite interval mapping has become the most popular method of choice among the scientific community [34]. In the early 1990s, Jansen [61], Zeng [62], and Zeng [37] proposed to combine conventional interval mapping, a method which uses a pair or two pairs of flanking markers at a time for mapping QTL on a chromosome, with multiple linear regression, to improve the accuracy of QTL prediction. The method was later named as "composite interval mapping" (CIM) [37]. This was a landmark development in QTL mapping followed by two more significant developments, the use of empirical threshold [63] and permutation tests [64] which reduced the chances of false-positive results and increased the QTL prediction efficiency [34]. These developments in interval mapping changed the way of QTL mapping that has prevailed to this day. However, where it is now easy to detect QTLs precisely, the recent developments in the field of marker technology, i.e., availability of GBS and array-based SNP markers, brought new opportunities for development of high-resolution maps along with the challenges of large data handling. Most of the computational tools that revolutionized the field of QTL mapping in the late twentieth and early twenty-first centuries such as MAPMAKER [65] proved ineffective after these developments. Thanks to the advancements and availability of open-source resources such as the statistical language R and many other Linux OS-supported programs, they are not only suitable for handling of large genotyping and phenotyping datasets that are common today but have also enabled new avenues for QTL discovery in different crop species in this omics era.

Bread wheat or common wheat, *Triticum aestivum* L., belongs to subtribe *Triticinae* and tribe *Triticeae* of the family *Poaceae*. Among crop plants, bread wheat (segmental allohexaploid) is the youngest (evolved ~10,000 years ago) polyploid species [66, 67] and has the largest genome size (~17 Gb), which is about sevenfold larger than the maize genome and ~37-fold larger than the rice genome. The wheat genome consists of ~80% repeated sequences and ~70% transposable elements. Bread wheat is an important crop for human consumption. In this chapter, we described a step-by-step experimental protocol of simple and robust "composite interval mapping" [37]. This method of QTL mapping uses a biparental homozygous DH mapping population and Illumina Infinium 90 K wheat SNP genotyping, without describing much of the statistics involved. Emphasis is given to use of open-source platforms for handling large genotyping and phenotyping datasets. A workflow of the same is presented in Fig. 1.

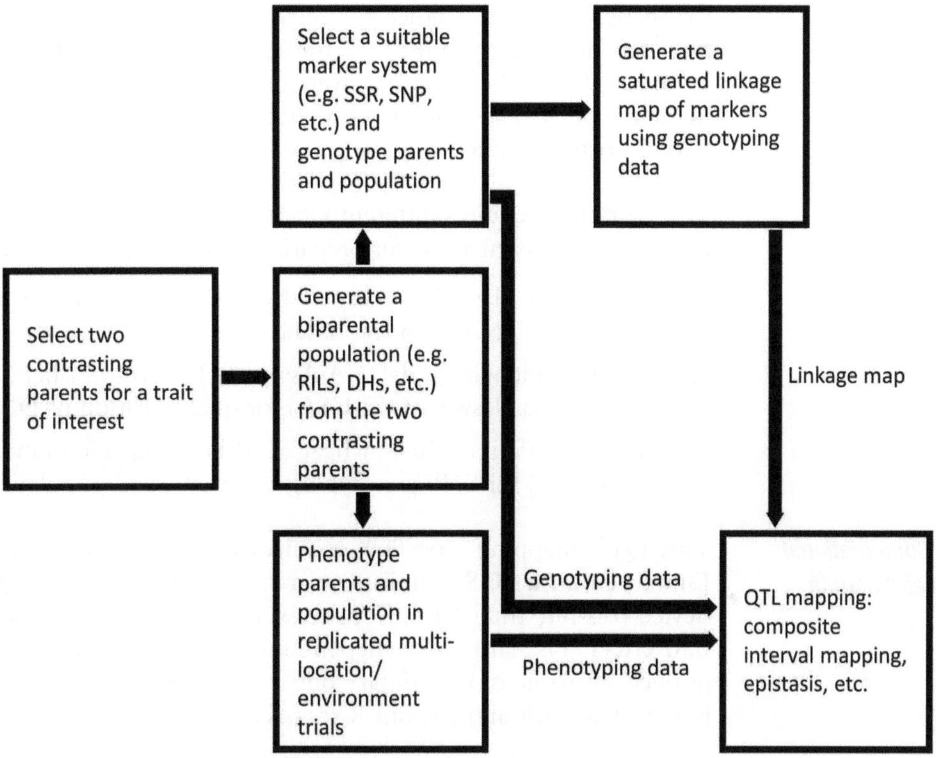


**Fig. 1** Overall workflow of quantitative trait loci (QTL) mapping. Firstly, contrasting parents are selected for a trait of interest. For example, if one wants to map disease resistance QTL(s) from a highly resistant line, this line should be crossed with a highly susceptible line to produce a large genetic mapping population containing a set of recombinant individuals such as recombinant inbred lines (RILs), doubled haploids (DHs), etc. The larger the number of lines that differ genotypically, the higher are the chances of generating a high-density map; however, both parental lines should be similar for most of the traits, except the target trait, for precise QTL mapping. Secondly, population and parents should be genotyped with a suitable marker system such as simple sequence repeats (SSRs), single nucleotide polymorphisms (SNPs), etc. followed by generation of a saturated linkage map using genotyping data of population. The third step involves phenotyping for a trait of interest in replicated trails in multiple environments (multilocation-year trials). Phenotyping can also be done for more than one trait of interest. Linkage mapping can be done while doing phenotyping since phenotyping takes a considerable amount of time. Finally, QTL mapping is carried out using the genotyping data, linkage map, and phenotyping data

## 2 Materials

### 2.1 Plant Material

A QTL mapping project typically requires the use of 50–300 individual lines derived from a biparental cross between two contrasting parents for a trait of interest [68]. However, a mapping population of 200–300 individuals is required for precise mapping of major effect QTLs and ~ 500 individuals for minor effect QTLs [69]. Different types of mapping populations such as second filial generation ($F_2$), backcross (BC), doubled haploids (DHs), recombinant inbred lines (RILs), and near isogenic lines (NILs) can be produced for mapping QTLs. Choice for selection of mapping population type depends mainly on the time frame of the project and resources available. Two other factors, project objectives and availability of marker type (i.e., dominant vs. codominant) for genotyping, are also important (*see* **Note 1**).

### 2.2 Kits and Reagents

This workflow requires different kits mentioned below. Follow the manufacturer instructions for preparing solutions (*see* **Note 2**):

1. DNeasy 96 Plant Kit (Qiagen Inc., Valencia, CA, USA) for isolation of DNA from leaf tissues.
2. Quant-iT PicoGreen dsDNA Assay Kit (Thermo Fisher Scientific Inc., Bartlesville, OK, USA) for quantification of DNA.
3. Illumina Infinium 90 K wheat SNP beadchip (Illumina, San Diego, CA) for SNP genotyping of parents and population.

### 2.3 Computational Tools and Scripts

This QTL mapping workflow requires the use of Windows and Linux (Ubuntu) OS running on a high-performance computer device (64-bit, Intel Core-i7 processor, at least 16GB RAM and 250-GB–1 TB hard drive, depending on the project). A set of publicly available or proprietary but free-to-use statistical and bioinformatics tools and custom R scripts are needed for the analysis:

1. GenomeStudio 2.0 (https://support.illumina.com/array/array_software/genomestudio/downloads.html) is a proprietary but free-to-use tool by Illumina to visualize and analyze SNP genotyping data generated by Illumina's platforms. Genotyping module of GenomeStudio allows data analysis of across five million markers and probes and can detect sample outliers.
2. MSTmap software (version 2.0) (http://mstmap.org/) is a software tool that is capable of constructing genetic linkage maps of ~100,000 markers efficiently using the minimum spanning tree algorithm [70].
3. MapDisto software (version 1.7.7.011) (http://mapdisto.free.fr/) is a flexible user-friendly program for mapping large numbers of markers such as those generated from GBS and array-based genotyping platforms. It runs on Windows and macOS

and has various features such as direct importation and conversion of Variant Call Format (VCF) files, data imputation on VCF files, and QTL detection, in addition to linkage mapping [71].

4. Windows QTL Cartographer (version 2.5) (https://brcwebportal.cos.ncsu.edu/qtlcart/index.php) is a user-friendly suite of programs to map QTLs on chromosomes. It performs various tasks, including simulation, reformatting, or analyzing data and visualizing the results [62, 37].
5. QTL Network (version 2.0) (http://ibi.zju.edu.cn/software/qtlnetwork/) is a user-friendly Windows-based software program for mapping QTLs with QxQ interaction (epistatic) effects, QxE interaction effects, and higher-order interactions in various populations, and for graphical presentation of QTL mapping results [72].
6. BLAST+ executables (https://www.ncbi.nlm.nih.gov/books/NBK279671/) is a suite of command-line tools to run BLAST. It consists of a number of utilities to perform BLAST searches, compare sequences, and identify homologous genomic sequences.
7. R (version 3.6.1) (https://www.r-project.org/) is a programming language and free software environment for statistical and graphic analyses supported by the R Foundation for Statistical Computing. It runs on various OS including Windows, macOS, and different UNIX platforms [73].
8. RStudio (version 1.2.5001) (https://rstudio.com/) is an integrated development environment (IDE) for R. It provides free and open-source tools for R.
9. R/qtl (version 1.46-2) (https://rqtl.org/) is an extensible, interactive environment for mapping QTLs in experimental crosses. It is implemented as an add-on R package. It is freely distributed as source code for Unix or compiled code for Windows or macOS.
10. OmicCircos (version 1.14.0) (http://bioconductor.org/packages/OmicCircos/) is an R application and Bioconductor package for generating high-quality circular plots for omics data [74]. It is used to generate genome-wide QTL plot in this protocol.
11. Shell and in-house R scripts are written to perform various tasks associated with linkage mapping (using MSTmap), QTL mapping (using R/qtl), and graphics generation (using OmicCircos).

## 3 Methods

The basic workflow for mapping QTLs (using composite interval mapping approach) and epistasis interaction in a biparental wheat population is presented in Fig. 1. This workflow can be divided into four main steps: selection of parents and generation of homozygous mapping population, phenotyping of parents and population, genotyping and linkage mapping, and QTL mapping.

### 3.1 Selection of Parents and Generation of Homozygous Mapping Population

For mapping of the QTL(s) from a wheat line or cultivar, one should cross a line/cultivar that possesses the trait of interest, e.g., disease resistance, with a line or cultivar that is lacking it. Generally, any parent can be selected as male or female (*see* **Note 3**). A suitable mapping population can be developed moving forward from the $F_1$ plants. While $F_2$ populations are constructed by selfing $F_{1s}$, BC populations are generated by crossing the $F_{1s}$ with one of the parents. Continuous selfing of individual $F_2$ plants and then their progenies individually using single seed descent (SSD; one seed from each line is used to grow the next generation) method for 7–8 generations allows construction of a series of homozygous RILs each containing a unique combination of chromosomal segments from the original parents [69]. DHs can be produced by chromosome doubling using colchicine treatment of haploids produced using a suitable technique such as anther culture [75], microspore culture [76, 77], crossing with *Hordeum bulbosum* (bulbosum technique) [78, 79], crossing with *Zea maize* L. (wheat-maize pollination technique) [80, 81], etc. However, the production of DH population needs expertise in tissue culture and more resources but considerably less time.

### 3.2 Phenotyping of Parents and Population

Both parents and whole population should be phenotyped in replicated (at least twice) trials/experiments in multiple (at least three) environments (year/location/both) for the target trait(s). It is always recommended to include some check cultivars. Parents and check cultivars should be repeated within replication. If the target trait is a disease or resistance to a disease, then phenotyping should be done in the disease nurseries with an appropriate disease pressure created after pathogen inoculations.

Phenotyping data can be scored as nominal (e.g., colors—green, white, yellow, purple, etc.), ordinal (e.g., a disease rating scale of 0–5, where 0 being immune and 5 highly susceptible, AUDPC, etc.), or metric (e.g., a percentage value such as grain protein percent, grain yield in kilograms, thousand kernel weight in grams, etc.) depending on the trait(s) of interest. Thus, ordinal and metric traits are always represented by numbers while nominal traits by alphanumeric strings. However, in most of the QTL mapping studies, data are recorded as metric. It is also necessary to record phenotypic data at an appropriate developmental stage. If a treatment such as drought or pathogen inoculation is given, it is necessary to wait until proper symptoms appear. Missing data can be shown by a period "." or a hyphen "-".

### 3.3 Genotyping and Linkage Mapping

Genotyping of parents and population can be done using different marker systems such as simple sequence repeats (SSRs), SNPs, etc. However, heterozygous populations such as $F_2$ and BCs should be genotyped with a codominant marker system such as SSRs. Homozygous populations such as RILs and DHs can be genotyped with any marker system. In this protocol, the genotyping method used by Dhariwal et al. [82] and Dhariwal et al. [83] is discussed. Genotyping example datasets can be downloaded from www.mdpi.com/2073-4425/9/1/19/s1.

#### 3.3.1 Genotyping

1. Isolate DNA from leaf samples of wheat plants (ideally 10 days old but can be collected at any developmental stage before physiological maturity of plants) using DNeasy 96 Plant Kit (Qiagen Inc., Valencia, CA, USA) following manufacturer instructions.
2. Quantify DNA samples using Quant-iT PicoGreen dsDNA Assay Kit (Thermo Fisher Scientific Inc., Bartlesville, OK, USA) following manufacturer instructions.
3. Genotype DNA samples using Illumina Infinium 90 K wheat SNP beadchip (Illumina, San Diego, CA) following manufacturer instructions. The following raw data files will be produced from genotyping: bead ID and sequence file "*.bpm," intensity file "*.idat," sample info file "*.csv," and cluster file "*.egt." These files should be saved on Windows workstation having GenomeStudio installed.

#### 3.3.2 Linkage Mapping

1. Download and install GenomeStudio 2.0 (Illumina, San Diego, USA) software. After installation, open the program and go to "File" menu to create a new genotyping project. Click on "New Project" option followed by "Genotyping." A project creation wizard will open; click "Next." Select a folder where you want to save your results and "Create" option. Give a name to your project using the "Create" option. Then again click "Next" followed by selecting the first option "Use sample sheet to load sample intensities." Click "Next" and then open the raw data files (bead ID and sequence file "*.bpm," intensity file "*.idat," sample info file "*.csv," and cluster file "*.egt") by browsing. Click "Next" followed by clicking on "Finish." GenomeStudio will create a GS workspace file ".bsc" from the raw data files specified in the previous step, and the project will be loaded. Select "No" option on pop-up window to save time. After data is loaded, click on "Sample Table" followed by

pressing the "calculate" button to calculate Call Rate. Then click on SNP Table. Browse the "Analysis" tab and click "Update SNP Statistics." To export high-quality SNP data, set the following logical expression filters for different single variable matrices: (i) Call Freq: >0.50, (ii) Minor Allele Freq: >0.03, (iii) AA Freq:! = 1, (iv) AB Freq:! = 1, and (v) BB Freq:! = 1. Press "control + S" to save project or save the project by navigating the "File" menu followed by "Save project." SNP data can be exported by navigating "Analysis" tab followed by "Reports" and "Report Wizard," in that order. On pop-up window, select "Final report" and click "Next." On the next screen, select "Standard," and then choose columns you want to have in SNP genotyping data file. Also select a grouping option from—"sample" or "SNP" and an output format from—"Tab" for tab delimited file or "Comma" for comma-delimited file. Click "Next" and browse a folder in the next window to save the report. Also change the name, if required, and click "Finish."

2. Data file exported from GenomeStudio will have SNP data in format similar to given in Table 1.

   You can remove extra columns between SNP names and genotype calls (shown by ".." in this example).

3. Open the data file in Microsoft Excel and sort SNP genotyping data by column of parent 1 genotype (i.e., P1.GType in this

**Table 1**
**Example format of genotyping data output file of GenomeStudio software**

| Index | Name | .. | P1. GType | P2. GType | L3. GType | L4. GType | L5. GType | L6. GType | L7. GType | .. | Ln. GType |
|---|---|---|---|---|---|---|---|---|---|---|---|
| 1556 | SNP1 | .. | AA | AA | AA | AB | AB | AB | AB | .. | AA |
| 2271 | SNP2 | .. | AB | AA | AA | AA | AA | AA | AA | .. | AA |
| 2373 | SNP3 | .. | AB | AA | AA | AB | AB | AB | AB | .. | AA |
| 3039 | SNP4 | .. | BB | AA | AA | AA | AA | AA | AA | .. | AB |
| 4468 | SNP5 | .. | AA | BB | AA | AA | AB | AB | AA | .. | AA |
| 5806 | SNP6 | .. | AB | AA | AA | AA | AB | AA | AB | .. | AA |
| 6309 | SNP7 | .. | AA | AB | AA | AB | AA | AA | NC | .. | AA |
| 6356 | SNP8 | .. | AA | AB | AA | AA | AA | AA | AA | .. | NC |
| 6513 | SNP9 | .. | NC | AA | AA | AA | AA | AA | AA | .. | AA |
| .. | .. | .. | .. | .. | .. | .. | .. | .. | .. | .. | .. |
| n | SNPn | .. | AA | BB | AA | AB | AB | AB | AB | .. | AA |

**Table 2**
**Example format of genotyping data sorted using parental profiles**

| Index | Name | .. | P1. GType | P2. GType | L3. GType | L4. GType | L5. GType | L6. GType | L7. GType | .. | Ln. GType |
|---|---|---|---|---|---|---|---|---|---|---|---|
| 1556 | SNP1 | .. | AA | AA | AA | AB | AB | AB | AB | .. | AA |
| 6309 | SNP7 | .. | AA | AB | AA | AB | AA | AA | NC | .. | AA |
| 6356 | SNP8 | .. | AA | AB | AA | AA | AA | AA | AA | .. | NC |
| 4468 | SNP5 | .. | AA | BB | AA | AA | AB | AB | AA | .. | AA |
| n | SNPn | .. | AA | BB | AA | AB | AB | AB | AB | .. | AA |
| 2271 | SNP2 | .. | AB | AA | AA | AA | AA | AA | AA | .. | AA |
| 2373 | SNP3 | .. | AB | AA | AA | AB | AB | AB | AB | .. | AA |
| 5806 | SNP6 | .. | AB | AA | AA | AA | AB | AA | AB | .. | AA |
| 3039 | SNP4 | .. | BB | AA | AA | AA | AA | AA | AA | .. | AB |
| 6513 | SNP9 | .. | NC | AA | AA | AA | AA | AA | AA | .. | AA |

example) and then by parent 2 (i.e., P2.GType in this example) using the sort function. After sorting, data will look similar to that presented in Table 2.

4. Remove the monomorphic SNPs by looking at both parental genotypes and deleting row(s) in data table. The first SNP is monomorphic in example dataset shown in Table 2.
5. Then select a set of SNP rows with parent 1 genotype of "AA" and parent 2 of "AB" (second and third SNPs in Table 2), and replace all "AA" genotype calls with "A" and "AB" with "B" using the replace function. Select the next set of SNPs with parent 1 genotype of "AA" and parent 2 of "BB" (fourth and fifth SNPs in Table 2), and replace all "AA" genotype calls with "A" and "BB" with "B" using replace function. If the population is heterozygous, then replace any other genotype (such as "AB" in this case) call with "H," otherwise by a hyphen (-). Complete this procedure for the rest of the SNPs, but always remember to assign "A" and "B" to parents 1 and 2, respectively. Replace all "NC" genotypes with a hyphen (-), but if "NC" allele is polymorphic between parents and segregating in the population in an expected ratio (e.g., 1:1 in case of DH population), then assign it to "A" or "B" allele, based on parental profile. After replacing all genotype calls, data will look similar to that presented in Table 3 and is ready for further analysis.
6. To check segregation distortions, download the MapDisto software from http://mapdisto.free.fr/Download_Soft/. Extract downloaded files and save in the documents folder;

**Table 3**
**Example format of genotyping data file suitable for linkage mapping using MapDisto software**

| Index | Name | .. | P1. GType | P2. GType | L3. GType | L4. GType | L5. GType | L6. GType | L7. GType | .. | Ln. GType |
|---|---|---|---|---|---|---|---|---|---|---|---|
| 6309 | SNP7 | .. | A | B | A | B | A | A | – | .. | A |
| 6356 | SNP8 | .. | A | B | A | A | A | A | A | .. | – |
| 4468 | SNP5 | .. | A | B | A | A | – | – | A | .. | A |
| n | SNPn | .. | A | B | A | – | – | – | – | .. | A |
| 2271 | SNP2 | .. | A | B | B | B | B | B | B | .. | B |
| 2373 | SNP3 | .. | A | B | B | A | A | A | A | .. | B |
| 5806 | SNP6 | .. | A | B | B | B | A | B | A | .. | B |
| 3039 | SNP4 | .. | A | B | B | B | B | B | B | .. | – |
| 6513 | SNP9 | .. | A | B | B | B | B | B | B | .. | B |

no installation is required. Open the MapDisto folder followed by MapDisto file (*.Win file), and press enable editing and contents, if required. Press the start button on the main screen followed by "Data" button on the second screen. On the data screen, press "Clear Data" button to make space for another dataset. Paste genotyping data in clear cells while names/IDs of SNPs (as rows) and genotypes (as columns) in yellow-highlighted cells (*see* **Note 4**). Fill genotyping data parameters (e.g., population type, DH, RILs, etc.; population size, number of individual lines, number of loci/markers and traits, if any) and the codes (homozygote parents 1 and 2, heterozygote, missing, etc.). In this example, homozygote parents 1 and 2 are represented by "A" and "B," respectively, heterozygote and missing by a hyphen (-). After completing these steps, read the data by pressing "Read data" button. Then press "Main menu" button to go to the first page followed by pressing "Segregation $\chi^2$s." Press "Use all loci" on pop-up window. $\chi^2$ results will be shown in a Table form; column 5 of this table contains "ns" or "*" to indicate whether segregation distortion is nonsignificant or significant, respectively. Significance level can be shown by >1*, depending on the level of segregation distortion. Save a copy of SNP data file and delete significantly distorted SNPs (whole rows) by looking at significance level. It is always better to keep just expectedly segregated (ns in this example) markers; however, if only very few markers are available, one can decide to keep markers showing slightly significant segregation distortion. After filtering SNPs that showed significant segregation distortion, data is ready for linkage mapping.

7. We recommend the use of MSTmap to construct the framework linkage map. Source code of MSTmap can be downloaded by visiting http://mstmap.org/download.html. Alternatively, execute the following command on your Linux machine (*see* **Note 5**):

```
git clone https://github.com/ucrbioinfo/MSTmap
```

To compile, extract the source code and then navigate to MSTmap directory. Execute the following commands (*see* **Note 6**):

```
cd MSTmap
make
```

To construct a linkage map using MSTmap, arrange all the marker/genotype data in a single Excel file in the format/example given in Table 4. To prepare a data file, first provide values to 12 different parameters in the first 12 rows; then after leaving a blank row, paste the genotyping data. The first row of genotyping data represents the numbers/IDs of the mapping genotypes/lines, while the first column represents the names/IDs of the genetic markers. IDs can be written as alphanumeric strings but without a space.

Parent 1 and 2 alleles in population should be written as "A" or "a" and "B" or "b," respectively. Heterozygotes, as in case of RILs, can be written with "X." Missing can be written either as a hyphen (-) or an "U." A complete description of different parameters and input file can be found on http://mstmap.org/download.html. After providing all the values, save a copy of the input file as a Tab-delimited text file (e.g., inputfile.txt) to the MSTmap directory/folder.

To construct the linkage map, descend into the MSTmap directory and execute the following command:

```
./MSTmap inputfile.txt outputmap.txt
```

In the above command, the input file name should be the same as provided for the genotyping data file. Output file name can be anything as shown in the above command. MSTmap can also be used online at http://www.mstmap.org/.

Output file of MSTmap is self-explanatory. Constructed linkage groups (LGs) are represented by header "; BEGINOFGROUP " and footer "; ENDOFGROUP." First and second columns of a LG represent marker IDs and distance in cM, respectively.

**Table 4**
**Example format of input data file for MSTmap**

| | | | | | | | | | | | |
|---|---|---|---|---|---|---|---|---|---|---|---|
| population_type | DH | | | | | | | | | | |
| population_name | Parent1 × Parent2 | | | | | | | | | | |
| distance_function | kosambi | | | | | | | | | | |
| cut_off_p_value | 0.000001 | | | | | | | | | | |
| no_map_dist | 15 | | | | | | | | | | |
| no_map_size | 2 | | | | | | | | | | |
| missing_threshold | 0.25 | | | | | | | | | | |
| estimation_before_clustering | no | | | | | | | | | | |
| detect_bad_data | yes | | | | | | | | | | |
| objective_function | ML | | | | | | | | | | |
| number_of_loci | 26 | | | | | | | | | | |
| number_of_individual | 11 | | | | | | | | | | |
| locus_name | L1 | L2 | L3 | L4 | L5 | L6 | L7 | L8 | L9 | L10 | Ln |
| SNP1 | A | B | A | B | A | A | A | B | A | B | A |
| SNP2 | B | A | B | B | A | B | A | B | A | A | B |
| SNP3 | B | B | A | B | A | A | A | B | A | B | A |
| SNP4 | A | B | A | B | A | A | A | B | A | B | A |
| SNP5 | B | B | A | B | A | A | A | B | A | B | A |
| SNP6 | B | B | A | B | A | A | A | B | A | B | A |
| SNP7 | B | B | A | B | B | A | A | A | B | A | B |
| SNP8 | A | B | A | B | B | A | A | A | B | A | B |
| SNP9 | B | B | A | B | B | A | A | A | B | A | B |
| SNP10 | B | B | A | B | B | A | A | A | B | A | B |
| SNP11 | B | B | A | B | B | A | A | A | B | A | B |
| SNP12 | A | B | – | B | B | A | A | A | B | A | B |
| SNP13 | B | B | B | A | A | A | A | B | B | B | A |
| SNP14 | B | B | A | B | B | A | A | A | B | A | B |
| SNP15 | B | B | A | B | B | A | A | A | B | A | B |
| SNP16 | B | B | A | B | B | A | A | A | B | A | B |
| SNP17 | B | B | A | B | B | A | A | A | B | A | B |
| SNP18 | A | A | B | B | B | A | A | A | B | A | B |
| SNP19 | B | A | B | B | B | A | A | A | B | A | B |

(continued)

**Table 4 (continued)**

| | | | | | | | | | | | |
|---|---|---|---|---|---|---|---|---|---|---|---|
| SNP20 | B | B | B | A | A | B | A | B | B | B | A |
| SNP21 | A | A | B | B | B | A | A | A | B | A | B |
| SNP22 | B | A | B | B | B | A | A | A | B | A | B |
| SNP23 | B | B | A | B | B | A | A | A | – | A | B |
| SNP24 | B | B | A | B | B | A | A | A | B | A | B |
| SNP25 | A | A | A | A | A | B | A | B | B | B | A |
| SNPn | A | B | A | B | B | A | A | A | B | A | B |

8. To assign a LG to a particular chromosome, one can search IDs of the markers in previously published map, if available. Otherwise, if markers are new such as those generated using GBS or sequencing of SSR-enriched libraries, one can BLAST search the marker sequences against the wheat genome. Both procedures are described separately step-by-step.

   *Using published map:* Open your map file, which you generated using MSTmap, in Excel and save in this format. Then download a recent wheat consensus map such as the one published by Wang et al. [84] from https://onlinelibrary.wiley.com/doi/full/10.1111/pbi.12183. After download, open the map file (Excel) and rename the Excel sheet tab to a short and simple name such as "90K" and save. Then right-click on renamed (e.g., 90K) tab and select move followed by selecting the name of MSTmap-generated file saved in Excel format from the drop-down menu and pressing "OK." This will move 90K sheet to the same Excel file/book which has the map you generated.

   Open the file that contains both maps and then the sheet/tab that contains the MSTmap-generated map. Then click in a cell such as the one in the third column against a marker of a first linkage group. Type the following formula: =VLOOKUP (A11,'90K'!C:D,2,FALSE) in that cell. This formula has five changeable parameters/expressions. A11 is a cell ID of the first marker of the first LG and can be changed to a different cell ID, depending on the cell that has a marker name to be searched in the reference/consensus map. MSTmap by default generates marker names in column A; thus it can be a cell number of column A. 90K is the tab name for the Excel sheet that has consensus map. In C:D, C represents the column that has marker names in 90K sheet, while D represents the column that has the chromosome names. 2 represents the number difference between the two selected columns, in this case C:D. As we want the exact match, FALSE should be selected.

Select the cell containing formula, and drag the formula down to the last marker of the last linkage group. It will generate/show the chromosome names of all the markers that are common between the two maps, MSTmap-generated and the published consensus map. In the next step, type the same formula in the next column and replace D with H and 2 with 6. H is a column in 90K file that contains rescaled marker positions (cM) that we want to compare with our map. Thus, 6 is the number difference between columns C and H. Also drag-down this formula until the last marker of the last linkage group. Now, in addition to MSTmap-generated map positions (cM) of all LGs, you will have chromosome names and map positions (cM) against every common marker in third and fourth columns, respectively. These chromosome names and positions are from consensus map. All the LGs are now assigned to a particular chromosome. One can compare map positions of MSTmap-generated LGs with LGs/chromosomes in consensus map. If the order of markers within MSTmap-generated LGs and consensus map LGs is similar, then this map can be saved for QTL analysis.

*Using similarity search*: This method utilizes BLAST+ program on Linux machine (Ubuntu in this case). To install the BLAST+ program, execute the following command (*see* **Note 7**):

```
sudo apt install ncbi-blast+
```

Or download BLAST+ suite from this ftp directory: https://ftp.ncbi.nlm.nih.gov/blast/executables/LATEST/. Right-click on the latest version (e.g., ncbi-blast-2.11.0+-x64-linux.tar.gz), and select "Save link as…" from the pop-up window. In the prompt, change to a preferred directory/folder, and click the "Save" button to save the selected archive to a preferred location on your Linux machine.

Extract the downloaded package using the following tar command:

```
tar zxvpf ncbi-blast-2.11.0+-x64-linux.tar.gz
```

or by a combination of following gunzip and tar commands:

```
gunzip -d ncbi-blast-2.11.0+-x64-linux.tar.gz
tar xvpf ncbi-blast-2.11.0+-x64-linux.tar
```

These commands will generate a new ncbi-blast-2.11.0+ directory under the working directory. A bin subdirectory under newly generated directory contains all BLAST+

programs. Configure the BLAST+ program by executing the following command in bash:

```
export PATH=$PATH:$HOME/ncbi-blast-2.11.0+/bin
```

Create a directory to store BLAST databases by executing the following command:

```
mkdir $HOME/blastdb
```

Execute the following command under bash to set the environment variable for created database storage directory:

```
export BLASTDB=$HOME/blastdb
```

After installation and configuration, BLAST+ can be used for similarity search analysis.

For assigning chromosomes to the generated LGs, first of all, prepare a text file containing sequences of all the mapped markers in the FASTA format and save it as the markers.fsa under "blastdb" directory.

Browse the https://urgi.versailles.inra.fr/download/iwgsc/IWGSC_RefSeq_Assemblies/v2.0/, and right-click on iwgsc_refseqv2.0_all_chromosomes.fa.zip file and select "Save link as. . ." from the pop-up window. In the prompt, change to "blastdb" directory and click the "Save" button to save the selected archive. Note that registration, user agreement, and sign-in may be necessary for downloading wheat IWGSC RefSeq v2.0 (*see* **Note 8**).

Then decompress these files using the following command:

```
unzip iwgsc_refseqv2.0_all_chromosomes.fa.zip
```

In the next step, create local database from FASTA file using the following command:

```
makeblastdb -in all_chromosomes.fa -dbtype nucl -out WheatDb -
title "IWGSC_RefSeq_v2.0" -parse_seqids
```

Then execute the following command to perform the BLAST search of marker sequences:

```
blastn -db WheatDb -word_size 7 -query markers.fsa -task
blastn -dust no -evalue 1e-6 -perc_identity 100 -outfmt 7 -
num_alignments 1 -out results.out
```

Open the results.out file in Microsoft Excel and sort the BLAST results according to subject ID and subject start site. As only one alignment will be allowed, file can easily open in Microsoft Excel. Delete all unnecessary information. Use VLOOKUP formula, as described in previous steps for published consensus map, to extract the chromosome name and genomic position of all the markers present in MSTmap-generated map and physically located on reference genome using BLAST search.

9. If the order of the marker within a LG is not similar (flipped or inaccurate) to the order in a published map/reference genome or there are more than one LGs for one chromosome, then we recommended to refine these LGs using MapDisto program (*see* **Note 9**). However, in this case, it is also recommended to recalculate all the LGs using the same program, i.e., MapDisto.

   We recommend using MapDisto for recalculating all LGs individually. To do the same, copy the genotyping data, along with marker and mapping population lines IDs, for all the markers belonging to a distorted LG or > 1 LGs related to a single chromosome. These can be copied on the basis of chromosome information obtained in the previous **step 8** using a published consensus map or similarity search. Paste the copied genotyping data onto data window in MapDisto, fill different parameters, and read the data as described in **step 6** of this section. After reading, go to main menu and press "Find linkage groups" button. Press "Use all loci" on the prompt followed by "OK" on the next prompt to calculate LGs at default parameters (LOD min: 3, r max: 0.3). Press "Yes" on the next step for filtering marker loci with lots of missing data. Fill the minimum 10% of missing data required to filter marker loci and hit "OK." Again, press "OK" to accept missing data adjustment made, if any. This will take some time for calculating the LG(s). If it still produces >1 LGs, gradually less stringent cutoff values (r max >0.3) should be tried. In case there is a large (cM max ≥50) gap between markers, which can be seen under the title of calculated LGs on the results page, they should be presented separately. Accepted calculated LG (s) should be added to main menu by pressing "Add groups." Go to main menu and press "AutoOrder" followed by the "AutoCheckInversions" and "AutoRipple," in that order but after completion of the previous step.

   Go to options on main menu by pressing "Options. . . ." On the prompt, change the "Format for map exportation" to "Win QTL Cart" and hit "OK." Press "Write map" on the main menu to export the LG(s). Press the "Export Map" button on the opened map page. Change the file name to an Excel format such as 1A_map.xlsx and press "Save" on the prompt. It will save the results to MapDisto directory by default but can be changed to any location before pressing "Save" on the previous prompt window. MapDisto map file (e.g., 1A_map.xlsx) contains three sheets/tabs; open "ChrDat" which contains the LG(s) maps. Compare the calculated LG(s) with a previously published consensus map or a reference genome, depending on the marker system utilized for genotyping. Comparison can be done following the procedure described in **step 8** of this section. Sometimes the mapping software generates inverted LGs which can be observed by looking at marker order as described in **step 8**. If the LG is generated in the inverted orientation, then go to the main menu of MapDisto, and provide the sequence (LG#) of LG in a cell against *Current working linkage group*. Then press the "Flip group..." and export the LG as described above. Recalculate all the LGs as described above.

10. Sometimes the mapping software generates spurious LGs (with unexpected marker order) due to factors like low recombination on a chromosome, etc. Marker orders can be guided by a physical map; however, it should be done only when parents of the genetic mapping populations are highly related to one which was used to produce a physical map, e.g., Chinese Spring. And, also if there is no other option available and one has to work with a particular chromosome. Genetic/linkage distances between markers ordered by physical map can be calculated later on using R/qtl.

    We recommend to follow the procedure given in **step 8** of this section (*Using similarity search*) to obtain the marker order based on the genome sequence. However, one should restrict to the physical map of a particular chromosome, rather than whole genome as described above, to save time. Using this procedure, genomic positions of all the markers will be obtained which should be utilized to sort the markers and genotyping data. Convert the physical positions (Mbp) to genetic positions (cM) assuming a constant recombination rate, or just enter physical map positions and re-estimate the marker distances using est.map() function of R/qtl. See the following example of the R/qtl input SNP data file (Table 5) sorted according to the physical map positions. Prepare a similar input file in Excel and save in the working directory as csv file (e.g., genodata.csv).

    The example of R/qtl input file in Table 5 has physical distances based SNP map and genotypic data of chromosome 1. While genotypes/individuals (ids) are represented as rows and markers as columns, the first row represents the marker IDs, the second chromosome number, and the third physical distances.

**Table 5**
**Example format of input marker data file for R/qtl**

| id | SNP1 | SNP2 | SNP3 | SNP4 | SNP5 | SNP6 | SNP7 | SNP8 | SNP9 | SNPn |
|---|---|---|---|---|---|---|---|---|---|---|
| | 1 | 1 | 1 | 1 | 1 | 1 | 1 | 1 | 1 | 1 |
| | 10.0 | 1230.0 | 1701.0 | 11021.0 | 17001.0 | 23452.0 | 430012.0 | 920345.0 | 1303534.0 | 1783673.0 |
| id1 | B | B | B | B | B | B | B | B | B | B |
| id2 | B | A | B | B | B | B | B | B | B | B |
| id3 | A | B | A | A | A | A | A | – | A | A |
| id4 | B | B | B | B | B | B | B | B | B | B |
| id5 | A | A | A | A | A | A | B | B | B | B |
| id6 | A | B | A | A | A | A | A | A | A | A |
| id7 | A | A | A | A | – | A | A | A | A | A |
| id8 | B | B | B | B | B | B | A | A | A | A |
| idn | A | A | A | A | A | A | B | B | B | B |

To estimate the genetic distances, we need to use the R/qtl package of R. Download the R installer from https://mirror.rcg.sfu.ca/mirror/CRAN/ or any other CRAN website and the RStudio Desktop installer from https://rstudio.com/products/rstudio/download/#download. Then install R first followed by RStudio Desktop by accepting default settings. Open the RStudio from the desktop icon, and click on "Packages" tab on the top of a small window on lower right. Click on "Install" button and write "qtl" on the "Packages" line. Also, click on "Install dependencies" and hit "Install" to install the R/qtl. Alternatively, use the following command in the console to install R/qtl package:

```
install.packages("qtl")
```

Change the working directory by navigating the "Session" tab and selecting "Set Working Directory" followed by "Choose Directory..." and a folder/directory on the prompt. Alternatively, use the following command in the console:

```
setwd("C:/Users/Path/to/Directory")
```

Load the R/qtl package by typing the following command:

```
library(qtl)
```

Read and attach the csv input data file (e.g., genodata.csv) by typing the following commands:

```
genodata <- read.cross("csv", ".", "genodata.csv") #to load a
datasets into R
attach(genodata) #to attach the data file
```

Check the data file using the following commands:

```
summary(genodata)
summary.cross(genodata)
```

Estimate the genetic distances between markers using the following command:

```
newmap <- est.map(genodata, error.prob=0.005)
```

Replace the old (physical map based) distances with genetic distances calculated in the last step by using the following command:

```
genodata <- replace.map(genodata, newmap)
```

Check the new map by using the following command:

```
summaryMap(genodata)
```

Plot the map by using the following command:

```
plotMap(genodata, show.marker.names=TRUE)
```

Save the re-estimated map using the following command:

```
write.csv(genodata,'map_reestimated.csv')
```

Linkage map generated in the above steps now can be used for QTL mapping.

### 3.4 QTL Mapping

*Main-effect QTL mapping*: Phenotypic and genotypic datasets along with the linkage maps generated in previous steps will be utilized for QTL mapping. We will utilize "Windows QTL Cartographer" and "QTL Network" for mapping main-effect and epistasis QTLs, respectively. First download these programs for Windows from https://brcwebportal.cos.ncsu.edu/qtlcart/WQTLCart.htm (for Windows QTL Cartographer) and http://ibi.zju.edu.cn/software/qtlnetwork/download.htm (for QTL Network). Both programs can be used without installation by extracting and copying the files to working directory. However, it is recommended to

**Table 6**
**Example format of input data file for QTL Cartographer**

| | | | | | | | | | | | | |
|---|---|---|---|---|---|---|---|---|---|---|---|---|
| #START# | | | | | | | | | | | | |
| Populations | 1 | | | | | | | | | | | |
| Samples | 81 | | | | | | | | | | | |
| Traits | 2 | | | | | | | | | | | |
| Chromosomes | 3 | | | | | | | | | | | |
| Markers | 4 | 4 | 2 | | | | | | | | | |
| MapFunction | Kosambi | | | | | | | | | | | |
| CrossType | Ri0 | | | | | | | | | | | |
| MisTraitValue | . | | | | | | | | | | | |
| TranslationTable | 2 | *1 | 0 | *2 | *3 | -1 | | | | | | |
| Other Traits | 0 | | | | | | | | | | | |
| #END# | | | | | | | | | | | | |
| MkID | SNP1 | SNP2 | SNP3 | SNP4 | SNP5 | SNP6 | SNP7 | SNP8 | SNP9 | SNP10 | Yr | Sr |
| Position | 0 | 1 | 2 | 10 | 0 | 3 | 4 | 5 | 7 | 30 | 1 | 2 |
| Chromosome1 | 1 | 1 | 1 | 1 | 2 | 2 | 2 | 2 | 3 | 3 | 1 | 2 |
| Chr Label | 1A | 1A | 1A | 1A | 1B | 1B | 1B | 1B | 1D | 1D | 1 | 2 |
| Ind-1 | 2 | 2 | 2 | 2 | 2 | 2 | 2 | 2 | 2 | 2 | 6.8 | 3.3 |
| Ind-2 | 0 | 0 | 0 | 0 | 0 | 0 | 0 | 0 | 0 | 0 | 8.7 | 7.1 |
| Ind-3 | 0 | 0 | 0 | 0 | 0 | 0 | 0 | 2 | 2 | 2 | 9.6 | 8.5 |
| Ind-4 | 0 | 0 | 0 | 0 | 0 | 0 | 0 | 0 | 0 | 0 | 9.2 | 6.8 |
| Ind-5 | 2 | 2 | 0 | 0 | 0 | 0 | 0 | 0 | 0 | 0 | 2.4 | 2.5 |
| Ind-6 | 0 | 0 | 2 | -1 | 0 | 0 | 2 | 0 | 2 | 2 | 9 | 6 |
| . | . | . | . | . | . | . | . | . | . | . | . | . |
| . | . | . | . | . | . | . | . | . | . | . | . | . |
| Ind-81 | 2 | 2 | 2 | 2 | 2 | 2 | 2 | 2 | 2 | 2 | 8.7 | 6.2 |

download the installer of Windows QTL Cartographer and install it for stability. After installation, prepare a single input data file in Microsoft Excel using the phenotypic and genotypic datasets, and the linkage maps in a format similar to one shown in Table 6 to map main-effect QTLs using QTL Cartographer. Save the file as single csv file (e.g., QTLdata.csv).

This file is quite self-explanatory. This example has ten parameter rows in starting (between #START# and #END# rows). Write # of populations, genotypes, traits, and chromosomes being studied in cells B2–B5, in that order. Write # of markers present on each chromosome in separate cells in row 6 in the same order as chromosomes arranged in genotyping data. Map function (Kosambi in this case), cross/population type (Ri0 for DHs in this case), and value for missing trait (.) should be given in cells B7–B9 as shown in Table 6. Assign different marker alleles in row 10 (cells B10–G10). Parent 1 allele "A" and parent 2 allele "B" are formatted as "2" and "0" in the genotypic dataset, while missing genotype as "−1." Other trait (nil in this case) is mentioned in cell B11. The last two columns in this example files are showing phenotypic data.

When your file is ready, open the Windows QTL Cartographer, and click on "Import" icon on the main menu. Select the "Microsoft CSV format (*.csv)" on the prompt followed by pressing "Next." On the next step, press the "CSV..." button to browse the input data file that you saved in the working directory. Select your input file (e.g., QTLdata.csv), and press "Open" followed by "Finish." Press "OK" on the next prompt. Hit "CIM" button on main menu followed by "Control..." button. Change the regression method to "3: Forwards & Backwards Method" and both (in and out) probability thresholds to 0.05 followed by "OK" on the prompt. Change default "Permutation Times" settings to 1000 and hit "OK." This may take few hours to days to complete, depending on the data size. After completion of permutation testing, click "START" button. A QTL plot window will open upon completion of analyses. Note the entire CIM interval above the threshold line as QTL confidence interval and the highest peak position as QTL position. Note the highest LOD score, $R^2$ (phenotypic variation) values and additive effect values of each QTL. $R^2$ plot (bottom plot) can be changed to additive effect plot by selecting "Show Additive Effect" option on the "Effects" tab on main menu of QTL plot window. After noting down all the information, navigate the "File" tab on main menu of QTL plot followed by selecting the "Save as CSV File..." option. Change the file name and location, if required, and press "Save" on the prompt to save the results.

A result file typically looks like the one presented in Table 7.

*Mapping of epistatic QTLs*: QTL Network requires two different files, one for linkage map and another for genotypic and phenotypic datasets, to map epistatic QTLs. Prepare the map file in Microsoft Excel in a format given in Table 8 and save as "population_name.map" Tab-delimited text file.

The main difference between this map file and the others used in the previous steps is that the marker positions in this file are shown as difference values of positions of adjacent markers. One can calculate this in Excel.

Prepare a single data file for genotypic and phenotypic data in Microsoft Excel in a format as given in Table 9, and save as Tab-delimited text file (e.g., data.txt).

**Table 7**
**Example format of output result file of QTL Cartographer**

| Trait | Chromosome | Marker | Position (cM) | LOD (H0: H1) | R2 (H0: H1) | TR2 (H0: H1) | Additive (H1) | S1 |
|---|---|---|---|---|---|---|---|---|
| 1 | 1 | 1 | 0.01 | 0.113803 | 0.001483 | 0.276559 | −0.10156 | 0.148981 |
| 1 | 1 | 1 | 1.01 | 0.146113 | 0.002021 | 0.277096 | −0.118615 | 0.165824 |
| 1 | 1 | 1 | 2.01 | 0.182692 | 0.002649 | 0.277724 | −0.135907 | 0.185812 |
| 1 | 1 | 1 | 3.01 | 0.22262 | 0.003337 | 0.278413 | −0.152679 | 0.208174 |
| 1 | 1 | 1 | 4.01 | 0.264566 | 0.00404 | 0.279116 | −0.168148 | 0.231541 |
| 1 | 1 | 1 | 5.01 | 0.306958 | 0.004704 | 0.27978 | −0.181636 | 0.254219 |
| 1 | 1 | 1 | 6.01 | 0.348194 | 0.005281 | 0.280357 | −0.19268 | 0.274589 |
| 1 | 1 | 1 | 7.01 | 0.386839 | 0.005733 | 0.280809 | −0.201024 | 0.291376 |
| 1 | 1 | 1 | 8.01 | 0.421752 | 0.00604 | 0.281115 | −0.20662 | 0.303783 |
| 1 | 1 | 1 | 9.01 | 0.452145 | 0.006192 | 0.281268 | −0.209539 | 0.311408 |
| 1 | 1 | 2 | 10.01 | 0.477663 | 0.006204 | 0.28128 | −0.210104 | 0.31428 |

This file contains three sections: (i) header, (ii) marker/genotyping data, and (iii) trait/phenotyping data. Though the file format is quite self-explanatory, some important features are discussed. Observations in header should have total number of observations taken for a trait of interest across environments, irrespective of missing data. Thus, these can be calculated by (#genotype)*(#Environment). In marker body, markers and genotypes are shown by rows and columns, respectively. The order of markers is the same as in chromosomes/LGs in map file; marker data of chromosome 1 is followed by chromosome 2 and then by chromosome 3.

Once the files are ready, open the QTL Network. Click on the "Project(s)" button on the main menu followed by "New(N)." Browse and select the map and data file created in the previous steps by using "Map File..." and "Data File..." buttons on the prompt. Hit "OK" following change made to results location, if any. Press start icon on the left side icon panel to start analysis. Press "OK" on the prompt to accept default settings. QTL Network will display the plots of identified QTLs and different interaction effects along with a result report upon analyses completion. A detailed analysis file (*.qnk) and a result file (*.pre) will be saved to the working directory (*see* **Note 10**).

A typical QTL Network result file (*.pre) looks like one presented in Table 10. Though all the results are important, we will briefly describe epistasis effect results from this analysis. QTL_i and QTL_j are the two QTLs involved in the detected epistatic

**Table 8**
**Example format of input map file for QTL Network**

| | | | |
|---|---|---|---|
| _MapFunction | K | | |
| _DistanceUnit | cM | | |
| _Chromosomes | 3 | | |
| _MarkerNumbers | 17 | 15 | 18 |
| *MapBegin* | | | |
| Marker# | ch1 | ch2 | ch3 |
| 1 | 0 | 0 | 0 |
| 2 | 19.236 | 12.9949 | 7.7618 |
| 3 | 16.2488 | 5.3402 | 13.2518 |
| 4 | 4.8552 | 22.2875 | 6.9239 |
| 5 | 4.8047 | 27.7327 | 9.8037 |
| 6 | 15.3881 | 6.3438 | 2.7929 |
| 7 | 15.5969 | 29.4517 | 17.5239 |
| 8 | 15.0048 | 10.2825 | 41.7545 |
| 9 | 3.8375 | 8.9339 | 37.3036 |
| 10 | 3.2747 | 12.824 | 15.8394 |
| 11 | 34.4392 | 8.4598 | 18.7639 |
| 12 | 2.5322 | 5.1683 | 2.5121 |
| 13 | 23.7979 | 10.1262 | 5.0168 |
| 14 | 8.2644 | 5.2896 | 28.9405 |
| 15 | 13.3483 | 13.2089 | 1.9109 |
| 16 | 33.5319 | | 22.7256 |
| 17 | 2.5622 | | 15.2455 |
| 18 | | | 32.48 |
| *MapEnd* | | | |

interaction. These are named with the relevant chromosome and marker intervals. The first (before "–") and second (after "–") parts of the name represent chr number and marker interval, respectively. Interval_i and interval_j are the flanking markers of QTL_i and QTL_j, respectively. Position_i and position_j are the distance between QTL_i/QTL_j and the first marker of the relevant chromosome. Range_i and range_j are the position support intervals of QTL_i and QTL_j, respectively. AA is the estimated additive x additive (epistatic) effect. SE is the standard error of estimated or predicted QTL effect.

**Table 9**
**Example format of input data (genotype and phenotype) file for QTL Network**

| | | | | | | | |
|---|---|---|---|---|---|---|---|
| _Population | DH | | | | | | |
| _Genotypes | 200 | | | | | | |
| _Observations | 400 | | | | | | |
| _Environments | Yes | | | | | | |
| _Replications | No | | | | | | |
| _TraitNumber | 2 | | | | | | |
| _TotalMarker | 50 | | | | | | |
| _MarkerCode | P1 = A | P2 = B | | | | | |
| *MarkerBegin* | | | | | | | |
| #MK | 1 | 2 | 3 | 4 | .. | 200 | ; |
| MK1 | A | A | A | A | .. | A | ; |
| MK2 | A | A | A | A | .. | A | ; |
| MK3 | B | A | A | A | .. | B | ; |
| MK4 | B | A | B | A | .. | A | ; |
| MK5 | A | B | B | A | .. | A | ; |
| MK6 | A | B | B | A | .. | A | ; |
| MK7 | A | A | B | A | .. | A | ; |
| MK8 | A | A | A | A | .. | A | ; |
| MK9 | A | A | A | A | .. | A | ; |
| MK10 | A | A | A | A | .. | A | ; |
| MK11 | A | A | A | A | .. | A | ; |
| MK12 | A | B | A | B | .. | A | ; |
| MK13 | A | B | . | B | .. | B | ; |
| MK14 | A | A | A | B | .. | B | ; |
| MK15 | A | A | A | A | .. | B | ; |
| MK16 | A | A | A | A | .. | B | ; |
| .. | .. | .. | .. | .. | .. | .. | ; |
| MK50 | A | A | A | A | .. | B | ; |
| *MarkerEnd* | | | | | | | |
| *TraitBegin* | | | | | | | |
| Env# | Geno# | TraitPheno1 | TraitPheno2 | ; | | | |
| 1 | 1 | 98.8331 | 90.4331 | ; | | | |

(continued)

**Table 9**
**(continued)**

| | | | | |
|---|---|---|---|---|
| 1 | 2 | 107.1500 | 102.1200 | ; |
| 1 | 3 | 105.6465 | 117.1100 | ; |
| . | . | . | . | . |
| . | . | . | . | . |
| 1 | 199 | 96.0623 | 93.1123 | ; |
| 1 | 200 | 95.2396 | 92.1122 | ; |
| 2 | 1 | 87.8537 | 85.8333 | ; |
| 2 | 2 | 110.2431 | 108.3431 | ; |
| . | . | . | . | . |
| . | . | . | . | . |
| 2 | 199 | 90.5981 | 88.4543 | ; |
| 2 | 200 | 106.7538 | 103.3434 | ; |
| *TraitEnd* | | | | |

*Results presentation*: Whole-genome QTL analysis results can be presented using R package OmicCircos. Create all the required files for the OmicCircos as suggested below.

1. First of all, create a Tab-delimited framework linkage map file in a format given in Table 11. If a LG is of 186.6 cM, such as chr1A in the example, it will be divided into 187 segments (to a whole number). Length of LGs can be obtained from the results saved at the end of **steps** 7, **9**, and **10** of Sect. 3.3. Please add all segments of all the LGs in this file and save it as "seg.frame.txt" Tab-delimited text file in the working directory.
2. Create a label file for names of mapped QTLs and their flanking markers as formatted in Table 12.

   Provide chromosome names (such as chr1A to chr7D) in column 1 (seg.name), marker/QTL position in column 2 (seg.po), and marker/QTL name in column 3 (marker). In this example, short names are provided for a QTL and its flanking markers. *See* **Note 11** for instructions for labeling QTLs and markers. Information about any QTL/marker that needs to be labeled should be provided in this file in rows. Although all marker/QTL names can be provided in this file, we recommend giving only flanking markers as all markers cannot be viewed on the QTL plot. Save this file as Tab-delimited text file in the working directory and rename it as "seg.labels.txt."

**Table 10**
**Example format of output result file of QTL Network**

| |
|---|
| _trait 1 TraitPheno1 |
| _variance_components (population mean: 100.1422) |
| V(G)/V(P) V(E)/V(P) V(GE)/V(P) V(e)/V(P) |
| 0.6034 0.0405 0.1403 0.2204 |
| V(A)/V(P) V(I)/V(P) V(A + AA)/V(P) V(AE)/V(P) V(IE)/V(P) V(AE + AAE)/V(P) |
| 0.4065 0.2103 0.6040 0.0654 0.0740 0.1403 |
| _QTL_effect |
| QTL interval position range A SE P-value AE1 SE P-value AE2 SE P-value; |
| 2–4 MK13-MK14 37.0 35.0–42.0 −4.1665 0.2331 0.000000 −0.0005 0.0171 0.977999 0.0005 0.0171 0.977426; |
| 3–6 MK27-MK28 51.0 45.0–58.0 −0.2405 0.2311 0.297897 −2.3118 0.3251 0.000000 2.2600 0.3251 0.000000; |
| _QTL_heritability |
| QTL h^2(a) h^2(ae) h^2(ae1) h^2(ae2) |
| 2–4 0.1689 0.0009 0.0009 0.0009 |
| 3–6 0.0035 0.0397 0.0397 0.0397 |
| _epistasis_effect |
| QTL_i interval_i position_i range_i QTL_j interval_j position_j range_j AA SE P-value AAE1 SE P-value AAE2 SE P-value; |
| 1–3 MK2-MK3 19.01 19.0–24.0 1–7 MK5-MK6 67.0 65.0–69.0 2.7192 0.2497 0.000000 0.0001 0.0171 0.995739 −0.0001 0.0171 0.995430; |
| _epistasis_heritability |
| QTL_i QTL_j h^2(aa) h^2(aae) h^2(aae1) h^2(aae2) |
| 1–3 1–7 0.0551 0.0011 0.0011 0.0011 |

(continued)

**Table 10 (continued)**

| | | | | | | |
|---|---|---|---|---|---|---|
| _genotype_value | | | | | | |
| Entry | G | G + GE1 | G + GE2 | | | |
| P1 | −0.9195 | −4.6809 | 2.8144 | | | |
| P2 | −2.0330 | 2.1823 | −6.1274 | | | |
| SuperiorLine(+) | 15.8043 | 16.2127 | 15.3787 | | | |
| SuperiorLine(−) | −15.8043 | −12.6620 | −18.8729 | | | |
| _superior_genotype | | | | | | |
| QTL | GSL(+) | GSL(−) | SL(+)1 | SL(+)2 | SL(−)1 | SL(−)2 |
| 1–3 | QQ | qq | QQ | QQ | qq | qq |
| 2–4 | qq | QQ | qq | qq | QQ | QQ |
| 3–6 | qq | QQ | qq | qq | QQ | QQ |
| 1–7 | QQ | QQ | QQ | QQ | QQ | QQ |
| Time used: 0:2:5 | | | | | | |

**Table 11**
**Example format of framework linkage map file for OmicCircos**

| seg.name | seg.Start | seg.End | the.v | NO |
|---|---|---|---|---|
| chr1A | 0 | 1 | NA | NA |
| chr1A | 1 | 2 | NA | NA |
| chr1A | 2 | 3 | NA | NA |
| . | . | . | . | . |
| . | . | . | . | . |
| chr1A | 185 | 186 | NA | NA |
| chr1A | 186 | 187 | NA | NA |
| chr1B | 0 | 1 | NA | NA |
| chr1B | 1 | 2 | NA | NA |
| chr1B | 2 | 3 | NA | NA |
| . | . | . | . | . |
| . | . | . | . | . |
| chr1B | 138 | 139 | NA | NA |
| chr1B | 139 | 140 | NA | NA |
| chr1D | 0 | 1 | NA | NA |
| chr1D | 1 | 2 | NA | NA |
| . | . | . | . | . |
| . | . | . | . | . |
| . | . | . | . | . |
| . | . | . | . | . |
| chr7D | 180 | 181 | NA | NA |
| chr7D | 181 | 182 | NA | NA |

3. Create a single LOD score file in the Microsoft Excel from the result file saved at the end of QTL mapping carried out using QTL Cartographer. Format of new file should be as given in Table 13. Give chromosome names in column 1, segment positions in column 2, and marker and QTL names in column 3. The segment positions (seg.po column) in this example contain same values as "Position(cM)" column in the QTL mapping result file. Columns 4–9 (named L1–L6) each contain LOD(H0:H1) score at every segment position as obtained from the QTL mapping result file, for each of the 6 different studied traits in this example. More than ten traits can be plotted easily. The last column contains a selected LOD

**Table 12**
**Example format of label file for OmicCircos**

| seg.name | seg.po | marker |
|---|---|---|
| chr1B | 34.90 | M1 |
| chr1B | 36.00 | Q1 |
| chr1B | 38.80 | M2 |

**Table 13**
**Example format of LOD score file for OmicCircos**

| seg.name | seg.po | marker | L1 | L2 | L3 | L4 | L5 | L6 | Th |
|---|---|---|---|---|---|---|---|---|---|
| chr1A | 0.01 | | 0.113803 | 0.200966 | 0.496227 | 0.099029 | 0.058747 | 0.333618 | 2.5 |
| chr1A | 1.01 | | 0.146113 | 0.159096 | 0.462578 | 0.07759 | 0.057959 | 0.292421 | 2.5 |
| chr1A | 2.01 | | 0.182692 | 0.368488 | 0.925778 | 0.075965 | 0.097638 | 0.271567 | 2.5 |
| chr1A | 3.01 | | 0.22262 | 0.272187 | 0.887817 | 0.06912 | 0.261406 | 0.456371 | 2.5 |
| . | . | . | . | . | . | . | . | . | . |
| . | . | . | . | . | . | . | . | . | . |
| chr1A | 185.01 | | 0.735864 | 0.562618 | 0.983225 | 0.228407 | 0.002031 | 0.361101 | 2.5 |
| chr1A | 186.01 | | 0.343331 | 0.567157 | 0.800971 | 0.176416 | 0.023354 | 0.501944 | 2.5 |
| chr1B | 0.01 | | 0.012083 | 0.118496 | 0.046226 | 0.424844 | 0.204101 | 1.108274 | 2.5 |
| chr1B | 1.01 | | 0.013021 | 0.121336 | 0.044868 | 0.399149 | 0.209057 | 1.066644 | 2.5 |
| . | . | . | . | . | . | . | . | . | . |
| . | . | . | . | . | . | . | . | . | . |
| chr1B | 34.90 | M1 | 6.829153 | 0.207018 | 0.264532 | 0.744456 | 0.175173 | 0.460523 | 2.5 |
| chr1B | 36.00 | Q1 | 7.02168 | 0.087382 | 0.171409 | 0.530252 | 0.111202 | 0.538538 | 2.5 |
| chr1B | 38.80 | M2 | 7.078191 | 0.09007 | 0.180853 | 0.50282 | 0.111625 | 0.543725 | 2.5 |
| . | . | . | . | . | . | . | . | . | . |
| . | . | . | . | . | . | . | . | . | . |
| chr7D | 179.01 | | 0.001222 | 0.254582 | 0.059368 | 0.002461 | 0.636679 | 0.28141 | 2.5 |
| chr7D | 180.01 | | 0.002991 | 0.287107 | 0.105647 | 0.000603 | 0.729838 | 0.333776 | 2.5 |
| chr7D | 181.01 | | 0.019826 | 0.31748 | 0.161038 | 0 | 0.813644 | 0.384557 | 2.5 |

threshold (2.5 in this example) to indicate significant QTLs. Save this file as Tab-delimited text file in the working directory and rename it as "seg.mapping1.txt."

4. Similarly, create a single phenotypic variation ($R^2$) score file in Microsoft Excel from the result file saved at the end of QTL mapping carried out using QTL Cartographer in a format

**Table 14**
**Example format of phenotypic variation ($R^2$) score file for OmicCircos**

| seg.name | seg.po | marker | R2_1 | R2_2 | R2_3 | R2_4 | R2_5 | R2_6 | Th |
|---|---|---|---|---|---|---|---|---|---|
| chr1A | 0.01 | | 0.001483 | 0.3215 | 0.7556 | 0.1291 | 0.0854 | 0.4778 | 10 |
| chr1A | 1.01 | | 0.002021 | 0.2546 | 0.7049 | 0.1014 | 0.0843 | 0.4197 | 10 |
| chr1A | 2.01 | | 0.002649 | 0.6068 | 1.4634 | 0.1003 | 0.1476 | 0.4077 | 10 |
| chr1A | 3.01 | | 0.003337 | 0.435 | 1.343 | 0.0901 | 0.3789 | 0.6518 | 10 |
| . | . | . | . | . | . | . | . | . | . |
| . | . | . | . | . | . | . | . | . | . |
| chr1A | 185.01 | | 1.2557 | 0.895 | 1.4875 | 0.301 | 0.003 | 0.5319 | 10 |
| chr1A | 186.01 | | 0.5886 | 0.9031 | 1.2172 | 0.2303 | 0.034 | 0.7211 | 10 |
| chr1B | 0.01 | | 0.0207 | 0.192 | 0.0716 | 0.6508 | 0.2982 | 1.5925 | 10 |
| chr1B | 1.01 | | 0.023 | 0.2021 | 0.0717 | 0.6264 | 0.3168 | 1.598 | 10 |
| . | . | . | . | . | . | . | . | . | . |
| . | . | . | . | . | . | . | . | . | . |
| chr1B | 34 | M1 | 5.3173 | 0.2001 | 0.3443 | 0.9016 | 0.2132 | 0.531 | 10 |
| chr1B | 35 | Q1 | 5.4573 | 0.0845 | 0.2232 | 0.6435 | 0.1353 | 0.6202 | 10 |
| chr1B | 36 | M2 | 5.7415 | 0.0885 | 0.2418 | 0.629 | 0.1391 | 0.6347 | 10 |
| . | . | . | . | . | . | . | . | . | . |
| . | . | . | . | . | . | . | . | . | . |
| chr7D | 179.01 | | 0.0023 | 0.4395 | 0.1002 | 0.0033 | 1.0324 | 0.4395 | 10 |
| chr7D | 180.01 | | 0.0055 | 0.4905 | 0.1757 | 0.0008 | 1.1641 | 0.5148 | 10 |
| chr7D | 182.01 | | 0.0356 | 0.5313 | 0.2605 | 0 | 1.2589 | 0.5793 | 10 |

given in Table 14. Columns 4–9 (named as R2_1 to R2_6) each contain R2(H0:H1) score at every segment position as obtained from the QTL mapping result file, for each of the 6 different traits studied in this example. The last column contains a selected $R^2$ threshold (10.0 in this example) to indicate major QTLs. Save this file as Tab-delimited text file in the working directory and rename it as "seg.mappingr.txt."

5. Similarly, create a single additive effect score file in Microsoft Excel from the result file saved at the end of QTL mapping carried out using QTL Cartographer in a format given in Table 15. Columns 4–9 (named as Add.1–Add.6) each contain Additive(H1) effect score at every segment position as obtained from the QTL mapping result file, for each of the six different traits studied in this example. This file does not

**Table 15**
**Example format of additive effect score file for OmicCircos**

| seg.name | seg.po | marker | Add.1 | Add.2 | Add.3 | Add.4 | Add.5 | Add.6 |
|---|---|---|---|---|---|---|---|---|
| chr1A | 0.01 | | −0.10156 | −0.101091 | −1.65938 | −0.514023 | −0.078447 | −0.163254 |
| chr1A | 1.01 | | −0.118615 | −0.089915 | −1.602575 | −0.457376 | −0.077644 | −0.15301 |
| chr1A | 2.01 | | −0.135907 | −0.138903 | −2.317272 | −0.456689 | −0.103168 | −0.150886 |
| chr1A | 3.01 | | −0.152679 | −0.117819 | −2.22713 | −0.433312 | −0.166294 | −0.191234 |
| . | . | . | . | . | . | . | . | . |
| . | . | . | . | . | . | . | . | . |
| chr1A | 185.01 | | 0.212934 | 0.169925 | 2.348564 | 0.784368 | 0.014702 | −0.175275 |
| chr1A | 186.01 | | 0.145766 | 0.17107 | 2.12794 | 0.685322 | −0.0499 | −0.203873 |
| chr1B | 0.01 | | −0.027464 | −0.080607 | −0.520661 | −1.166941 | −0.148118 | −0.302246 |
| chr1B | 1.01 | | −0.028873 | −0.082528 | −0.5202 | −1.143697 | −0.152346 | −0.302257 |
| . | . | . | . | . | . | . | . | . |
| . | . | . | . | . | . | . | . | . |
| chr1B | 34 | M1 | −2.268102 | 0.177347 | 0.121715 | −0.169622 | −0.973886 | 0.911057 |
| chr1B | 35 | Q1 | −2.340911 | 0.115278 | 0.098057 | −0.14337 | −0.776809 | 0.986655 |
| chr1B | 36 | M2 | −2.392835 | 0.117942 | 0.101948 | −0.14179 | −0.78752 | 0.995788 |
| . | . | . | . | . | . | . | . | . |
| . | . | . | . | . | . | . | . | . |
| chr7D | 179.01 | | −0.008908 | 0.118372 | 0.608281 | −0.082379 | 0.270614 | 0.157263 |
| chr7D | 180.01 | | 0.013863 | 0.125114 | 0.805406 | −0.04067 | 0.287445 | 0.170203 |
| chr7D | 181.01 | | 0.035254 | 0.130293 | 0.980595 | −0.000123 | 0.299046 | 0.18055 |

need a threshold column at the end. Save this file as Tab-delimited text file in the working directory and rename it as “seg.mappinga.txt.”

6. Prepare a Tab-delimited file for plotting the QTL confidence interval as bars beneath LOD score peaks as formatted in Table 16. It contains three columns. Provide names of the chromosomes for each detected QTL in column 1 (seg. name), start position of the confidence intervals in column 2 (seg.Start), and end position of the confidence intervals in column 3 (seg.End). This information can be obtained from graphical output of main-effect QTL analysis carried out using QTL Cartographer in previous steps. It can also be viewed by looking on LOD score values above threshold in the raw file of detailed QTL analysis results obtained using QTL Cartographer. Save this file as Tab-delimited text file in the working directory and rename it as “arcdata.txt.”

**Table 16**
**Example format of input data of QTL confidence intervals for OmicCircos**

| seg.name | seg.Start | seg.End |
|---|---|---|
| chr1B | 33.0 | 38.5 |
| chr2B | 64.2 | 71.8 |
| chr2B | 64.9 | 66.7 |
| chr2D | 37.9 | 41.9 |
| chr2D | 40.2 | 63.7 |
| . | . | . |
| . | . | . |
| . | . | . |
| chr7D | 77.8 | 91.5 |
| chr7D | 113.0 | 118.0 |

**Table 17**
**Example format of input data of epistatic QTLs for OmicCircos**

| seg1 | po1 | name1 | seg2 | po2 | name2 | name3 |
|---|---|---|---|---|---|---|
| chr1A | 17.7 | DI | chr1B | 138 | DI | DI |

7. Prepare a Tab-delimited link file for epistatic QTLs in the format given in Table 17. This file will utilize epistasis results information obtained using QTL Network software. It contains two rows (one for header, another for data) and seven columns. Provide the chromosome name of first epistatic QTL (QTL_i) in column 1 (seg1) and second (QTL_j) in column 4 (seg2). In this example the first epistatic QTL is present on chromosome 1A, while the second is on chromosome 1B. Provide the genetic map positions of QTL_i and QTL_j in columns 2 (po1) and 5 (po2), respectively. Write the name of related trait in columns 3 (name1), 6 (name2), and 7 (name3).

   Information about multiple pairs of epistatic loci can be added in this file, but we recommended to prepare a separate file for each pair of epistatic loci identified. Separate files provide more opportunities to make different desirable changes on the QTL plot. Save this file in the working directory and rename it as "seg.link1.txt." Similarly, prepare files for the rest of each identified pair of epistatic loci, and save them as a series of "seg.link2.txt, seg.link3.txt, seg.link4.txt, etc." in working directory.

8. To generate a whole-genome QTL plot, download and install package OmicCircos by executing the following commands in console of RStudio:

```
if (!requireNamespace("BiocManager", quietly = TRUE))
 install.packages("BiocManager")
BiocManager::install("OmicCircos")
```

9. Change the working directory by navigating the "Session" tab and selecting "Set Working Directory" followed by "Choose Directory..." and a folder/directory on the prompt. Alternatively, use the following command in the console:

```
setwd("C:/Users/Path/to/Directory")
```

10. Execute the following commands to load and run the OmicCircos:

```
library(OmicCircos)
options(stringsAsFactors = FALSE)
set.seed(1234)
```

11. Read, attach, and check the files created in previous steps by executing the following commands:

```
seg.f <- read.table("seg.frame.txt", header=TRUE, sep="\t")
attach(seg.f)
head(seg.f)
seg.l <- read.table("seg.labels.txt", header=TRUE, sep="\t")
attach(seg.l)
head(seg.l)
 seg.v <- read.table("seg.mappingl.txt", header=TRUE,
sep="\t")
attach(seg.v)
head(seg.v)
 seg.v1 <- read.table("seg.mappingr.txt", header=TRUE,
sep="\t")
attach(seg.v1)
head(seg.v1)
 seg.v2 <- read.table("seg.mappinga.txt", header=TRUE,
sep="\t")
attach(seg.v2)
head(seg.v2)
arc.d1 <- read.table("arcdata.txt", header=TRUE, sep="\t")
attach(arc.d1)
head(arc.d1)
link.v1 <- read.table("seg.link1.txt", header=TRUE, sep="\t")
```

```
attach(link.v1)
head(link.v1)
```

Read, attach, and check the additional link files, if any, as shown in the example below using the following commands:

```
link.v2 <- read.table("seg.link2.txt", header=TRUE, sep="\t")
attach(link.v2)
head(link.v2)
link.v3 <- read.table("seg.link3.txt", header=TRUE, sep="\t")
attach(link.v3)
head(link.v3)
link.v4 <- read.table("seg.link4.txt", header=TRUE, sep="\t")
attach(link.v4)
head(link.v4)
```

For this example dataset, we attached the 16 additional "seg. link.txt" files as used by Dhariwal et al. [83].

Execute the following commands to start making whole-genome scan QTL plot:

```
names <- levels(as.factor(seg.f$seg.name))
seg.num <- length(names)
```

Set the color option using the following command:

```
colors <- rainbow(1, alpha=0)
```

Generate the db and plot using the following commands (*see* **Note 12**):

```
db <-segAnglePo(seg.f, names)
pdffile <- "130120-9.pdf";
pdf(pdffile, 5.5, 5.5)
par(mar=c(0, 0, 0, 0))
plot(c(-30,850), c(-30,850), type="n", axes=FALSE, xlab="",
ylab="", main="")
```

Plot the linkage map and chromosome labels using the following commands:

```
circos(R=326, cir=db, type="chr", col=colors, print.chr.lab=-
TRUE, W=8, scale=FALSE) #Will plot chromosome labels.#
circos(R=320, cir=db, type="chr", col="firebrick3", print.chr.
lab=FALSE, W=8, scale=TRUE) #Will plot framework linkage map.#
```

Plot the marker and QTL labels using the following commands:

```
circos(R=370, cir=db, W=8, mapping=seg.l, col.v=3, type="la-
bel", B=FALSE, side="out", col=c("blue", "firebrick3","blue",
"blue", "firebrick3","blue"), cex=0.7) #You can set a differ-
ent color for every marker/QTL label by making changes under
col=c().
```

Colors for only 6 marker/QTL labels are shown in the above command to save space and time but colors were specified for each label to regenerate QTL plot shown in this protocol.

Plot the LOD score using the following command:

```
circos(R=240, cir=db, W=90, mapping=seg.v, col.v=4, type="ml",
B=FALSE, col=c("forestgreen", "violetred", "lightskyblue",
"greenyellow", "cyan3", "palevioletred1", "blue"), lwd=1, sca-
le=TRUE)
```

Every trait is assigned a different color.

Plot the QTL confidence interval bars beneath LOD curve peaks using the following command:

```
circos(R=230, cir=db, W=40, mapping=arc.d1, type="arc2",
B=FALSE, col=colors, lwd=6, cutoff=0)
```

Plot the phenotypic variation ($R^2$) score using the following command:

```
circos(R=190, cir=db, W=50, mapping=seg.v1, col.v=4,
type="ml", B=FALSE, col=c("forestgreen", "violetred", "light-
skyblue", "greenyellow", "cyan3", "palevioletred1", "blue"),
lwd=1, scale=TRUE)
```

Plot the additive effect score using the following command:

```
circos(R=150, cir=db, W=50, mapping=seg.v2, col.v=4,
type="ml", B=FALSE, col=c("forestgreen", "violetred", "light-
skyblue", "greenyellow", "cyan3", "palevioletred1"), lwd=1,
scale=TRUE)
```

Plot the links between each pair of epistatic loci separately using the different link.v data frames by executing the commands pasted below (*see* **Note 13**):

```
circos(R=150, cir=db, W=10, mapping=link.v13, type="link2",
lwd=7.3977, col="cyan3")
circos(R=150, cir=db, W=10, mapping=link.v1, type="link2",
lwd=1.4385, col="lightskyblue")
```

If the AxA effect is a negative value, plot this type of interaction using the trick below. Run the same command twice, but make two changes when you are ready to run a second time, change the color to white and "lwd" equal to 0.5. In this way, the first line will be shown as hollow but have desired color and width. See command examples below:

```
circos(R=150, cir=db, W=10, mapping=link.v2, type="link2",
lwd=1.4096, col="lightskyblue")
circos(R=150, cir=db, W=10, mapping=link.v2, type="link2",
lwd=0.5, col="white")
circos(R=150, cir=db, W=10, mapping=link.v4, type="link2",
lwd=3.898, col="cyan3")
circos(R=150, cir=db, W=10, mapping=link.v4, type="link2",
lwd=0.5, col="white") #Run this command for every pair of
epistatic loci. Only a few are shown here to save space.#
```

Plot the figure legends by executing the commands pasted below:

For effect legend.

```
legend(653, 168, legend=c("PHT", "DTA", "DI", "DS", "VRI",
"DON", "th."),
 col = c("forestgreen", "violetred", "lightskyblue", "green-
yellow", "cyan3", "palevioletred1", "blue"), lty = c(1, 1,
1, 1, 1, 1, 1), lwd = 0.9, cex=0.7,
 title=expression(bold("Traits")), title.col="orange", text.
font=1, box.lty=0)
```

For legend of interaction types.

```
legend(772, 168, legend=c("-AA", "+AA"),
 col = c("grey", "grey"), lty = c(1, 1), lwd = 2, cex=0.7,
 title=expression(bold("Interactions")), title.col="grey",
text.font=1, box.lty=0)
legend(772, 168, legend=c("-AA", "+AA"),
 col = c("white", "grey"), lty = c(1, 1), bty = "n",lwd = 0.5,
cex=0.7,
 title=expression(bold("Interactions")), title.col="orange",
text.font=1, box.lty=0)
```

For AA interaction scale.

```
legend(770, 88, legend=c("0.1", "1.0", "2.5", "5.0", "7.5"),
 col = c("grey", "grey", "grey", "grey", "grey"), lty = c(1, 1,
1, 1, 1), bty = "n",lwd = c(0.1, 1.0, 2.5, 5.0, 7.5), cex=0.7,
 title=expression(bold("AA scale")), title.col="orange", text.
font=1, box.lty=0)
```

Save the results and exit using dev.off() function:

```
dev.off()
```

A representative whole-genome QTL plot based on CIM analysis method used in this workflow is reproduced from Dhariwal et al. [83] and presented in Fig. 2.

## 4 Notes

1. $F_2$ populations should be used if codominant markers such as simple sequence repeat (SSR) are available for genotyping. As $F_{2s}$ are highly heterozygous, only codominant markers can reveal the genetic makeup of all genotypes.
2. Prepare all buffers and solutions using doubled distilled water or other ultrapure water such as milli Q Type I water. Readers are referred to Green and Sambrook [85] for preparation of all standard buffers and solutions. Prepare and store all buffers and solutions at room temperature except RNase A solution and isopropanol or as otherwise indicated. Carefully follow all local guidelines and regulations for disposal of waste.
3. Population generated from highly contrast parents will increase the chances of mapping of a large number of polymorphic markers well distributed across chromosomes; however, if both lines/parents are similar for most of the traits, except the target trait, it will increase precision of QTL mapping for target trait.
4. Parental genotypes are not required for further analysis. Do not paste genotype calls of parents in MapDisto. Also, do not paste any other information that is present in other columns present in genotyping data file.
5. If you encounter an error, execute the following command to install git:

```
sudo apt install git
```

6. g++ is required to compile MSTmap. If you are running Linux OS on a 64-bit device and make command is not working, then execute the following commands:

```
sudo apt-get install gcc-multilib
sudo apt-get install build-essential
```

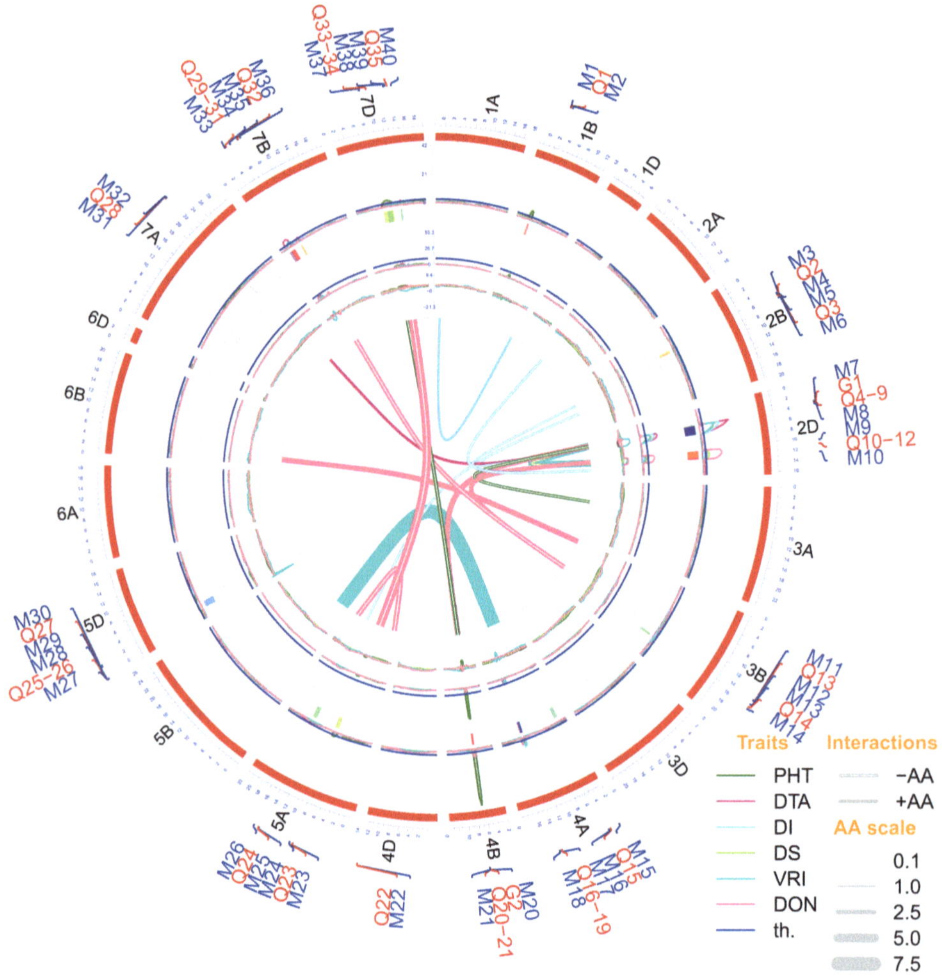


**Fig. 2** A representative figure of whole-genome QTL plot based on CIM analysis and reproduced from Dhariwal et al. [83]. The outermost circle/track shows the 21 chromosomes (1A–7D) arranged in a clockwise direction with the selected QTLs (Q1–Q35), candidate genes (G1–G2), and linked/flanking markers (M1–M40) in 0.1X scale (cM). Three inner circles and line connections in the middle represent the mean LOD score (second track from outside), % phenotypic variation ($R^2$) explained (third track from outside), additive effect (AE) (fourth track from outside) of individual QTL, and epistatic effect (AxA or AA) of digenic QTL x QTL interactions (line connections in the middle) for different measured traits. In the second and third tracks from outside, blue lines show an LOD threshold of 2.5 and $R^2$ threshold of 10%, respectively. QTL confidence intervals are shown in different colors beneath the QTL scans in the second track from outside. LOD, $R^2$, and AE peaks of different traits are represented in different colors as shown in the trait legend in the lower-right corner of the diagram. Three most significant and stable QTL regions identified are on chromosomes 2D and 4B. AA interactions between QTL pairs of the same or different linkage groups for each trait are represented in the middle of the diagram by line connections of different colors, as shown in the trait legend in the lower-right corner of the diagram. Negative and positive AA interactions are represented by hollow and solid lines, respectively. The width of the line connections represents the strength of AA effect, as shown in the AA interaction scale in the lower-right corner of the diagram

7. Alternatively, one can install and configure NCBI-BLAST using the following commands (needs access to root):

```
sudo apt-get install ncbi-blast+
mkdir $HOME/blastdb
export BLASTDB=$HOME/blastdb
```

8. Alternatively, one can download the sequences by executing the following command with user-supplied user name and password:

```
wget --user your_user_name --password your_password https://
urgi.versailles.inra.fr/download/iwgsc/IWGSC_RefSeq_Assem-
blies/v2.0/iwgsc_refseqv2.0_all_chromosomes.fa.zip
```

or with ask password option to supply password later:

```
wget --user=your_user_name --ask-password https://urgi.ver-
sailles.inra.fr/download/iwgsc/IWGSC_RefSeq_Assemblies/v2.0/
iwgsc_refseqv2.0_all_chromosomes.fa.zip
```

9. It is obvious that LGs may have some markers that mapped to another chromosome in a previous study or on reference genome. One should not worry about these unless a really long or significant proportion of LG contain markers from other chromosome(s). These may be the result of a translocation or a mapping error. To assess whether these resulted due to error, recalculate that particular LG using MapDisto at lower threshold values.
10. Sometimes QTL Network stops upon completion of the analyses. It happens mainly when data is huge and complex. In that case, it does not generate a result file (*.pre), but detailed analysis file (*.qnk) is always generated. This file also contains the same results in tabular form, as of *.pre file, at the end of each trait. These tables can be copied.
11. Markers, candidate genes, and QTLs should be labeled following guidelines given in Catalogue of Gene Symbols for Wheat by McIntosh et al. [86].
12. However, picture formats require significantly more time than *.pdf format; one can plot results to a different file format such as *.png using the following command:

```
png("QTLplot.png", width=8, height=8, units="in", res=1200)
```

13. Set or scale the "lwd" values equal to values of AxA effect to reflect the epistasis interaction effect as line width. Write a

different color name for every different interaction in the last argument (col="") in separate commands. Epistatic loci with the highest AxA effect should be plotted first as executed in this example. As the width of connected line between first two epistatic loci is highest, it can be viewed easily under other lines.

## References


1. Jiménez-Gómez JM (2011) Next generation quantitative genetics in plants. Front Plant Sci 2:77–77. https://doi.org/10.3389/fpls.2011.00077
2. Heino M (2014) Chapter four—quantitative traits. In: Cadrin SX, Kerr LA, Mariani S (eds) Stock identification methods, 2nd edn. Academic Press, San Diego, pp 59–76. https://doi.org/10.1016/B978-0-12-397003-9.00004-7
3. Kearsey MJ, Farquhar AGL (1998) QTL analysis in plants; where are we now? Heredity 80 (2):137–142. https://doi.org/10.1046/j.1365-2540.1998.00500.x
4. Mendel G (1866) Experiments in plant hybridization (1865). In: Bennett JH (ed) Proceedings of the Natural History Society of Brünn. Oliver and Boyd, Edinburgh, pp 3–47
5. Galton F (1876) A theory of heredity. J Anthropol Inst 5:329–348
6. Galton F (1877) Typical laws of heredity. Nature 15:492–495. 512–514, 532–533
7. Galton F (1889) Natural inheritance. Macmillan, London
8. Gleason K (2017) Sex limited inheritance in drosophila (1910) by Thomas Hunt Morgan. http://embryo.asu.edu/handle/10776/11509
9. Boveri T (1902) Über mehrpolige Mitosen als Mittel zur Analyse des Zellkerns (On multipolar mitosis as a means to analyze the cell nucleus). In: Verhandlungen der physicalisch-medizinischen Gesselschaft zu Würzburg (Proceedings of the physical-medical company at Wurzburg). pp 67–90
10. Sutton WS (1902) The chromosomes in heredity. Biol Bull 4:231–250
11. Johannsen W (1909) Elemente der xakten Erblichkeitsliehre. Fisher, Jena
12. Nilsson-Ehle H (1909) Kreuzungsuntersuchungen an Hafer und Weizen. Lunds Universitets Arsskrift 7:280–291
13. East EM (1916) Studies on size inheritance in Nicotiana. Genetics 1:164–176
14. Lander ES, Botstein D (1989) Mapping mendelian factors underlying quantitative traits using RFLP linkage maps. Genetics 121 (1):185–199
15. Zhang T, Jia X, Xu Z (2018) An alternative analysis on Nilsson-Ehle's hybridization experiment in wheat—theory of dual multiple factors and three Normal distributions on quantitative inheritance (continuation). Appl Math 9:1005–1014. https://doi.org/10.4236/am.2018.98068
16. Morgan TH (1910) Chromosomes and heredity. Am Nat 44(524):449–496
17. Morgan TH (1910) Sex limited inheritance in drosophila. Science 32(812):120–122
18. Morgan TH (1911) Chromosomes ansd associative inheritance. Science 34:384
19. Morgan TH (1911) Random segregation versus coupling in Mendelian inheritance. Science XXXIV:1911
20. Sturtevant AH (1913) The linear arrangement of six sex-linked factors in Drosophila, as shown by their mode of association. J Exp Zool 14:43–59
21. Fisher RA (1919) XV.—the correlation between relatives on the supposition of Mendelian inheritance. Trans R Soc Edinburgh 52 (2):399–433. https://doi.org/10.1017/S0080456800012163
22. Nelson RM, Pettersson ME, Carlborg Ö (2013) A century after Fisher: time for a new paradigm in quantitative genetics. Trends Genet 29(12):669–676. https://doi.org/10.1016/j.tig.2013.09.006
23. Sax K (1923) The association of size differences with seed-coat pattern and pigmentation in PHASEOLUS VULGARIS. Genetics 8 (6):552–560
24. Rasmusson JM (1935) Studies in the inheritance of quantitative characters in Pisum: I. preliminary notes on the genetics of flowering. Hereditas 20:161–180
25. Everson EH, Schaller CW (1955) The genetics of yield differences associated with awn barbing in the barley hybrid (lion x Atlas10) x atlas. Agron J 47:276
26. Burnham CR (1966) Cytogenetics in plant improvements. In: Frey KJ (ed) Plant breeding

symposium. Iowa State University, Ames, Iowa, pp 139–187

27. Law CN (1967) The location of genetic factors controlling a number of quantitative characters in wheat. Genetics 56:445–461
28. Geldermann H (1975) Investigations on inheritance of quantitative characters in animals by gene markers I. methods. Theor Appl Genet 46 (7):319–330. https://doi.org/10.1007/BF00281673
29. Mather K (1949) Biometrical genetics, 1st edn. Methuen, London
30. Dudley JW (1993) Molecular markers in plant improvement: manipulation of genes affecting quantitative traits. 33 (4):cropsci1993.0011183X003300040003x. doi: https://doi.org/10.2135/cropsci1993.0011183X003300040003x
31. Bailey DW (1971) Recombinant inbred strains. Transplantation 11:325–327
32. Burr B, Burr FA, Thompson KH, Albertson MC, Stuber CW (1988) Gene mapping with recombinant inbreds in maize. Genetics 118 (3):519–526
33. Mackay TFC, Stone EA, Ayroles JF (2009) The genetics of quantitative traits: challenges and prospects. Nat Rev Genet 10(8):565–577. https://doi.org/10.1038/nrg2612
34. Kulwal PL (2018) Trait mapping approaches through linkage mapping in plants. In: Varshney RK, Pandey MK, Chitikineni A (eds) Plant genetics and molecular biology. Springer International Publishing, Cham, pp 53–82. https://doi.org/10.1007/10_2017_49
35. Liu BH (1998) Statistical genomics: linkage, mapping, and QTL analysis. CRC Press, Boca Raton, FL. https://doi.org/10.1201/9780203738658
36. Collard BCY, Jahufer MZZ, Brouwer JB, Pang ECK (2005) An introduction to markers, quantitative trait loci (QTL) mapping and marker-assisted selection for crop improvement: the basic concepts. Euphytica 142 (1):169–196. https://doi.org/10.1007/s10681-005-1681-5
37. Zeng ZB (1994) Precision mapping of quantitative trait loci. Genetics 136(4):1457–1468
38. Li H, Ye G, Wang J (2007) A modified algorithm for the improvement of composite interval mapping. Genetics 175(1):361–374. https://doi.org/10.1534/genetics.106.066811
39. Hackett CA, Meyer RC, Thomas WTB (2001) Multi-trait QTL mapping in barley using multivariate regression. Genet Res 77(1):95–106. https://doi.org/10.1017/S0016672300004869
40. Kao C-H, Zeng Z-B, Teasdale RD (1999) Multiple interval mapping for quantitative trait loci. Genetics 152(3):1203–1216
41. Satagopan JM, Yandell BS, Newton MA, Osborn TC (1996) A bayesian approach to detect quantitative trait loci using Markov chain Monte Carlo. Genetics 144(2):805–816
42. Jansen RC, Nap J-P (2001) Genetical genomics: the added value from segregation. Trends Genet 17(7):388–391. https://doi.org/10.1016/S0168-9525(01)02310-1
43. Breitling R, Li Y, Tesson BM, Fu J, Wu C, Wiltshire T, Gerrits A, Bystrykh LV, de Haan G, Su AI, Jansen RC (2008) Genetical genomics: spotlight on QTL hotspots. PLoS Genet 4(10):e1000232. https://doi.org/10.1371/journal.pgen.1000232
44. Jansen RC, Tesson BM, Fu J, Yang Y, McIntyre LM (2009) Defining gene and QTL networks. Curr Opin Plant Biol 12(2):241–246. https://doi.org/10.1016/j.pbi.2009.01.003
45. Keurentjes JJB, Koornneef M, Vreugdenhil D (2008) Quantitative genetics in the age of omics. Curr Opin Plant Biol 11(2):123–128. https://doi.org/10.1016/j.pbi.2008.01.006
46. Joosen RVL, Ligterink W, Hilhorst HWM, Keurentjes JJB (2009) Advances in genetical genomics of plants. Curr Genomics 10 (8):540–549. https://doi.org/10.2174/138920209789503914
47. Tanksley SD (1993) Mapping polygenes. Annu Rev Genet 27(1):205–233. https://doi.org/10.1146/annurev.ge.27.120193.001225
48. Mackay TFC (2001) The genetic architecture of quantitative traits. 35(1):303–339. https://doi.org/10.1146/annurev.genet.35.102401.090633
49. Mauricio R (2001) Mapping quantitative trait loci in plants: uses and caveats for evolutionary biology. Nat Rev Genet 2(5):370–381. https://doi.org/10.1038/35072085
50. Asíns MJ (2002) Present and future of quantitative trait locus analysis in plant breeding. 121 (4):281–291. https://doi.org/10.1046/j.1439-0523.2002.730285.x
51. Doerge RW (2002) Mapping and analysis of quantitative trait loci in experimental populations. Nat Rev Genet 3(1):43–52. https://doi.org/10.1038/nrg703
52. Hackett CA (2002) Statistical methods for QTL mapping in cereals. Plant Mol Biol 48 (5):585–599. https://doi.org/10.1023/A:1014896712447
53. Gupta PK, Kulwal PL, Mir RR (2013) QTL mapping: methodology and applications in cereal breeding. In: Gupta PK, Varshney RK (eds) Cereal genomics II. Springer

Netherlands, Dordrecht, pp 275–318. https://doi.org/10.1007/978-94-007-6401-9_11

54. Frommlet F, Bogdan M, Ramsey D (2016) Statistical methods of QTL mapping for experimental populations. In: Frommlet F, Bogdan M, Ramsey D (eds) Phenotypes and genotypes: the search for influential genes. Springer London, London, pp 73–104. https://doi.org/10.1007/978-1-4471-5310-8_4
55. Soto-Cerda BJ, Vasudevan A, Laroche A, Ragupathy R (2021) Genome-wide association studies (GWAS) in cereals. In: Bilichak A, Laurie JD (eds) Accelerated breeding for cereal crops. Springer Science+Business Media, LLC, New York, NY
56. Wu R, Zeng ZB (2001) Joint linkage and linkage disequilibrium mapping in natural populations. Genetics 157(2):899–909
57. Gupta PK, Rustgi S, Kulwal PL (2005) Linkage disequilibrium and association studies in higher plants: present status and future prospects. Plant Mol Biol 57(4):461–485. https://doi.org/10.1007/s11103-005-0257-z
58. Zhu C, Gore M, Buckler ES, Yu J (2008) Status and prospects of association mapping in plants. 1(1):5–20. https://doi.org/10.3835/plantgenome2008.02.0089
59. Würschum T, Liu W, Gowda M, Maurer HP, Fischer S, Schechert A, Reif JC (2012) Comparison of biometrical models for joint linkage association mapping. Heredity 108 (3):332–340. https://doi.org/10.1038/hdy.2011.78
60. Gupta PK, Kulwal PL, Jaiswal V (2014) Association mapping in crop plants: opportunities and challenges. Adv Genet 85:109–147. https://doi.org/10.1016/b978-0-12-800271-1.00002-0
61. Jansen RC (1993) Interval mapping of multiple quantitative trait loci. Genetics 135 (1):205–211
62. Zeng ZB (1993) Theoretical basis for separation of multiple linked gene effects in mapping quantitative trait loci. Proc Natl Acad Sci 90 (23):10972–10976. https://doi.org/10.1073/pnas.90.23.10972
63. Churchill GA, Doerge RW (1994) Empirical threshold values for quantitative trait mapping. Genetics 138(3):963–971
64. Doerge RW, Churchill GA (1996) Permutation tests for multiple loci affecting a quantitative character. Genetics 142(1):285–294
65. Lander ES, Green P, Abrahamson J, Barlow A, Daly MJ, Lincoln SE, Newburg L (1987) MAPMAKER: an interactive computer package for constructing primary genetic linkage maps of experimental and natural populations. Genomics 1(2):174–181. https://doi.org/10.1016/0888-7543(87)90010-3
66. Feldman M (2001) The origin of cultivated wheat. In: Bonjean A, Angus W (eds) The wheat book. Lavoisier Tech & Doc, Paris, pp 1–56
67. Feldman M, Lupton FGH, Miller TE (1995) Wheats. In: Smartt J, Simmonds NW (eds) Evolution of crop plants, 2nd edn. Longman Scientific, London, pp 184–192
68. Niemirowicz-Szczytt K (1997) Excessive homozygosity in doubled haploids—advantages and disadvantages for plant breeding and fundamental research. Acta Physiol Plant 19(2):155–167. https://doi.org/10.1007/s11738-997-0032-7
69. Dhingani RM, Umrania VV, Tomar RS, Parakhia MV, Golakiya BA (2015) Introduction to QTL mapping in plants. Ann Plant Sci 4 (4):1072–1079
70. Wu Y, Bhat PR, Close TJ, Lonardi S (2008) Efficient and accurate construction of genetic linkage maps from the minimum spanning tree of a graph. PLoS Genet 4(10):e1000212. https://doi.org/10.1371/journal.pgen.1000212
71. Heffelfinger C, Fragoso CA, Lorieux M (2017) Constructing linkage maps in the genomics era with MapDisto 2.0. Bioinformatics 33 (14):2224–2225. https://doi.org/10.1093/bioinformatics/btx177
72. Yang J, Hu C, Hu H, Yu R, Xia Z, Ye X, Zhu J (2008) QTLNetwork: mapping and visualizing genetic architecture of complex traits in experimental populations. Bioinformatics 24 (5):721–723. https://doi.org/10.1093/bioinformatics/btm494
73. Core Team R (2013) R: a language and environment for statistical computing. R Foundation for Statistical Computing, Vienna
74. Hu Y, Yan C (2015) OmicCircos: high-quality circular visualization of omics data. R package version 1.16
75. Hu D, Tang Y, Yuan Z, Wang J (1983) The induction of pollen sporophytes of winter wheat and the development of the new variety Jinghua no. 1. Sci Agric Sin 1(1983):29–35
76. Datta SK, Wenzel G (1987) Isolated microspore derived plant formation via embryogenesis in Triticum aestivum L. Plant Sci 48 (1):49–54. https://doi.org/10.1016/0168-9452(87)90069-0
77. Tuvesson D, Kirstine I, Öhlund V, Charlotte R (1993) Plant regeneration through culture of isolated microspores of Triticum aestivum L. Plant Cell Tissue Organ Cult 34

(2):163–167. https://doi.org/10.1007/BF00036097

78. Barclay IR (1975) High frequencies of haploid production in wheat (Triticum aestivum) by chromosome elimination. Nature 256 (5516):410–411. https://doi.org/10.1038/256410a0
79. Inagaki M (1985) Chromosome doubling of the wheat haploids obtained from crosses with Hordeum bulbosum L. Jpn J Breed 35 (2):193–195. https://doi.org/10.1270/jsbbs1951.35.193
80. Laurie DA, Bennett MD (1988) The production of haploid wheat plants from wheat x maize crosses. Theor Appl Genet 76 (3):393–397. https://doi.org/10.1007/BF00265339
81. Suenaga K, Nakajima K (1989) Efficient production of haploid wheat (Triticum aestivum) through crosses between Japanese wheat and maize (Zea mays). Plant Cell Rep 8 (5):263–266. https://doi.org/10.1007/bf00274125
82. Dhariwal R, Fedak G, Dion Y, Pozniak C, Laroche A, Eudes F, Randhawa H (2018) High density single nucleotide polymorphism (SNP) mapping and quantitative trait loci (QTL) analysis in a biparental spring triticale population localized major and minor effect fusarium head blight resistance and associated traits QTL. Genes 9(1):19
83. Dhariwal R, Henriquez MA, Hiebert C, McCartney CA, Randhawa HS (2020) Mapping of major fusarium head blight resistance from Canadian wheat cv. AAC tenacious. Int J Mol Sci 21(12):4497. https://doi.org/10.3390/ijms21124497
84. Wang S, Wong D, Forrest K, Allen A, Chao S, Huang BE, Maccaferri M, Salvi S, Milner SG, Cattivelli L, Mastrangelo AM, Whan A, Stephen S, Barker G, Wieseke R, Plieske J, Lillemo M, Mather D, Appels R, Dolferus R, Brown-Guedira G, Korol A, Akhunova AR, Feuillet C, Salse J, Morgante M, Pozniak C, Luo M-C, Dvorak J, Morell M, Dubcovsky J, Ganal M, Tuberosa R, Lawley C, Mikoulitch I, Cavanagh C, Edwards KJ, Hayden M, Akhunov E (2014) Characterization of polyploid wheat genomic diversity using a high-density 90 000 single nucleotide polymorphism array. Plant Biotechnol J 12(6):787–796. https://doi.org/10.1111/pbi.12183
85. Green MR, Sambrook J (2014) Molecular cloning, a laboratory manual. Cold Spring Harbor Laboratory Press, Cold Spring Harbor, NY
86. McIntosh RA, Hart GE, Devos KM, Gale MD, Rogers WJ (1998) Catalogue of gene symbols for wheat. doi:https://wheat.pw.usda.gov/ggpages/wgc/98/

# Chapter 3

# Introduction to Marker-Assisted Selection in Wheat Breeding

**Sudhakar Pandurangan, Clare Workman, Kirby Nilsen, and Santosh Kumar**

## Abstract

Marker-assisted selection (MAS) is a selection method for improving agriculturally essential traits in wheat. MAS allows efficient screening of difficult-to-phenotype traits, introgression of genomic complements from donor germplasm into elite breeding lines, and gene/trait pyramiding for quantitative traits. The discovery of a large number of QTL-specific molecular markers for qualitative and quantitative traits has accelerated MAS in recent years. Markers for loci related to disease resistance and agronomic and quality traits are available in wheat. The convenience of detecting and following the inheritance pattern of major genes/QTL at a low cost has improved genotyping and germplasm selection in the breeding programs. The development of high-throughput sequencing, genetic loci probing technologies, precision phenotyping systems, crop molecular physiology, and computational tools further extends the possible uses of MAS in wheat. The pan-genomic resources and the annotated genome sequence increase the effectiveness of MAS in hexaploid wheat by allowing for primer design, in silico validation of primers for specific binding, QTL walking, and genome-anchored fine mapping. This chapter sheds light on the use of MAS in shortening the breeding cycle, improving biotic and abiotic stress resistance, and sustaining the yield potential of wheat.

**Key words** Marker-assisted selection, Genotyping, Phenotyping, Stress resistance, Yield, Wheat

## 1 Introduction

### 1.1 Breeding for Field-Ready Cultivars and its Challenges

Bread wheat (*Triticum aestivum* L.) is an important cereal crop and is consumed all over the world. It is the staple food for one-third of the world's population. In 2018, the global production of wheat was 734 million tons [1]. The current wheat yield is insufficient to meet the projected demands of the rising world population [2]. Therefore, it is critical to increase the yield potential of wheat varieties. Breeding field-ready cultivars is an integral part of attaining food security for the developing and the underdeveloped parts of the world. An ideal field-ready wheat cultivar has a high yield, biotic and abiotic stress tolerance, and optimum end-use quality. Wheat yield and production have steadily risen worldwide since the

Andriy Bilichak and John D. Laurie (eds.), *Accelerated Breeding of Cereal Crops*, Springer Protocols Handbooks, https://doi.org/10.1007/978-1-0716-1526-3_3, © Springer Science+Business Media, LLC, part of Springer Nature 2022

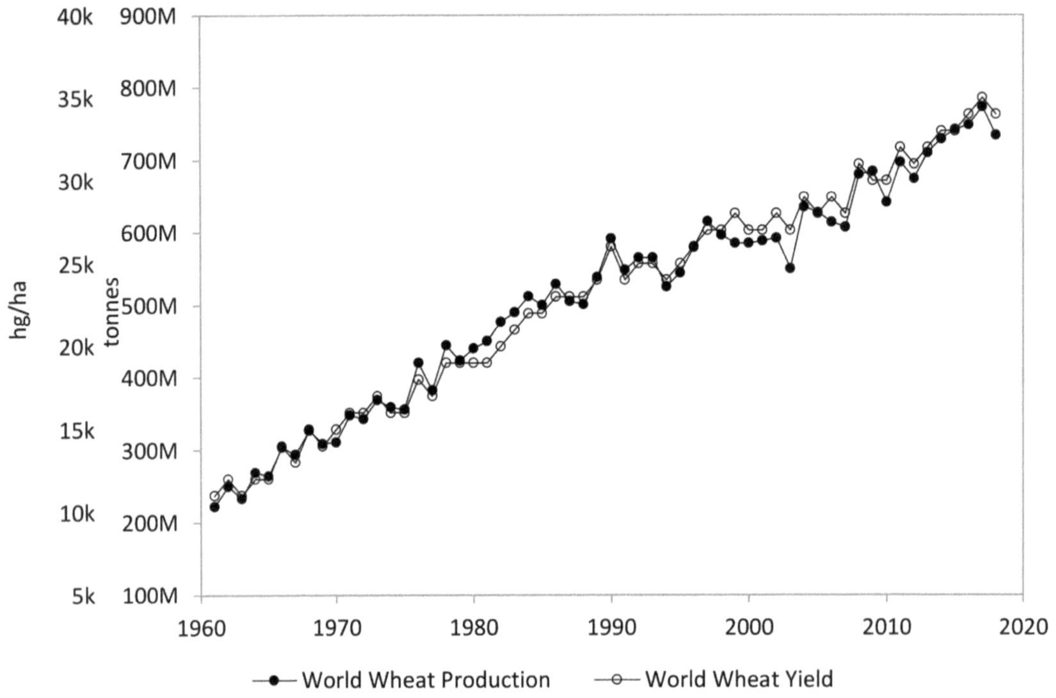


**Fig. 1** World wheat production and yield from 1961 to 2018

Green Revolution (Fig. 1) [1]. However, the low average annual genetic gain of premium quality wheat at 0.6–0.7% [3, 4] remains a crucial challenge to wheat production. Climate change and increased urbanization will impact sustainable agriculture due to increased carbon dioxide concentration, rising sea levels, reduced agricultural land, and increased input costs [5]. Wheat yield and production must continue to increase in order to overcome these challenges. MAS is an essential breeding tool that can make it easier and faster for breeders to select traits of interest to increase productivity in the field-ready wheat cultivars. In this chapter, we present an overview of the recent developments in MAS for wheat breeding.

### 1.2 MAS as an Assistive Technology

Practical plant breeding relies on rigorous selection for desirable traits. Phenotypic selection, coupled with traditional breeding, has been utilized successfully in developing crop cultivars. Despite the significant contributions, traditional breeding has several limitations. For instance, selection based entirely on phenotype is inefficient for complex traits that are less heritable and have high environmental influence and requires an extended period (10–12 years) to develop varieties [6]. Phenotypic selection procedures are often costly and extremely challenging, for example, the selection for tolerance to abiotic stresses and quarantined diseases. As assistive technology, MAS helps overcome the limitations of traditional phenotype-based breeding and significantly improves plant breeding efficiency primarily for incorporation of multi-allelic traits to generate high-yielding cultivars [7].

MAS is a method of screening plants based on the heritability of a defined sequence of nucleic acids with the trait of interest, where the molecular marker linked to the trait is selected [8]. MAS is useful in situations when:

(a) Phenotypic selection is restricted (race-specific screening for Ug99).

(b) Selection depends on specific environmental conditions or developmental stages that influence the expression of the target trait (preharvest sprouting (PHS), seedling/adult plant resistance for rust pathogens).

(c) Maintenance of recessive alleles in backcrossing programs and to accelerate the recurrent genome recovery.

(d) Pyramiding multiple monogenic traits (resistance to various diseases, baking quality traits) or traits controlled by major QTL such as Fusarium head blight (FHB).

(e) Phenotypic selection is simply too costly and time-consuming (root architecture). MAS has its limitations as it is relatively efficient for simple traits with high heritability. Whereas for complex traits governed by many minor effect QTLs with low heritability, MAS is less efficient compared to conventional phenotypic selection [9, 10]. As a complementary approach, genomic selection (GS) employing genome-wide markers may be more effective for selecting complex traits like yield [10, 11].

### 1.3 Cost-Effective Implementation of MAS

MAS could be implemented at different stages of the breeding cycle (selection of parents for developing breeding populations, early-generation and advanced-generation selections). Using MAS on segregating populations, the desirable allele frequency can be improved during early-generation selection. Application of MAS in the $F_2$ can expedite identifying superior allele combinations [12]. However, stringent selection at early generation based on a few markers may reduce genetic diversity and may have detrimental effects on adaptability and resistance to biotic and abiotic stresses [13]. Sustaining a minimum number of plants to maintain sufficient allelic variability at the unselected loci is advantageous for developing elite material adapted to various local needs.

To adopt MAS into a breeding program, it should be cost-effective. The reduced labor and field equipment cost of conventional breeding must justify the cost of high-throughput MAS. The cost of implementing MAS into a conventional breeding program depends on the application of MAS. For example, MAS combined with phenotypic screening was more effective than phenotypic screening alone for FHB resistance [14]. MAS, used in conjunction with other advanced breeding techniques such as doubled haploids (DHs), is more efficient and reduces the cost by 40% over traditional breeding [15]. Implementing MAS using simple sequence repeats (SSRs) and single-nucleotide polymorphisms (SNPs) had a moderate start-up cost, but low cost per analysis [16]. The cost of genotypic screening for a marker associated with a disease resistance reduced when multiple markers combined into a multiplex assay [12]. Besides, the costs associated with equipment, consumables, marker development/validation, and skilled persons contribute to breeding expenditures. To reduce the cost for MAS to be feasible:

(a) Conduct genotypic assays only on progenies possessing the relevant marker allele based on their pedigree.
(b) Screen out plants based on undesirable plant ideotype, thus allowing MAS on the most promising materials.
(c) Use MAS for high-priority traits that are difficult, time-consuming, or expensive to measure.
(d) Use of high-throughput scalable markers such as markers based on SNP.
(e) Use of multiplexed assays and high-throughput genotyping systems.
(f) Choice of targeted [kompetitive allele-specific PCR (KASP) assay] vs whole-genome scan [array- or next-generation sequencing (NGS)-based] depending on the number of markers and population size.
(g) Automation vs manual assays based on throughput and needs of the program (potential for automation increases the investment costs while the manual methods lower the cost of MAS but compromise the throughput).
(h) Availability of on-demand genotypic service providers with good track record and low genotyping error rates.

Implementing MAS in different crop breeding programs demonstrated significant cost and time savings in wheat [15], maize [17], rice [7, 18], potato [12], and soybean breeding [19]. With the continued improvements in crop genomics and computational analysis, the design and assay costs may fall, further facilitating MAS in plant breeding.

### 1.4 Trait Consideration and Marker Implementation

Both qualitative and quantitative traits are selected using markers [20]. MAS's utility for manipulating qualitative/simple/single-gene traits is straightforward and has been well documented [7, 8, 12]. These traits include (a) *disease/pest resistance*, including resistance against various rusts, FHB, common bunt, orange wheat blossom midge, wheat stem sawfly, Hessian fly, and Russian wheat aphid; (b) *agronomic traits*, plant height, photoperiod response, and thousand-kernel weight; and (c) *grain quality traits*, including protein content, gluten strength, color hardness, and tolerance to PHS. MAS for the improvement of polygenic traits is possible. In general, MAS seems to be most effective for traits with high heritability, with much of the variation controlled by a few genes with large effects [20]. When using MAS to target a specific QTL, the effects should be stable across environments [21] and explain at least 10–20% of the phenotypic variance [22]. In wheat, the major FHB resistance QTL on 3BS, referred to as *Fhb1*, accounts for 20–40% of the phenotypic variation for type II FHB resistance in different genetic backgrounds widely used in MAS [22]. MAS to pyramid recently cloned stem rust R genes *Sr22*, *Sr45*, and *Sr50* is necessary to provide durable resistance against the Ug99 family of stem rust [23]. With the recent development of molecular tools and genetic maps, MAS has become more attractive and practical to develop broad-spectrum resistance to diseases and insects. Abiotic stress tolerance traits like drought, heat, salinity, waterlogging, and metal toxicity are the future targets for wheat improvement using MAS.

The MASWheat consortiums in the USA and Mexico (CIMMYT) have made remarkable progress in implementing MAS strategies for cultivar development. MAS is advantageous for active selection in early-generation and passive selection in advanced-generation germplasm. In active selection, progenies positive for a marker are selected while the rest (progenies that do not possess the marker of interest or possess a marker with adverse effects) are discarded. Using active selection to only select progenies homozygous for a gene/marker reduces the population size, thus reducing the genetic variability required to select superior gene combinations that may be more adaptable. Selecting progenies heterozygous for a marker might segregate in the successive generations leading to a considerable proportion of lines with undesired traits. Passive selection is useful when genes are segregating, and the breeder wants to keep the population large enough to observe the transgressive segregants and phenotypic traits of interest. Passive selection involves performing MAS but may not result in discarding lines purely based on the markers. MAS information can be used passively to discard lines in combination with phenotypic, agronomic, or grain quality information obtained at different generations of the breeding cycle. Molecular markers are also employed to trace the presence of target genes/QTL introgressed from the donor parent (foreground selection) as well as for accelerating the recovery of the recurrent parent genome (background selection) in backcross programs [20, 24].

### 1.5 Specialized Knowledge and Instrumentation

The crucial aspects of integrating MAS in conventional breeding are the awareness on breeding priorities, availability of appropriate germplasm, screening efficiencies of segregating traits, and the knowledge of marker-trait associations [20]. The availability of qualified expertise and financial prudence are also critical considerations. DNA extraction, marker identification, validation on germplasm, genotyping, and data analysis require knowledge on molecular biology, breeding, genetics, and genomics. Therefore, a multidisciplinary team's interaction is vital in using MAS effectively [7]. In particular, the role of computational methods (including bioinformatics) is the leading force behind realizing the full potential of DNA marker-assisted approaches [25]. Overall, closing the knowledge gap among specialists in their field would harness MAS's full potential in crop breeding [7].

DNA marker technology and genotyping techniques have developed rapidly in the past two decades. SSR markers have been used extensively, primarily due to compatibility with existing lab equipment [26]. A new generation of molecular markers based on SNP promises high-throughput assays and multiplexing [27]. The development of a 384-well DNA extraction protocol (NucleoMag 384 Plant, Takara Bio USA, Inc.) and multiplexed high-throughput genotyping equipment resulted in shortening the processing time and made a significant impact on progressing deserving germplasm through the breeding pipeline [27].

### 1.6 Efficiencies of MAS over Traditional Breeding

MAS refers to the use of molecular markers to assist phenotypic selections in crop improvement. Molecular breeding or MAS offers an opportunity to accelerate traditional breeding. Traditional breeding complicates selecting suitable genotypes as most agronomic traits are quantitative, and selection is based only on phenotype. Various phenotypic markers are commonly implemented by the breeders depending on the germplasm, growing conditions, and the expression of the phenotypic markers (Table 1). Also, conventional breeding involves repeated testing of breeding lines under natural or artificial inoculations. This process is time-consuming, costly, and prone to influence by the environment. Genotype-by-environment interaction (GEI) is a common problem, and it further confounds the line selection and phenotyping procedures. Thus, it is relevant to supplement phenotypic selection with MAS. By employing molecular markers, desirable genes can be pyramided or stacked and fixed in the early generations. Moreover, molecular markers are unaffected by environmental conditions and are detectable in all stages of plant growth. The use of MAS in crop improvement may reduce the cost of developing new wheat varieties and increase the precision and efficiency of selection in the breeding program.

As a method of selection, MAS requires establishing a correlation between the desired trait and molecular marker(s). Correlation is established by phenotyping a genetic mapping population followed by QTL analysis. DNA markers associated with QTL/genes for major economic traits are being deployed through MAS in wheat breeding programs. MAS should be employed with caution

**Table 1**
**Phenotypic traits as breeding target for wheat improvement**

| Phenotype | Gene(s) | Comment |
|---|---|---|
| Growth condition/habit | *Vrn1, Vrn2, Vrn3, Vrn4* | Spring vs winter |
| Earliness | Photoperiod-insensitive alleles of *Ppd-1* (*Ppd-A1a*, *Ppd-B1a*, and *Ppd-D1a*) | Flower early regardless of photoperiod |
| Longer coleoptiles | *Lcol-A1* | Suitable for deep sowing (opposite of *Rht* genes). Promotes good emergence and establishment |
| Plant height | *Rht1*, *Rht2*, *Rht8*, *Rht18*, *Rht24*, *Rht25* | Semidwarf plant type |
| Leaf tip necrosis | *Ltn1/Lr34/Yr18/Sr57/PM38* *Lr67/Yr46* | Tolerance to wheat leaf rust, powdery mildew, stem rust, and stripe rust |
| Erect leaf posture | Spot blotch genes (*Sb1*, *Sb2*, *Sb3*), *Ltn1*, *Lr34* | Selection for moderate resistance to spot blotch |
| Leaf rolling and chlorotic streaking | Russian wheat aphid (RWA), *Diuraphis noxia* (Mordvilko) | Susceptible to RWA |
| Flag leaf size (under moderate drought stress) | Unknown | Increased photosynthetic capacity and positive role in drought resistance |
| Stem melanism or "pseudo-black chaff" | *Sr2/Lok1* | Resistance to wheat stem rust (*Fhb1* may be excluded as *Sr2* and *Fhb1* are in tight repulsion) |
| Anther extrusion | *Rht-D1a*, *APETELA2 (AP2-D)* | Selection for enhanced FHB resistance. AE also favors efficient cross-pollination in hybrid production |
| Paired spikelet | *Teosinte Branched1 (TB1)*, *Photoperiod-1 (Ppd-1)*, *FLOWERING LOCUS T (FT)* | Increased number of grains per spikelet and crop yield |
| Kernel color | "R" genes *(chalcone synthase (CHS), chalcone isomerase (CHI), flavanone 3-hydroxylase (F3H),* and *dihydroflavonol 4-reductase (DFR)* | Hard red/white kernel color |
| Dormant kernels, red kernel | *R-1 gene (R-A1b, R-B1b, and R-D1b), Myb-type transcription factor (TaMyb10)* | Tolerance to preharvest sprouting |
| Kernel size/weight | *Sucrose synthase (Sus2-2B), trehalose 6-phosphate phosphatase* (*TaTPP-6AL1a*) | High TKW and variation in flour water absorption tolerance |
| Hard kernel | Puroindoline genes (*Pina, Pinb*) | Better milling, water absorption capacity of flour, and baking quality |
| Delay in senescence | *No apical meristem* (*NAM-B1*) allele | Decreased wheat protein, Fe and Zn content |

as enhancing an allele may have a negative effect due to linkage drag (*Gpc-B1* is associated with yield penalty), eliminate the genes linked in repulsion (*Fhb1* and *Sr2*), and/or co-transfer deleterious genes or unfavorable traits (negative correlation between *Fhb-5Ac* and plant height). In wheat breeding, MAS helped accelerate the introgression of many genes that improve disease resistance and agronomic and quality traits. Wheat breeding programs targeting improvement in these traits have been in progress in several countries, including Canada, the USA, Europe, China, Australia, and Mexico (CIMMYT). Other traits targeted for wheat improvement using MAS are drought, heat, salinity, waterlogging, and metal toxicity tolerance (e.g., boron, aluminum, and arsenic). Overall, MAS offers an enormous scope of accelerated progress in selecting stress-tolerant crop plants with increased precision.

## 2 Marker Technologies

### 2.1 A Brief History of Breeding Markers

Genetic markers are critical developments in the field of plant breeding. Markers are heritable polymorphisms and are associated with a QTL/trait in a specific gene pool and act as a sign or flag [28]. Numerous marker technologies have been developed since the application of molecular markers in the 1980s for genome mapping [29, 30]. DNA markers largely adopted to wheat breeding are restriction fragment length polymorphism (RFLP) in 1989 [30], random amplified polymorphic DNA (RAPD) in 1992 [31], sequence-tagged site (STS) in 1994 [32], microsatellite or SSR in 1995 [33, 34], sequence-characterized amplified region (SCAR) in 1996 [35], amplified fragment length polymorphism (AFLP) in 1998 [36], inter-simple sequence repeat (ISSR) in 1998 [37], cleaved amplified polymorphic sequence (CAPS) in 2000 [38], expressed sequence tag (EST) in 2004 [39], SNP in 2003–2004 [40, 41], and diversity arrays technology (DArT) markers in 2006 [42]. These technologies are used in MAS with varying degrees of efficiency.

PCR-based markers in wheat are well established, with their advantages and limitations. The markers can be dominant or codominant. A codominant marker can differentiate the heterozygote (Aa) from the two homozygotes (AA or aa), while the heterozygote for a dominant marker will be identical to the homozygote showing "presence" of the marker [24]. RFLP, CAPS, SSR, and SNP are examples of codominant markers. RAPD, AFLP, SCAR, ISSR, and DArT are examples of dominant markers. Codominant markers are preferred in MAS selection schemes because they allow the differentiation between homozygous and heterozygous loci. At $F_2$, codominant markers segregate in a 1:2:1 ratio, and dominant markers segregate in a 3:1 ratio. The segregation ratio is critical to assess the marker loci and to decide appropriate statistical analysis [24]. RAPD, AFLP, and ISSR, in general, are more limited in describing genetic variation due to their dominant nature [24]. In recent years, SSR and SNP markers have become widely applied to genetic linkage map constructions, QTL mapping, genetic diversity analyses, and marker-assisted breeding (MAB). Abundance throughout the genome, reliability in detection, and suitability for automation make SNP the preferred marker of choice over SSRs. Besides, the implementation of SNPs in high-throughput genotyping methods, including DNA array chips, allele-specific PCR, and genotyping-by-sequencing (GBS), makes SNPs the favorite marker in wheat.

The identification and use of SNPs as markers increased significantly with progress in high-throughput sequencing technologies [28, 43]. Although each marker system has its advantages and disadvantages, the choice of marker system is dictated by the intended application, convenience, and the cost involved. In Canadian wheat breeding, SNP-KASP markers are used for high-throughput genotyping. SSR markers are used when a validated SNP marker on Canadian germplasm is not available [44]. Targeted progress was made in developing a SNP-KASP marker for FHB-5AS QTL, where a difficult-to-score SSR was the only marker available [45].

### 2.2 Molecular Markers

Based on the method of analysis or detection techniques, molecular markers are grouped into two categories: (a) hybridization or non-PCR-based and (b) PCR-based (Fig. 2). Since the discovery of first-generation RFLP markers [29] to the current third-generation SNP markers, MAS has seen remarkable progress. The advancements in sequencing technologies expedite the marker identification and increase the number of informative markers. DNA markers associated with important agronomic traits played a significant role in the crop improvement of many species [46–53].

Hybridization-based RFLP markers have been replaced with PCR-based markers that are easy to handle and require less DNA. PCR- and sequence-based DNA markers are used mostly due to their abundance. They have enormous potential to improve conventional plant breeding precision and efficiency via MAS. The commonly used DNA-based marker systems (DArT, SCAR, EST, SSR, and SNP) in hexaploid wheat are briefly addressed here.

#### 2.2.1 DArT

DArT is used for high-throughput characterization of a given genomic representation of a genotype based on hybridization. DArT markers generate whole-genome fingerprints by scoring the presence versus absence of DNA fragments in genomic DNA [42, 54]. For its efficiency and high genome coverage, DArT has been applied to wheat in numerous QTL mapping studies [55, 56],

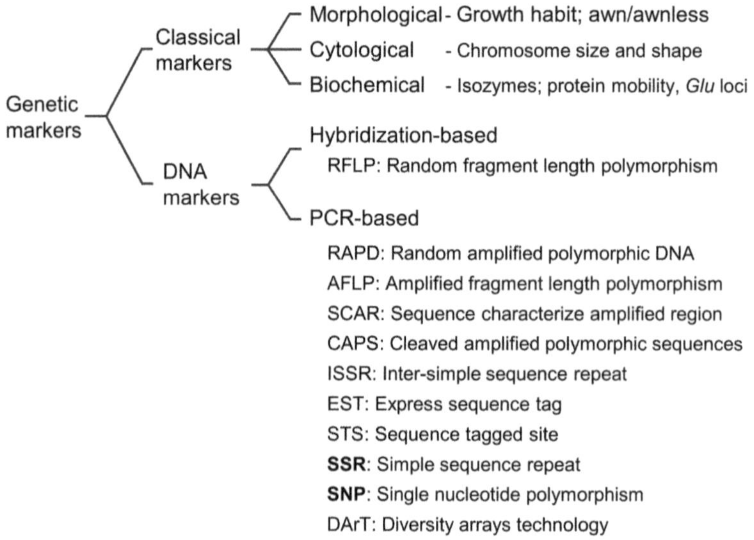


**Fig. 2** Classification of genetic markers. Highlighted in bold are markers widely used in MAS in wheat

genome-wide association studies (GWAS) [57, 58], and GBS [59]. It is a high-throughput array-based system that has no prerequisite for sequence information. However, as DArT exclusively generates dominant markers, other genotyping methods are often preferred. Also, the development of DArT for each species requires significant effort and time.

#### 2.2.2 Scar

A SCAR is a PCR-based monolocus codominant marker that requires the use of two specific primers (20–25 bp) designed from the nucleotide sequence established in cloned RAPD fragments. SCAR markers were first developed to tag downy mildew in lettuce [60]. The conversion of RAPD into more user-friendly SCAR represents the development of codominant markers from dominant markers to avoid the lengthy RAPD procedures. Developing SCAR markers for wheat can be difficult. Nevertheless, SCAR offers several advantages over RAPD and is easy, reliable, less sensitive to reaction conditions, and highly reproducible [60, 61]. SCARs are a variety of STS markers.

In wheat, SCAR markers linked to resistance genes against fungal pathogens, viruses, and insect pests have been characterized and used in the identification of the *Pm2* gene conferring resistance to powdery mildew [62], *Bipolaris sorokiniana* causing spot blotch of wheat [63], chemotypes of divergent *F. graminearum* populations [64], *Tilletia controversa* Kuhn, the causal fungus of wheat dwarf bunt [65], barley yellow dwarf virus (BYDV) resistance gene *Bdv2* [61], and Russian wheat aphid resistance gene *Dn2* [66].

#### 2.2.3 ESTs

ESTs or single-pass cDNA sequences are fragments of expressed genes generated by the sequencing of mRNA (cDNA libraries). ESTs provide a quick and inexpensive option for genotyping [67], obtaining data on gene expression and regulation [68, 69], gene identification [70], and constructing genetic/consensus map [71]. EST databases are sources to develop functional markers from the open reading frames (ORFs) or 5′- or 3′-untranslated regions (UTRs). EST-SSRs also provide a new source for genetic and evolutionary studies based on homology searches of putative SSR functions. A list of 15,605 wheat ESTs with both 5′ and 3′ sequences generated and their GenBank accession is available (https://wheat.pw.usda.gov/NSF/curator/wheat_est_index.html).

#### 2.2.4 SSRs

SSR or microsatellites are stretches of DNA containing a variable number of short tandem repeats of an underlying motif of <6 bp. They are generally codominant, highly polymorphic, abundant, and reliable, and can be readily developed from existing genomic sequences (g-SSRs) or EST-SSRs. The codominant nature of SSRs allows for the direct measurement of heterozygosity and makes them the marker of choice for a variety of purposes, including MAS in practical plant breeding [26]. Genes/QTLs associated with various traits like yield, quality, and biotic and abiotic stress tolerance can be used in a breeding program using their flanking SSR markers [72]. Röder et al. [73] developed 230 polymorphic primer sets containing 279 microsatellites covering the seven homoeologous chromosome groups in wheat with the designation "gwm" (Gatersleben wheat microsatellite). Microsatellite markers from other research groups include the Wheat Microsatellite Consortium, GDM, CFA, CFD, and BARC. A large number of SSRs for wheat are publicly available in GrainGenes (https://wheat.pw.usda.gov/GG3/) and Gramene (http://www.gramene.org) databases. The online hexaploid wheat SSR genomic resource, TaSSRDb [74] (http://webtom.cabgrid.res.in/wheatssr/), contains 476,169 putative SSR markers reported so far, including 268 SSRs associated with 11 traits.

SSRs are the most commonly shared markers between wheat genetic maps to enable the construction of consensus maps or comparative maps to increase marker density in specific regions. The first high-density microsatellite consensus map for bread wheat was constructed by joining four independent genetic maps. A total of 1235 microsatellite loci were mapped, covering 2569 cM, giving an average interval of 2.2 cM [75]. These SSR markers have been used in wheat for the construction of a DNA fingerprinting database for genetic diversity assessment [76], varietal improvement and management [74], seed purity testing and differentiation of hybrids [77], evolutionary studies [78], association studies [79], QTL mapping and gene identification [80], haplotype determination [81], and MAS [44]. However, in recent years, the domination of SSRs in wheat breeding was reduced by the increased deployment of SNP markers. Many of the important SSRs are anchored on genetic and physical maps. This allows for the identification and development of breeder-friendly SNP-based marker assays.

#### 2.2.5 SNPs

SNPs are the most common type of genetic variation. Each SNP represents a single-nucleotide base difference at a locus due to natural mutation [82]. SNPs are usually scored as biallelic markers and classified (a) *based on nucleotide substitutions*, as either transition (C/T or G/A) or transversion (C/G, A/T, C/A, or T/G); (b) *based on genomic location*, as noncoding SNPs (ncSNPs) and exonic SNPs or coding SNPs; and (c) *based on phenotype*, as functional SNPs or candidate SNPs that affect the function of a gene and give rise to phenotypic effect and anonymous SNPs that do not affect the gene function. The ncSNPs located in introns are known as intronic SNPs. An exonic SNP that does not lead to a change in the amino acid sequence is called a synonymous SNP (synSNP), while a nonsynonymous SNP (nsSNP) alters the amino acid sequence and also called as diagnostic SNP. Most SNPs belong to the anonymous SNP category [24]. The SNPs in polyploid species are classified as true or simple SNPs and hemi-SNPs or homoeo-SNPs. A simple or true SNP detects allelic variation between homologous loci of the same genome, whereas hemi-SNPs or homoeo-SNPs detect homoeologous/paralogous loci in the two or more genomes [83]. SNPs are incredibly abundant, easy to detect, and ideal for automation and high throughput, and have relatively low mutation rates. SNPs are also codominant markers and have emerged as preferred genetic markers in genetic studies and plant breeding. When SNPs are used in dominant assays, the heterozygotes are indistinguishable from one of the homozygotes [84]. In practical use, a dominant KASP assay for the null allele *Wx-B1* favored the selection of partial waxy mutants in wheat [43].

The success of wheat breeding strongly depends on the availability of a valuable diversity within breeding populations [85]. SNPs provide maximum beneficial diversity, and typical frequencies range from 1 SNP every 100–300 bps [86, 87]. Using whole-genome resequencing data from 18 wheat accessions, 36.4 million SNPs were detected at the whole-genome level [88]. In wheat, SNPs have been extensively used as molecular markers to characterize genetic resources, marker-assisted backcrossing (MABC), MAS, linkage mapping, GWAS, and GS. A convenient and low-cost method for genotyping a specific SNP employed in small-scale experiments is RFLP (SNP-RFLP) or by using the cleaved amplified polymorphic sequence (CAPS) marker technique [89]. Application of CAPS markers for genotyping is the best option when a diagnostic SNP associated with a trait is not suitable for primer design using the parameter of a good primer provided the SNP occurs within the recognition site of a restriction enzyme. If one allele contains a recognition site for a restriction enzyme while the other does not, the digestion will produce fragments of different sizes. Modern SNP genotyping platforms capable of high-throughput genotyping (i.e., TaqMan or KASP) support automated allele calling and generate high-quality data.

#### *2.2.6 Significance of KASP Assay and Primer Design*

With the development of DNA sequencing technology, the third-generation molecular markers or SNP markers augmented the development of molecular breeding. They have shown advantages such as high quantity, high density, high genetic stability, and easy automatic detection. Thus, SNPs are rapidly replacing traditional markers (RAPD, AFLP, CAPS, and SSR) and are now widely used in gene discovery and molecular breeding. Several technologies have been developed for high-throughput SNP genotyping in wheat [90]. Promising genotyping platforms such as RNase H2 enzyme-based amplification (rhAmp) [91], semi-thermal asymmetric reverse PCR (STARP) [92], and the Amplifluor SNP genotyping system [93] were also developed, followed by the development of KASP and TaqMan. However, KASP genotyping has been widely employed in wheat breeding as it offers flexible criteria for designing assay and easy to run at an affordable cost. The quantity and detection methods of molecular markers play a pivotal role in the MAS procedure. KASP detection technology based on both PCR and fluorescence detection meets low-, medium-, and high-throughput genotyping requirements. Additionally, KASP assays can accommodate SNPs, insertions and deletions (InDels), and CAPS markers and provide a flexible, relatively inexpensive solution for genotyping compared to SNP arrays or GBS. KASP technology is efficient at identifying polymorphic loci and haplotype analysis. Recently, the allelic differentiation of *Lr34* loci was further differentiated using haplotype analysis, and diagnostic KASP markers were developed to identify *Lr34a*, *Lr34b*, *Lr34c*, *Lr34d*, and *Lr34e* alleles [94].

For KASP assays, the allele-specific primers are designed carrying the standard FAM (5′-GAAGGTGACCAAGTTCATGCT-3′) and HEX (5′-GAAGGTCGGAGTCAACGGATT-3′) tails and with the targeted SNP at the 3′ end. A common genome-specific primer is designed, and the total amplicon length ranges from 50 to 125 bp. Primers can be manually designed using software like Primer3, designed from an automated bioinformatics pipeline for SNP assay development (http://www.polymarker.info/ or obtained from the list of predesigned KASP primers available in CerealsDB (https://www.cerealsdb.uk.net/cerealgenomics/) and MASWheat (https://maswheat.ucdavis.edu/). The primer mixture comprises 46 μl $ddH_2O$, 30 μl common primer (100 μM), and 12 μl of each tailed primer (100 μM). KASP assays are tested in 96-well plates in a 10 μl reaction volume (5 μl of 2 × KASP master mixture, 0.14 μl of primer mix, 3 μl DNA at 25 ng/μl, and 1.86 μl $ddH_20$). Assays are also performed in a 384-well plate in 5 μl reaction volume for high throughput. PCR cycling is performed using the following protocol: hot start at 94 °C for 15 min, followed by 10 touchdown cycles (94 °C for 20 s; touchdown at 65 °C initially and decreasing by −0.8 °C per cycle for 60 s), followed by 32 additional cycles (94 °C for 20 s; 57 °C for 60 s). The PCR plates are read with a microplate reader, and clusters are visualized using an SNP allele calling software. The genotypes in each cluster are analyzed to see its association with phenotypic data.

#### Copy Number Variation (CNV)

CNV is an essential source of genetic variation that has been largely untapped. In recent years, whole-genome resequencing, pan-genomics, and developing computational methods have allowed genome-wide characterization of CNVs in diverse species. The detection of CNVs for a relatively small number of genes can be analyzed by quantitative PCR (qPCR) [95]. For variation detection at the genomic level, tools like CNVnator are available [96]. Various studies in hexaploid wheat detected CNVs containing gene copies and their role in plant development. Examples include:

(a) In hexaploid wheat, *Photoperiod-1* (*Ppd-1*) loci are essential regulators of heading time, with *Ppd-B1* conferring photoperiod insensitivity through CNVs. The higher copy number of *Ppd-B1* resulted in an earlier heading [97].

(b) The interaction between the genes *Vernalization 1* (*Vrn-A1*) and *Frost resistance 2* (*Fr-A2*) primarily affect the frost tolerance of hexaploid wheat. Increased copy number of recessive *vrn-A1* allele and *CBF* (*C-repeat Binding Factor*) genes at the frost resistance locus, *Fr-A2*, improves frost tolerance and facilitates adaptation in different winter conditions [95, 98].

(c) Solid stem is a gene with CNV within the stem-solidness QTL *SSt1*. Increased copy number of the gene increases the resistance to the wheat stem sawfly (WSS) (Nilsen et al. 2020, In press).

### 2.3 Targeted Versus Genome-Wide Genotyping

There is a need to identify specific sequence variants in many DNA samples (single loci query/targeted genotyping) or genotype a large number of markers rapidly across a large sample (genome-wide and ultra-high-throughput). Several high-throughput single-plex and multiplex marker platforms in wheat [43, 99, 100], rice [101], legumes [102], and other crops have been proposed for MAS and GBS.

Single loci query is dominated by the use of traditional genotyping methods involving scalable single PCR-based assays such as TaqMan (Applied Biosystems, Foster City, CA), KASP ([84]; LGC, Herts, UK), and High-Resolution Melt (HRM) assays. Use of PCR tetrad machines (handling 384 PCR reactions at a time), capillary electrophoresis, and automation provide realistic throughput and cost-effectiveness in identifying single marker in thousands of samples. The deployment of SNP markers in forward breeding, where only a few specific markers are tracked, entails flexible, cost-effective, small-scale high-throughput uniplex assays. Allele-specific hybridization, coupled with Taq polymerase activity during PCR, forms the basis of the TaqMan chemistry. One pair of PCR primers and two different probes to one SNP site are used. Fluorescence occurs when one of the probes matches an SNP allele, which leads to the separation of the quencher and the fluorescent dye. Though the TaqMan assay is robust and reliable in genotyping, due to cost and less flexibility in assay design, KASP was developed as an alternative to TaqMan. HRM uses intercalating fluorescent dyes to monitor the melting profile (unmelted to melted) of PCR products. Currently, the KASP assay is widely preferred considering the ease of use, flexibility in assay development, efficiency, and cost per sample or data point. It has now evolved into a global benchmark technology for genotyping [103]. Nonetheless, in wheat, many traits are typically influenced by multiple genes, and targeted genotyping often results in the poor detection of minor genes by functional markers. Therefore, it is vital to establish simple, accurate, whole-genome high-throughput genotyping platforms for MAS.

Multiplexing and automation enhance the effectiveness of SNP genotyping. SNP detection with the iPLEX Assay and the MassARRAY System from Agena Bioscience (San Diego, CA) and SNPlex technologies from Applied Biosystems (Foster City, CA, USA) offers genotyping for a few markers. A locus-specific PCR, single-base extension using mass-modified dideoxynucleotide terminators and matrix-assisted laser desorption/ionization time-of-flight (MALDI-TOF) mass spectrometry (MS) detection constitute the iPLEX assay [104]. The iPLEX genotyping on the MassARRAY System facilitates a multiplexing level of ≥40-plex, offering a high level of flexibility and a low cost per genotype. SNPlex technology [105] allows the investigation of one to several hundred samples with sets of 48 SNP markers. Both iPLEX and SNPlex are a good compromise between single loci query and whole-genome genotyping.

Whole-genome genotyping is performed at a large scale using array-based (DArT and high-density Infinium/Axiom) and NGS-based (GBS) technologies. DArT was one of the initial array-based technologies developed as a microarray hybridization-based alternative to identify random single-copy sequences using biallelic dominant markers. DArT is a sequence-independent system that simultaneously scores several thousand loci in a single assay originally developed for rice [54]. Based on this technology, hundreds of markers were generated in wheat [42]. DArT markers were successfully used for genotyping diversity studies and genetic mapping in many crop species and among species of Triticeae [106–108]. DArTseq method deploys sequencing of the genome representations on the NGS platforms to create even more scalable and cost-effective than the original DArT method [107]. Illumina GoldenGate and BeadXpress multiplexing platforms were used for germplasm characterization in wheat [109, 110]. NGS technologies created the basis for the development of high-density SNP genotyping platforms. SNP arrays significantly reduced the cost and increased the genotyping capacity [111]. Currently, high-density SNP arrays are widely used in the breeding of major crops, including wheat. Illumina's BeadArray based on single-base extension or allele-specific primer extension (www.illumina.com) and Affymetrix's GeneChip based on molecular inversion probe hybridization (www.affymetrix.com) are the most widely used platforms for large-scale SNP genotyping in crop plants [112]. Several SNP arrays for wheat are available, such as 9 K [113], 90 K [114], 820 K [115], 35 K [116], and 660 K [117]. Different custom arrays are derived from the high-density SNP arrays, such as 9 K and 15 K [118] (from 90 K), 35 K (from 820 K), and 55 K (from 660 K) (Jia Jizeng; personal communication) for specifically tailored use. Other available SNP arrays include 18 K [119], 50 K [47], 280 K [90], and 420 K (Etienne Paux; personal communication). Fixed genotyping chips were often preferred over the GBS due to relatively low cost a few years ago. However, GBS is commonly used now to generate data sets of common sequence variants at low cost, but there is an additional cost of bioinformatics analysis when using GBS. The most promising application of these arrays is the GS for the improvement of economically important traits. The main drawback of the fixed arrays is that they do not allow the discovery of new SNPs, and the per sample cost is several times greater than less expensive options such as GBS.

#### *2.3.1 GBS*

First described by Elshire et al. [120], GBS is a low-cost NGS approach based on reduced representation sequencing for simultaneous SNP discovery and genotyping. GBS-based platforms are applicable for various crops regardless of prior knowledge of the genome sequence. This technique was developed to investigate the high-resolution association in maize under the Illumina NGS platform. The GBS relies on the digestion of genomic DNA with restriction enzymes to generate a reduced representation of the genome followed by barcode adapter ligation, PCR amplification, and sequencing of the amplified fragments on short-read sequencing platforms. GBS requires less DNA than the original restriction site-associated DNA (RAD) sequencing method and lacks a size selection procedure. Alignment of the generated reads and the complications due to low coverage and extensive missing data require the establishment of demanding bioinformatics pipelines and imputation algorithms. GBS was applied to bread wheat, resulting in the incorporation of thousands of markers in the bread wheat map [121]. A GBS-based map containing 28,644 markers harbors broad-spectrum pleiotropic adult plant resistance (APR) genes (*Lr27/Yr30/Sr2*, *Lr34/Yr18/Sr57*, *Lr46/Yr29/Sr58*) and 15 other published QTLs for wheat rusts [122]. GBS is a powerful approach in producing a large number of SNPs for genotyping and genetic analysis for implementing phylogeny relationships [123], genetic diversity and population structure analysis [124, 125], GWAS [126], GS [127], genome profiling and high-resolution QTL mapping [128], marker and gene discovery [129], and MAB [130]. QTL mapping through GBS using high-density linkage maps has been studied for various traits like FHB [131], adult plant resistance to leaf rust and stripe rust [132], flag leaf traits [133], coleoptile length [134], plant architecture [135], nitrogen use efficiency [57, 136], heat tolerance [126, 137], and yield traits [135, 138, 139] in wheat. Moreover, Elbasyoni et al. [140] using GBS and a 90 K SNP array in winter wheat showed that GBS-SNPs led to higher genomic prediction accuracy compared to array SNPs.

It is expensive to perform whole-genome sequencing to depths sufficient for confident variant calling, particularly in wheat with large genome sizes. The sequencing of multiple individuals at low coverage combined with imputation is often referred to as skim sequencing [141]. In skim sequencing, heterozygous alleles are often eliminated by sequencing recombinant inbred line (RIL) or DH populations [142]. For example, whole-genome skim sequencing of 12 RIL lines from the Paragon × Chinese Spring population identified the QTL for altered gene conversion and *RecQ-7* gene, an important target for wheat breeders to increase recombination frequencies [143]. Increasing recombination rates can reduce linkage drag and generate new genetic combinations [143]. Likewise, skim genotyping-by-sequencing (SkimGBS) involves resequencing of multiple individuals at low depth (usually <1x) followed by alignment of the reads to the reference sequence. SkimGBS helps genotyping large populations and significantly reducing cost per sample [142].

## 3 Marker Development Pipeline

### 3.1 Identify Genomic Loci from Biparental and GWAS

Understanding the genetic architecture of complex traits and the associated genomic region is crucial to improve the integration of molecular markers into wheat breeding. Saturated genetic maps are essential for identifying genomic regions associated with genes and QTL that control agronomic traits. Three are two main approaches

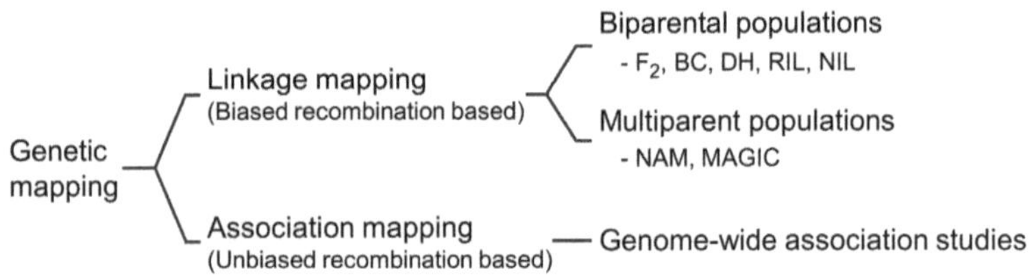


**Fig. 3** Genetic mapping approach for the identification of QTL

for genetic mapping in plants: linkage mapping and association mapping or linkage disequilibrium (LD) mapping (Fig. 3). Mapping studies use a segregating population derived from biparental, multiparent, or natural populations across diversified environments [144]. Röder et al. [34] compiled the first SSR-based wheat map, which included 279 loci. A high-density microsatellite consensus map for bread wheat was constructed by joining four independent genetic maps [75]. A novel genetic map consisting of 591 loci for yield, distributed across 21 wheat chromosomes, was constructed using SSR, STS, ISSR, DArT, and sequence-related amplified polymorphism (SRAP) markers [145]. Several biparental mapping and GWAS have identified genomic loci to dissect complex traits using SNP arrays or NGS and construction of ultra-high-density maps [146–149].

Biparental mapping has proven to be useful in identification of QTL targeted toward crop breeding objectives [150]. Biparental mapping is typically done using DH lines derived from the $F_1$, or RILs derived from the $F_6$ or backcross generations. The addition of markers to populations beyond a density of about one per 15 cM has a limited effect on precision [151]. In wheat, mapping populations has been used extensively to map many traits [152]. Multiparent mapping populations have been constructed to overcome the limitations of fewer phenotypic diversity and resolution of biparental populations. The genetic diversity of multiple parents leads to increased phenotypic diversity, making it suitable for high-resolution mapping. Two popular experimental designs of multiparent populations are nested association mapping (NAM) and multi-parent advanced-generation intercrosses (MAGIC).

GWAS has been proposed as a much more robust approach compared to linkage mapping. The use of a wide range of genetic backgrounds in GWAS analyses increases the probability of detecting QTL regions associated with traits of interest, compared to the limited genetic variation of a biparental mapping population. However, the limitations of GWAS, such as population structure, can lead to spurious associations between a candidate marker and a specific phenotypic trait. To eliminate such association, the mixed linear model incorporates structure data (Q) and relative kinship effects (K), resulting in the reduction of false-positive associations [153]. GWAS can also examine large gene pools representative of diverse breeding reservoirs. Several significant QTLs for traits related to abiotic and biotic stresses have been identified in wheat using GWAS, followed by fine mapping of candidate genes (Table 2). Meta-QTL analysis studies are carried out to identify redundant or consistent QTL among assorted studies. A total of 19 meta-QTL conferring resistance to tan spot was most recently identified from 104 initial QTLs detected in 15 previous QTL mapping studies [171].

The growing number of SNPs linked to the identified QTL moves forward the design of markers to support efficient selection in improved cultivar development. MAS for traits including FHB, rust diseases, and other agronomic and quality traits is widely practiced in Canadian wheat breeding combined with conventional field and quality testing. However, the breeder-friendly and validated SNP-KASP markers are still not available for many traits. Relatively large-size QTL spanning hundreds of markers/genes make the marker validation cumbersome and time-consuming. The knowledge of these QTL precise locations would lead to tightly linked markers [172] and a reduction of linkage drag from the donor/recurrent parent [173]. Information on the specificity of resistance loci to different pathogen strains and their effects under different climatic conditions would facilitate pyramiding the most promising QTL for durable resistance. In Canada, *Fhb1*, *Fhb2*, and *Fhb-5AS* from the Chinese variety Sumai-3 are the resistance genes most often used by breeding programs to combat FHB [45].

### 3.2 Marker Validation on Breeding Populations

Once genetic markers associated with traits are identified, their prediction accuracy should be validated before they can be applied routinely in MAS [174, 175]. Marker validation is the process of evaluating a marker for its effectiveness in determining the associated polymorphism in different genetic backgrounds and separate from the one in which the marker-trait linkage was identified initially [176]. Moreover, the marker alleles must be in strong linkage disequilibrium with alleles of interest of the trait locus, and the quantified positive or negative effects are consistently assigned to the same marker alleles. A list of validated KASP markers for hexaploid spring wheat is available [43, 44, 177]. MAS has been a routine operation in the wheat breeding programs, targeting agronomic, disease, and grain quality traits. However, the conversion of essential and novel QTL into trait-predictive DNA-based marker assays remains a bottleneck. SNP marker validation involves (a) identification of potential SNPs associated with the QTL or trait of interest, (b) examination of the effectiveness of the marker-trait linkage in the breeding materials identified, and (c) critical analysis of the results. An approach for validation of an SNP-KASP marker for *Fhb-5AS* was carried out as described (Fig. 4).

**Table 2**
**Examples of marker-trait associations identified using GWAS for agronomic, disease, and quality traits in wheat**

| Trait | Population | Marker-trait associations | Marker | Reference |
|---|---|---|---|---|
| FHB (type II) | 171 winter wheat cultivars | Four stable FHB resistance QTLs identified on 1B, 4A, 5D, and 7A. QTL on 4AL and 5DL are novel | Illumina 90 K SNP array | Hu et al. [154] |
| FHB spread and DON accumulation | 213 lines (152 spring and 61 winter wheat) | Six stable loci identified for FHB spread on 1A, 2BL, 3A, & 3B; and 3 for DON on 2BL, 3A, and 7A | Illumina 90 K SNP array | Wu et al. [155] |
| Leaf rust—Seedling and adult plant resistance (APR) | 338 spring wheat breeding lines | Identified 10 major QTLs for each seedling and field resistance, and 2 novel QTL on 1AL and 4AS with APR. also detected resistance effects in the known APR genes (*Lr34*, and possibly *Lr68*) region | Illumina 90 K SNP array | Gao et al. [156] |
| Leaf rust | 331 worldwide collection (297 spring and 34 winter wheat) | Detected 5 potential novel QTLs for LR resistance on 1AL, 4AS, 5AS, 5AL, and 7AS. Six QTLs were identified in the vicinity of known genes or QTL including the APR gene *Lr46* | 9K-SNP array | Sapkota et al. [157] |
| Leaf rust, stripe rust, and tan spot | 646 spring wheat advanced breeding lines (CIMMYT) | Identified markers and candidate genes associated with seedling resistance to LR, YR, and TS and also APR to YR (on 2DL, 3B, and 7DS). The disease resistance gene-rich segment on 2AS is very promising | GBS | Juliana et al. [158] |
| Stripe rust APR | 93 Chinese wheat landraces (SW + WW) | Six novel APR loci for stripe rust identified on 4A, 6A, and 7D | DArT and SSR | Yao et al. [57] |
| Stripe rust (ASR + APR) | 616 US spring wheat cultivars and breeding lines | Ten new QTLs for stripe rust identified on 1AS, 1DL, 2B, 4AL, 4BS, 5BL, 7AL, & 7DS | GMS plus 23 *Yr* gene-tagging markers | Liu et al. [79] |
| Powdery mildew | 329 worldwide wheat varieties | Four QTLs on chromosomes 3A, 3B, 6D, and 7D are potentially new QTL | Illumina 90 K SNP array | Kang et al. [159] |

(continued)

**Table 2 (continued)**

| Trait | Population | Marker-trait associations | Marker | Reference |
|---|---|---|---|---|
| Black point (BP) | 406 bread wheat and advanced breeding lines | 386 SNPs distributed across all chromosomes were significantly related to BP resistance. A novel and superior QTL was identified on 6D | 90 K and 660 K SNP arrays | Lv et al. [160] |
| Nitrogen use efficiency (NUE) | 93 European winter wheat | 183 MTAs identified for 11 traits spreading over almost the entire genome. Most of the MTAs were environmental-dependent and revealed new genomic regions | GBS and silico -DArT | Monostori et al. [58] |
| Potassium use efficiency (KUE) | 150 Pakistani spring wheat varieties | Of the 11 stable QTLs, 10 were novel associations for potassium-related traits | Illumina 90 K SNP array | Safdar et al. [161] |
| Coleoptile length (CL) | 707 Chinese wheat landraces | Identified 17 QTLs associated with CL, among which *QCl. sicau-6B.2*, was recognized as a novel major locus | GBS | Ma et al. [134] |
| Terminal drought and related agronomic traits | 339 pre-breeding lines from three-way crosses | Significant stable associations of Y on 4A, TKW on 3B, SL on 2D, and number of grains/ spikes on 3A & 6B were identified | DArTseq | Shokat et al. [162] |
| Heat-responsive physiological traits | 236 soft red winter wheat lines | Ten MTAs with pleiotropic effects mapping eight stable loci associated with physiological and yield traits were detected under higher temperature stress | GBS | Pradhan et al. [137] |
| Preharvest sprouting and low falling numbers | 469 white winter wheat cultivars and elite breeding lines | Fourteen potentially novel QTLs for PHS and 5 for falling numbers were identified | Illumina 90 K SNP array | Martinez et al. [163] |
| Grain protein content (GPC) and wet gluten content (WGC) | 163 bread wheat cultivars from China | The MTA on 6AS was novel and the most significant loci for GPC and WGC. The disease resistance RPP13-like protein 1 (*TaRPP13L1-B1*) gene was associated with flour color | Illumina 90 K SNP array | Chen et al. [164] |

(continued)

**Table 2 (continued)**

| Trait | Population | Marker-trait associations | Marker | Reference |
|---|---|---|---|---|
| GPC, TKW, and SDS sedimentation volume (SSV) | 192 bread wheat lines (25 synthetic hexaploid lines, 80 landraces, and 87 cultivars) | Significant SNPs associated with GPC (27), TKW (30), and SSV (34) were detected. RNA-Seq analysis showed the association of *Traes_7AL_1A21E8798* with increased GPC and an ubiquitin-specific protease (*TaUBP24*) with PH | Illumina 90 K SNP array | Liu et al. [165] |
| Gluten strength, color, and milling quality | 243 durum wheat cultivars and inbred lines ($F_{5:9}$) | The major QTL on 5B & 7B for high grain protein, for semolina extraction on 2B & 6B, and for polyphenol oxidase on 3A & 3B are novel loci | Illumina 90 K SNP array | Johnson et al. [166] |
| Flour-related traits | 469 soft white winter wheat cultivars and advanced-generation breeding lines | Potentially new loci for total flour yield, lactic acid solvent retention capacity (SRC), flour SSV, and flour swelling volume were detected. The MTA on chromosome 7BS is potentially a novel locus controlling flour SSV and lactic acid SRC | Illumina 90 K SNP array | Jernigan et al. [167] |
| Thousand-kernel weight (TKW) | 105 elite winter wheat varieties and advanced lines | 24 MTAs identified for nine yield-related traits, including 4 major loci: 1 for RLW (6B), 1 for TKW (7A), and 2 loci for PH (7B) | Illumina 90 K SNP array | Wang et al. [168] |
| Spikelet sterility and yield-related traits | 96 ancient hexaploid wheat accessions | 14 SNPs within the most significant regions at 1B, 2A, 2D, and 5B were used to define the candidate genes. Two genes (*metal ion transport* and *Gibberellin 2-oxidases* (*GA2oxs*) genes showed strong involvement in controlling the spikelet sterility | Illumina 15 K SNP array | Alqudah et al. [169] |
| Yield-related traits | 384 wheat cultivars | Several superior alleles with significant effects on TKW, GWS, TSNS, PH, and SNPP may be novel | Illumina 55K SNP array | Li et al. [170] |

(continued)

**Table 2 (continued)**

| Trait | Population | Marker-trait associations | Marker | Reference |
|---|---|---|---|---|
| Grain yield and yield stability | 4302 advanced spring bread wheat lines | Two major QTLs 3B and 6B showed large effects on GY, and 4B and 7B identified as "hot spot" regions for yield stability | GBS | Sehgal et al. [139] |

*MTA* marker-trait associations, *LR* leaf rust, *YR* stripe rust, *DON* deoxynivalenol, *RLW* ratio of kernel length/kernel width, *TKW* thousand-kernel weight, *Y* yield, *SL* spike length, *GWS* grain weight per spike, *TSNS* total spikelet number per spike, *SNPP* spike number per plant, *GMS* genotyping by multiplexed sequencing

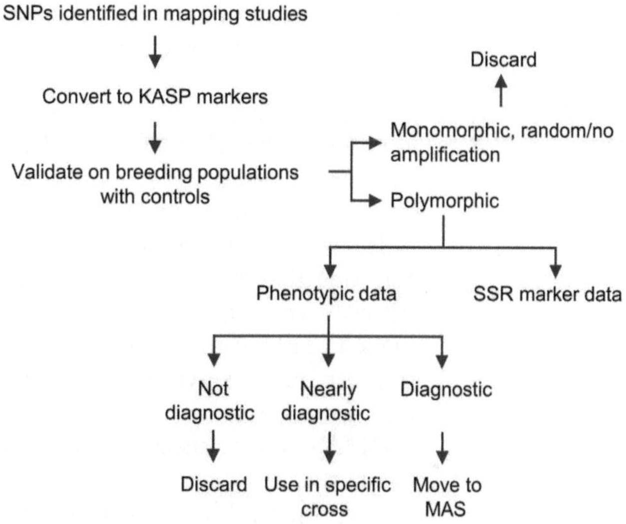


**Fig. 4** Validation of SNPs associated with a trait using KASP assay for MAS

For validation, genotyping data from the KASP assays on an unbiased population should be compared with phenotypic data and gel-based PCR markers, if available. Markers with the lowest false-positive (known negative genotypes incorrectly classified as positive) or false-negative (known positive genotypes incorrectly classified as negative) rates on a diverse genetic background could be used with caution in specific crosses where it can differentiate the parents involved in the cross or together with the phenotypic data [178]. Markers that pass the validation and diagnostics for a trait are added to the list of routinely used markers for MAS.

## 4 Implementation of MAS on Breeding Populations and Trait Considerations in each Population

### 4.1 Parent Panel

A parent panel is a suite of potential breeding parents with desirable traits and genes available for breeding. Each breeder may develop, maintain, and tailor their genotype panel to their breeding goals. The parent panel is tested for the marker's association with its target QTL. The assessed panel represents the near complete allelic composition present in a breeding program, and the breeding strategy focuses on increasing the frequency of favorable haplotypes [176]. The germplasm in a parent panel may be registered cultivars, or advanced breeding lines from a breeder's collection. For example, CIMMYT has germplasm commercially available and over 150,000 wheat accessions. A cross from a breeder's collection may only have one or a few desirable traits, and these traits can be introduced into new crosses with the assistance of MAS.

### 4.2 Early Generation

MAS can be a powerful tool in identifying traits in the early generations of a breeding program. Using MAS in the early generations can narrow down the scope of the entire cross so that only plants with targeted traits are carried forward. For example, MAS could only select plants containing reduced height genes (e.g., *Rht24*). Carrying that cross forward, the population should largely express the reduced height trait in further generations. Identifying genetic markers early in the population allows a breeder to focus on selection for traits like yield, which are screened phenotypically in the later generations.

MAS can be used on the $F_1$ generation of a complex cross (more than two parents) to identify the presence of favorable alleles in the progeny. If more than two parents are crossed, and each has a desirable trait associated with a genetic marker, then the final $F_1$ progeny of the complex cross can be tested to select desirable progeny. Conversely, this same concept can be used to only progress plants that do not have undesirable traits. MAS is useful in backcrossing important traits from wild relatives into elite breeding populations. This approach minimizes time and cost of maintaining large populations, can reduce phenotyping requirements, and increases accuracy of trait integration without a large linkage drag.

Screening for genetic markers at the early generations has an advantage of making genetic gains with little to no phenotypic data. A few disadvantages of using MAS at the early generation are selection targeted to a few traits, reduced allelic diversity, and resulting narrow germplasm diversity. MAS at early generation should be used on narrow crosses where the germplasm is segregating for a few traits, and MAS severely limits the possibility of unique gene combination resulting from unbiased recombinations.

#### 4.3 Advanced Generation

MAS can be used as a selective tool on advanced breeding populations to fine-tune the allelic composition of the lines that are reaching homozygosity. Markers that regulate GEI are used to select for best-suited germplasm that is stable over a vast geographical area and growing conditions. MAS allows identification of the breeding lines that are near misses from becoming a variety and are recycled into the crossing panel for making new breeding populations. MAS can be useful when issues arise in the field, making phenotypic selections difficult (e.g., natural disasters), and the advanced-generation lines need to be moved further into the breeding pipeline. MAS ties up the various agronomic, disease, detailed grain, and quality data obtained from advanced breeding generations with the right set of markers, thus associating the genotype and the phenotype of the select few advanced germplasms. This adds another layer of confidence in the advanced germplasm coming from a breeding pipeline and can clarify manual errors that may occur in pedigree-based breeding.

#### 4.4 DH Population

DH populations are produced by doubling the chromosomes from haploid plants, which results in complete homozygous fertile plants in one generation [179]. DHs (derived from $F_1$) can be propagated as true-breeding lines and reintroduced into a breeding scheme (Fig. 5) as advanced-generation lines, thus shortening the breeding cycle [180]. MAS assists in screening of somaclonal variations and elimination of lines that have recessive alleles for the trait of interest. Somaclonal variation is the heritable variation generated in cells that are subjected to tissue culture and in the plants regenerated from them [24]. Molecular markers are used in assessing the basis of somaclonal variation and genetic differences between regenerants and the corresponding parents involved in the cross. The production and use of DHs followed by MAS have been a viable strategy in wheat breeding for decades [181, 182]. McKenzie, a hard red spring wheat, was the first DH wheat cultivar registered in Canada in 1997 [183]. AC Transcend, the first durum variety created using this technology, was released in 2011 [184] and has reached 43% of the durum acreage in Canada.

DNA markers at the haploid stage allow the direct selection for the desired allele for traits of interest such as disease resistance (e.g., *Fhb1*, *Lr34*), insect resistance (e.g., *Sm1*, *SSt1*), abiotic stress tolerance (*Cdu1* regulating grain cadmium concentration), and/or end-use quality (e.g., *Gpc-B1*). DH plants can also be subjected to marker evaluation. Moreover, the genetic composition or marker data need not be regenerated in successive generations as necessary in a segregating population. Examples of DH cultivars of the Canadian Western red spring (CWRS) class developed using MAS include Burnside and Lillian with *Gpc-B1/Yr36* for improved grain protein content [185, 186], Shaw with the gene *Sm1* for resistance to the orange wheat blossom midge [181], AC Carberry

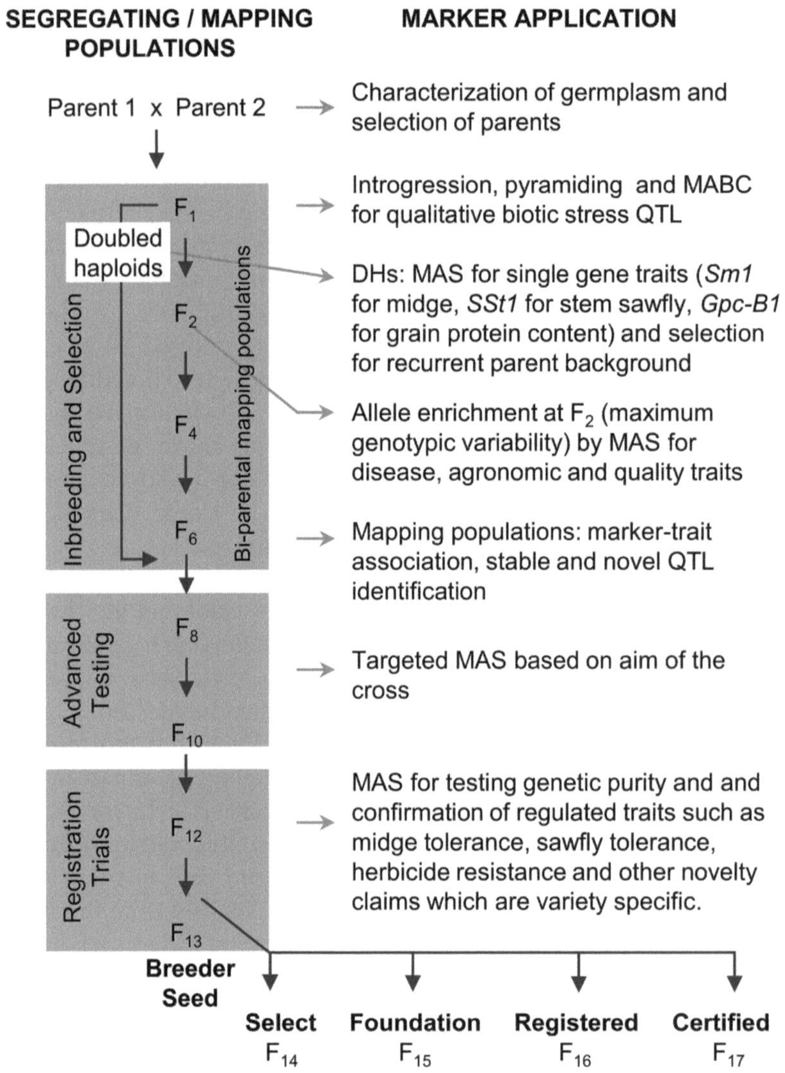


**Fig. 5** Wheat breeding scheme and use of MAS in the CWRS program at the Brandon Research and Development Centre, Canada. $F_3$, $F_5$, and $F_7$s are grown in off-season nursery in New Zealand, and selection is entirely based on phenotype. *MAS* marker-assisted selection, *MLT* multilocation trials, *GS* genomic selection

with *Fhb1* for improved resistance to FHB [182], and AAC Magnet with the broad-spectrum rust resistance gene *Lr34* [187]. DH populations are also preferred for QTL mapping as they are genetically stable and provide more accurate and effective phenotyping. DH populations have recently been used to identify and map QTL for traits such as those conferring resistance against common bunt [188], leaf rust [94], stem rust [189], and FHB [190].

## 5 Genetic, Genomic, and Agronomic Considerations

Several factors pose risks to wheat farming leading to yield reduction. The crop genetics (G), diverse growing conditions experienced in the vast grain crop region in a given year (E), and management practices (M) may cause instability in yield. Incorporation of yield stability factors (G × E × M) helps stabilize the yield over a large geographical area under different growing conditions. MAS can assist in selecting genotype resistance to prevalent diseases, acclimatized to photoperiod, day length, and growing degree day sensitivity as well as adapt to various soil/nutrient availability.

### 5.1 Agronomic Traits

Important agronomic traits that have been exploited in wheat improvement programs to aid cultivar development are early flowering and maturity, plant height, harvest index, total biomass, kernel weight, number of grains per spike, spike fertility, number of productive tillers, leaf morphology, photosynthetic efficiency, and root-related traits [191]. Most of the important agronomic traits are quantitative in nature [126]. Quantitative traits are easily affected by the environment, and therefore QTL analysis is typically performed using repeated field tests in various environments [16]. Many important agronomic traits have corresponding genomic structural variations like presence/absence and copy number variation [47]. The combination of environmental impact and the effects of multiple genes can make predicting the outcome of a plant difficult. However, several molecular markers for agronomic traits have been developed and are used in wheat breeding programs [44]. Historical varieties were taller, had thinner stems, and allocated more biomass to the stem than semidwarf varieties [192]. The successful introduction of dwarfing genes [193] and the availability of diagnostic markers [43, 44, 177] allow selection for plants carrying the alleles *Rht1-B1b* with reduced height, greater response to fertilizer, and higher yields [194]. The photoperiod-insensitive allele *Ppd-D1a* promoting early flowering and short growth, as well as the gibberellin-insensitive alleles *Rht-B1b* or *Rht-D1b* reducing plant height, is well represented in wheat adapted to temperate zones [195, 196]. Including molecular markers resulting in GS [197] and other associated advancements in breeding and agronomy continues to translate into yield improvements for wheat at the farm level [198].

### 5.2 Disease

Leaf and spike diseases can be devastating to a wheat crop. Under favorable conditions for infection, disease damage can result in the crop's total loss [199]. The most effective, cost-efficient, and environmentally friendly way to control the wheat disease is by breeding for disease resistance cultivars [199]. Rusts (stem, leaf, and stripe), FHB, loose smut, bunt, leaf spot, and other fungal diseases have been widely studied, and molecular markers have been developed for stable QTL. Markers linked to the known disease resistance genes of *FHB* (*Fhb1*, *Fhb2*, *Fhb-2DL*, *Fhb-4B*, and *Fhb-5AS*), leaf rust (*Lr14a*, *Lr16*, *Lr21*, *Lr22a*, *Lr23*, *Lr34*, *Lr37*, and *LrCen*), stem rust (*Sr2, Sr6, Sr7a, Sr8a, Sr9b, Sr11, Sr17, Sr23, Sr38, Sr57*, and *SrCad*), stripe rust (*Yr1*, *Yr5*, *Yr15*, *Yr18*, *Yr36*, and *YrSP*), loose smut (*Ut6*), common bunt (*Bt-Mck*), and tan spot (*TSN1*) are routinely used in MAS [44, 177].

With evolving pathogens, the breeding programs have to be vigilant and include new sources of resistance to different diseases. Molecular pathologists continue to identify new sources of resistance, and numerous databases are available identifying the QTL and genes linked to the new sources of disease resistance. For instance, there are dozens of genes associated with rusts (https://wheat.pw.usda.gov/GG3/node/657). Gene pyramiding (gene stacking) is crucial for improving and sustaining resistance to various diseases. For example, a recent study found that germplasm with *Yr15*, *Yr62*, and *Yr65* had complete resistance to stripe rust [200]. A combination of disease resistance genes (*Lr34/Yr18* and *Lr46/Yr29*) was introduced into wheat germplasm to improve disease resistance to a broad spectrum of pathogens. However, it is costly and time-consuming to confirm the presence of multiple genes/QTL using phenotyping alone. MAS makes it possible to stack genes and confirm their presence when phenotypic results can be misleading and inaccurate [201]. A key consideration while gene pyramiding is the physiological cost of adding resource hungry genes which do provide long-lasting resistance but come at the cost of yield and grain quality. For example, the presence of *Lr34/Yr18* has been shown to add resistance for stripe and leaf rust, but has also been linked to taller plants, yield loss, and lower test weights [202].

Disease races evolve and travel worldwide, making it advantageous to have broad-spectrum resistance against new and old disease-causing pathogens. Monitoring programs are essential to track races locally and worldwide. The emergence of new races can cause lines to lose their resistance, as seen with challenges from various stripe rust races in China [200, 203]. The changes in disease races in a location can occur in a matter of years, resulting in a cultivar's ineffectiveness shortly after its release [203]. MAS allows identification of genes/QTL in older germplasm that vary for reaction to new races of pathogens. It is much preferred to use an older cultivar for gene pyramiding than bringing a new source of resistance from a wild relative. MAS allows for regaining resistance to a new race of a pathogen in a relatively short amount of time.

### 5.3 Abiotic Stresses

In wheat, abiotic stresses such as drought, high temperature, and salinity cause significant production losses than biotic stress [204]. Plants use specialization, generalization, and phenotypic plasticity to manage abiotic stress [16]. With globally changing climate conditions, land-use change, and plant species invasiveness, phenotypic plasticity becomes increasingly crucial for plants to thrive [205]. Several wheat genes (Ta*DREB1*, *TaABRE3*, *TaWRKY44*, *TaNAC29*, *TaPIMP1*, *TabZIP1*, and *TabZIP60*) change expression levels due to drought stress, suggesting their role for improving stress tolerance in wheat [206, 207]. Fructan exohydrolase (FEH) gene (*1-FEH w3* on 6B) involved in stem water-soluble carbohydrate (WSC) remobilization was associated with drought response [208]. The allelic-specific functional marker of *Dreb-B1* on 3B and a CAPS marker of *1-FEH w3* on 6B were used to improve drought tolerance [206, 208]. MAS encouraged the introgression of 1BL.1RS translocation with yield advantage and impact on canopy water status [43, 209]. Also be noted, 1BL.1RS has different effects on grain yield and quality in different wheat classes and genetic backgrounds [210]. Heavy metal toxicity is a prime concern in premium quality wheat. Using the markers *Usw47* and *ScOPC20*, MAS was used in the development of durum cultivars (AC Strongfield and AAC Durafield) with low grain cadmium concentration [177, 211].

Numerous studies in wheat using crosses involving the source of abiotic stress resistance genotypes identified QTL for various traits implicated in abiotic stress. QTLs are identified for abscisic acid content [212], photosynthesis [213], seminal root angle and seminal root number [214], root-to-shoot weight ratio [215], SPAD chlorophyll content (SCC) [216], traits related to parameters of chlorophyll fluorescence kinetics (PCFKs) [217], canopy temperature [218], waterlogging [219], cadmium [220], and aluminum toxicity [221]. Despite considerable progress in molecular techniques and identification of QTL, the overall influence of MAS on the breeding of stress-tolerant cultivars has not been significant [222]. The challenges that limit the use of markers include polygenic nature of the trait, QTL × QTL and QTL × QTL × environment interactions, necessity for fine mapping, lack of standardized phenotyping techniques, and the structural complexity of the wheat genome.

### 5.4 Quality Traits

Breeding for quality to meet the demands of the wheat value chain is a challenging task. Kernel characteristics, protein content, flour characteristics, milling performance, dough properties, and baking quality are essential indicators of wheat quality [223]. Grain protein content (GPC) has gained special consideration among end-use quality traits to indicate the quality performance of wheat products such as bread, cake, noodles, and pasta [224]. Breeding for an increase in GPC is difficult due to the negative association that exists between grain protein content and yield [225]. Many quality traits cannot be determined phenotypically, and processing samples for quality traits can be time-consuming and expensive and require a large amount of grain. Therefore, MAS can be an essential tool to select lines with improved grain protein- and starch-related quality traits.

The *Gpc-B1* locus is one of the key regulators of total grain protein content [226]. The molecular marker available to detect *Gpc-B1* is a valuable resource for MAS [227]. MAS for protein can be done as early as $F_2$, whereas near-infrared (NIR) spectroscopy techniques are typically used to measure protein in advanced generations. Gluten (composed of glutenins and gliadins) strength is an important breeding target and is the basis of classification of Canadian wheat types. Gluten strength should be adequate for high-volume pan-bread production. Gluten strength is determined through a farinograph, extensograph, or functional bake tests that are time-consuming and require a large amount of grains. MAS overcomes these limitations and enriches favorable alleles in the early generations. A 660 K SNP array [117] has succeeded in identifying QTL for bread-making quality [228]. Markers for genes controlling grain hardness (*PinA* and *PinB*) are available to select cultivar for the hard or soft wheat market.

Wheat flour color is an important consideration in various Asian products including noodles, with a crucial factor being color retention and the appearance of noodle sheets. The genes like *polyphenol oxidase (PPO)* [229], *lipoxygenase (Ta-LoxB1)* [230], *phytoene synthase (Psy1)* (He [231], *zeta-carotene desaturase (TaZds-D1)* [232], and *peroxidase (TaPod-A1)* [233] have been shown to affect wheat flour color and processing quality. Allele-specific markers for these genes are available. The usage of these markers for the identification of genotypes is in practice within many breeding programs.

The tolerance to PHS is an important breeding objective, and reliable phenotyping is time-consuming. Susceptibility to PHS leads to poor falling number leading to poor dough strength. Use of markers for PHS can reduce phenotyping load and avoid escape of susceptible germplasm which conditions are not favorable for PHS. Validated KASP markers for a major QTL (PHS1) related to PHS, and functional markers developed for the genes involved in PHS response, namely, *Viviparous (TaVp1)*, seed dormancy gene (*TaSdr1*), and a major *Myb* class of gene for grain color (*TaMyb10*), are available in wheat [43, 44].

## 6 Conclusions

MAS has been explored to increase the precision and efficiency of various economic traits in wheat breeding. MAS offers several advantages as selection can be carried out in early-generation germplasm for increasing the allele frequency of desirable alleles. This reduces the number of individuals carried forward in the breeding cycle, thus reducing overall breeding costs. MAS has enormous potential for targeting traits of interest that are costly and cumbersome to phenotype. MAS increases the efficiency of gene pyramiding by combining minor and major QTL where the traits are governed by many major QTL. MAS allows distinction of resistance vs escape for various biotic stresses and is not dependent on growing conditions.

The discovery of DNA markers that are strongly associated with the desired phenotypes and different genotyping platforms has competed for the process of introgression and tracking of a gene/QTL and increased selection efficiency and genetic gain.

Mapping and fine mapping QTL emerged as viable solutions for breeding crops tolerant to biotic and abiotic stresses. Many QTLs are identified and cloned in wheat, and the majority of the QTL for biotic stresses have been integrated in the breeding programs. MAS for quantitative traits such as yield and abiotic stress tolerances are the focus areas for the next generation of markers based on genomics data. The explosion in genomics resources such as the 10+ Wheat Genomes Project and low-cost genome resequencing now enables identifying tightly linked SNP markers to capture copy number variation, structural variation, and genes from a single reference genome. The enhanced resolution of the wheat reference genome should also help accelerate the cloning of yield-related genes, develop functional markers, and identify favorable haplotypes to develop wheat cultivars with improved end-use quality and yield.

Cereal breeding programs have delivered many superior varieties with improved grain yield and quality. Nonetheless, the progress is slow due to lengthy breeding cycles, which often takes 10–15 years from crossing to cultivar release. Speed breeding has been refined and, in association with genome-wide selection, can achieve up to six generations of wheat per year. The efficiency of speed breeding to develop "field-ready cultivars" is under investigation by many breeding programs due to the inability to study genotype by environment interactions under the speed breeding process. To maintain and increase the annual rates of genetic gains, targeted crossings, rapid cycling of segregating populations, application of molecular markers, and phenotyping platforms for critical traits in conjunction with yield trials should be emphasized.

Advances in cereal genomics have enabled breeders to improve the prediction of phenotypes from genotypes in wheat breeding. Efficient breeding strategies combining precision breeding tools could lead to a step change in the rate of genetic gain. MAS would be crucial to develop wheat varieties with a higher yield, abiotic and biotic stress resistance, and increased nutritional value while contributing to reaching the objectives for sustainable food security.

## Acknowledgments

The authors wish to acknowledge the invaluable discussions with the Canadian cereal breeders and the professionals in cereal genomics, cytogenetics, pathology, physiology, and informatics. We acknowledge the funding provided by the Canadian Agricultural Partnership comprising Agriculture and Agri-Food Canada and the Canadian National Research Coalition.

### References

1. FAOSTAT (2019) Food and Agriculture Organization of the United Nations. In: FAOSTAT statistical database. Rome, Accessed July 13, 2020
2. Ray DK, Mueller ND, West PC et al (2013) Yield trends are insufficient to double global crop production by 2050. PLoS One 8: e66428
3. Bassi FM, Nachit MM (2019) Genetic gain for yield and allelic diversity over 35 years of durum wheat breeding at ICARDA. Crop Breed Genet Genom 1:e190004
4. Thomas JB, Graf RJ (2014) Rates of yield gain of hard red spring wheat in western Canada. Can J Plant Sci 94:1–13
5. Searchinger T, Waite R, Hanson C et al (2019) Adapt to climate change. In: Matthews E (ed) World resources report: creating a sustainable food future (final report), Washington, DC, pp 209–219
6. Tuberosa R (2012) Phenotyping for drought tolerance of crops in the genomics era. Front Physiol 3:347
7. Collard BC, Mackill DJ (2008) Marker-assisted selection: an approach for precision plant breeding in the twenty-first century. Philos Trans R Soc Lond Ser B Biol Sci 363:557–572
8. Hospital F (2009) Challenges for effective marker-assisted selection in plants. Genetica 136:303–310
9. Zhao Y, Mette MF, Gowda M et al (2014) Bridging the gap between marker-assisted and genomic selection of heading time and plant height in hybrid wheat. Heredity 112:638–645
10. Bassi FM, Bentley AR, Charmet G et al (2016) Breeding schemes for the implementation of genomic selection in wheat (*Triticum* spp.). Plant Sci 242:23–36
11. Bhat JA, Ali S, Salgotra RK et al (2016) Genomic selection in the era of next generation sequencing for complex traits in plant breeding. Front Genet 7:221
12. Slater AT, Cogan NOI, Forster JW (2013) Cost analysis of the application of marker-assisted selection in potato breeding. Mol Breed 32:299–310
13. Van Frank G, Rivière P, Pin S et al (2020) Genetic diversity and stability of performance of wheat population varieties developed by participatory breeding. Sustainability 12:384
14. Zhou WC, Kolb FL, Bai GH et al (2003) Validation of a major QTL for scab resistance with SSR markers and use of marker-assisted selection in wheat. Plant Breed 122:40–46
15. Kuchel H, Ye G, Fox R et al (2005) Genetic and economic analysis of a targeted marker-assisted wheat breeding strategy. Mol Breed 16:67–78
16. Priyadarshan PM (2019) Breeding for abiotic stress adaptation. In: Priyadarshan PM (ed) Plant breeding: classical to modern. Springer, Singapore, pp 413–455
17. Heffner EL, Lorenz AJ, Jannink J-L et al (2010) Plant breeding with genomic selection: gain per unit time and cost. Crop Sci 50:1681–1690
18. Cobb JN, Biswas PS, Platten JD (2019) Back to the future: revisiting MAS as a tool for modern plant breeding. Theor Appl Genet 132:647–667
19. Ma Y, Reif JC, Jiang Y et al (2016) Potential of marker selection to increase prediction accuracy of genomic selection in soybean (*Glycine max* L.). Mol Breed 36:113
20. Boopathi NM (2020) Marker-assisted selection (MAS). In: Genetic mapping and marker-assisted selection. Springer, Singapore
21. Goudemand E, Laurent V, Duchalais L et al (2013) Association mapping and meta-analysis: two complementary approaches for the detection of reliable *Septoria tritici* blotch quantitative resistance in bread wheat (*Triticum aestivum* L.). Mol Breed 32:563–584
22. Miedaner T, Korzun V (2012) Marker-assisted selection for disease resistance in

wheat and barley breeding. Phytopathology 102:560–566

23. Larochea A, Frick M, Graf RJ et al (2019) Pyramiding disease resistance genes in elite winter wheat germplasm for Western Canada. Crop J 7:739–749
24. Singh BD, Singh AK (2015) Sequence-based markers. In: Marker-assisted plant breeding: principles and practices. Springer, India, pp 77–124
25. Varshney RK, Graner A, Sorrells ME (2005) Genomics-assisted breeding for crop improvement. Trends Plant Sci 10:621–630
26. Gupta PK, Varshney RK (2000) The development and use of microsatellite markers for genetic analysis and plant breeding with emphasis on bread wheat. Euphytica 113:163–185
27. Dubcovsky J (2004) Marker-assisted selection in public breeding programs: the wheat experience. Crop Sci 44:1895–1898
28. Davey JW, Hohenlohe PA, Etter PD et al (2011) Genome-wide genetic marker discovery and genotyping using next-generation sequencing. Nat Rev Genet 12:499–510
29. Botstein D, White RL, Skolnick M et al (1980) Construction of a genetic linkage map in man using restriction fragment length polymorphisms. Am J Hum Genet 32:314–331
30. Chao S, Sharp PJ, Worland AJ et al (1989) RFLP-based genetic maps of wheat homoeologous group 7 chromosomes. Theor Appl Genet 78:495–504
31. Devos KM, Gale MD (1992) The use of random amplified polymorphic DNA markers in wheat. Theor Appl Genet 84:567–572
32. Schachermayr GM, Messmer MM, Feuillet C et al (1995) Identification of molecular markers linked to the *Agropyron elongatum*-derived leaf rust resistance gene *Lr24* in wheat. Theor Appl Genet 90:982–990
33. Devos KM, Bryan GJ, Collins AJ et al (1995) Application of two microsatellite sequences in wheat storage proteins as molecular markers. Theor Appl Genet 90:247–252
34. Röder MS, Korzun V, Gill BS et al (1998) The physical mapping of microsatellite markers in wheat. Genome 41:278–283
35. Dedryver F, Jubier M-F, Thouvenin J et al (1996) Molecular markers linked to the leaf rust resistance gene *Lr24* in different wheat cultivars. Genome 39:830–835
36. Ma Z-Q, Lapitan N (1998) A comparison of amplified and restriction fragment length polymorphism in wheat. Cereal Res Commun 26:7–13
37. Kojima T, Nagaoka T, Noda K et al (1998) Genetic linkage map of ISSR and RAPD markers in einkorn wheat in relation to that of RFLP markers. Theor Appl Genet 96:37–45
38. Khan IA, Procunier JD, Humphreys DG et al (2000) Development of PCR-based markers for a high grain protein content gene from *Triticum turgidum* ssp. *dicoccoides* transferred to bread wheat. Crop Sci 40:518–524
39. Lazo GR, Chao S, Hummel DD et al (2004) Development of an expressed sequence tag (EST) resource for wheat (*Triticum aestivum* L.): EST generation, unigene analysis, probe selection and bioinformatics for a 16,000-locus bin-delineated map. Genetics 168:585–593
40. Yanagisawa T, Kiribuchi-Otobe C, Hirano H et al (2003) Detection of single nucleotide polymorphism (SNP) controlling the waxy character in wheat by using a derived cleaved amplified polymorphic sequence (dCAPS) marker. Theor Appl Genet 107:84–88
41. Guillaumie S, Charmet G, Linossier L et al (2004) Colocation between a gene encoding the bZip factor SPA and an eQTL for a high-molecular-weight glutenin subunit in wheat (*Triticum aestivum*). Genome 47:705–713
42. Akbari M, Wenzl P, Caig V et al (2006) Diversity arrays technology (DArT) for high-throughput profiling of the hexaploid wheat genome. Theor Appl Genet 113:1409–1420
43. Rasheed A, Wen W, Gao F et al (2016) Development and validation of KASP assays for genes underpinning key economic traits in bread wheat. Theor Appl Genet 129:1843–1860
44. Toth J, Pandurangan S, Burt A et al (2019) Marker-assisted breeding of hexaploid spring wheat in the Canadian prairies. Can J Plant Sci 99:111–127
45. Pandurangan S, Nilsen KT, Kumar S (2020) Validation of a SNP-KASP marker for the fusarium head blight resistance QTL on chromosome 5AS. Can J Plant Sci. https://doi.org/10.1139/CJPS-2020-0099
46. Eagles HA, Barianab HS, Ogbonnaya FC et al (2001) Implementation of markers in Australian wheat breeding. Aust J Agric Res 52:1349–1356
47. Rasheed A, Xia X (2019) From markers to genome-based breeding in wheat. Theor Appl Genet 132:767–784
48. Paux E, Sourdille P, Mackay I et al (2012) Sequence-based marker development in wheat: advances and applications to breeding. Biotechnol Adv 30:1071–1088

49. Mackill D, Nguyen H, Zhang J (1999) Use of molecular markers in plant improvement programs for rainfed lowland rice. Field Crops Res 64:177–185
50. Prasanna BM, Pixley K, Warburton ML et al (2010) Molecular marker-assisted breeding options for maize improvement in Asia. (special issue: quantitative genetics and plant genomics.). Mol Breed 26:339–356
51. Kumar J, Choudhary AK, Solanki RK et al (2011) Towards marker-assisted selection in pulses: a review. Plant Breed 130:297–313
52. Spasibionek S, Mikołajczyk K, Ćwiek-Kupczyńska H et al (2020) Marker-assisted selection of new high oleic and low linolenic winter oilseed rape (*Brassica napus* L.) inbred lines revealing good agricultural value. PLoS One 15:e0233959
53. Foolad MR, Panthee DR (2012) Marker-assisted selection in tomato breeding. Crit Rev Plant Sci 31:93–123
54. Jaccoud D (2001) Diversity arrays: a solid state technology for sequence information independent genotyping. Nucleic Acids Res 29:25e-25
55. Mergoum M, Harilal VE, Simsek S et al (2013) Agronomic and quality QTL mapping in spring wheat. J Plant Breed Genet 1:19–33
56. Cabrera A, Guttieri M, Smith N et al (2015) Identification of milling and baking quality QTL in multiple soft wheat mapping populations. Theor Appl Genet 128:2227–2242
57. Yao F, Zhang X, Ye X et al (2019) Characterization of molecular diversity and genome-wide association study of stripe rust resistance at the adult plant stage in northern Chinese wheat landraces. BMC Genet 20:38
58. Monostori I, Szira F, Tondelli A et al (2017) Genome-wide association study and genetic diversity analysis on nitrogen use efficiency in a central European winter wheat (*Triticum aestivum* L.) collection. PLoS One 12: e0189265
59. Heffner EL, Jannink J-L, Iwata H et al (2011) Genomic selection accuracy for grain quality traits in biparental wheat populations. Crop Sci 51:2597–2606
60. Paran I, Michelmore RW (1993) Development of reliable PCR-based markers linked to downy mildew resistance genes in lettuce. Theor Appl Genet 85:985–993
61. Zhang Z, Xu J, Xu Q et al (2004) Development of novel PCR markers linked to the BYDV resistance gene Bdv2 useful in wheat for marker-assisted selection. Theor Appl Genet 109:433–439
62. Ma P, Xu H, Li L et al (2016) Characterization of a new Pm2 allele conferring powdery mildew resistance in the wheat germplasm line FG-1. Front Plant Sci 7:546
63. Aggarwal R, Gupta S, Banerjee S et al (2011) Development of a SCAR marker for detection of Bipolaris sorokiniana causing spot blotch of wheat. Can J Microbiol 57:934–942
64. Župunski V, Jevtić R, Lalošević M et al (2019) Diversity of trichothecene genotypes of *Fusarium graminearum* sensu stricto from winter wheat in Serbia. Eur J Plant Pathol 155:461–473
65. Gao L, Yu H, Han W et al (2014) Development of a SCAR marker for molecular detection and diagnosis of Tilletia controversa Kuhn, the causal fungus of wheat dwarf bunt. World J Microbiol Biotechnol 30:3185–3195
66. Myburg AA, Cawood M, Wingfield BD et al (1998) Development of RAPD and SCAR markers linked to the Russian wheat aphid resistance gene *Dn2* in wheat. Theor Appl Genet 96:1162–1169
67. Yumurtaci A, Sipahi H, Al-Abdallat A et al (2017) Construction of new EST-SSRs for fusarium resistant wheat breeding. Comput Biol Chem 68:22–28
68. Dabab Nahas L, Al-Husein N, Lababidi G et al (2019) In-silico prediction of novel genes responsive to drought and salinity stress tolerance in bread wheat (*Triticum aestivum*). PLoS One 14:e0223962
69. Oiestad AJ, Martin JM, Cook J et al (2017) Identification of candidate genes responsible for stem pith production using expression analysis in solid-stemmed wheat. Plant Genome 10:1–8
70. Shen X, Ma L, Zhong S et al (2015) Identification and genetic mapping of the putative *Thinopyrum intermedium*-derived dominant powdery mildew resistance gene *PmL962* on wheat chromosome arm 2BS. Theor Appl Genet 128:517–528
71. Wen W, He Z, Gao F et al (2017) A high-density consensus map of common wheat integrating four mapping populations scanned by the 90K SNP array. Front Plant Sci 8:1389
72. Twyford AD, Ennos RA (2012) Next-generation hybridization and introgression. Heredity (Edinb) 108:179–189
73. Röder MS, Korzun V, Wendehake K et al (1998) A microsatellite map of wheat. Genetics 149:2007–2023

74. Jaiswal S, Sheoran S, Arora V et al (2017) Putative microsatellite DNA marker-based wheat genomic resource for varietal improvement and management. Front Plant Sci 8:2009
75. Somers DJ, Isaac P, Edwards K (2004) A high-density microsatellite consensus map for bread wheat (*Triticum aestivum* L.). Theor Appl Genet 109:1105–1114
76. Kumar S, Kumar V, Kumari P et al (2016) DNA fingerprinting and genetic diversity studies in wheat genotypes using SSR markers. J Environ Biol 37:319–326
77. Wang L, Liu L, Li H et al (2014) Detecting seed purity of wheat varieties using microsatellite markers based on eliminating the influence of non-homozygous loci. Seed Sci Technol 42:393–413
78. Han B, Wang C, Tang Z et al (2015) Genome-wide analysis of microsatellite markers based on sequenced database in Chinese spring wheat (*Triticum aestivum* L.). PLoS One 10:e0141540
79. Liu L, Wang M, Zhang Z et al (2020) Identification of stripe rust resistance loci in US spring wheat cultivars and breeding lines using genome-wide association mapping and *Yr* gene markers. Plant Dis. https://doi.org/10.1094/PDIS-11-19-2402-RE
80. Zhou X, Wang M, Chen X et al (2014) Identification of *Yr59* conferring high-temperature adult-plant resistance to stripe rust in wheat germplasm PI 178759. Theor Appl Genet 127:935–945
81. Wang Y, Hao C, Zheng J et al (2015) A haplotype block associated with thousand-kernel weight on chromosome 5DS in common wheat *(Triticum aestivum* L.). J Integr Plant Biol 57:662–672
82. Liao PY, Lee KH (2010) From SNPs to functional polymorphism: the insight into biotechnology applications. Biochem Eng J 49:149–158
83. Deschamps S, Campbell MA (2010) Utilization of next-generation sequencing platforms in plant genomics and genetic variant discovery. Mol Breed 25:553–570
84. Allen AM, Barker GLA, Wilkinson P et al (2013) Discovery and development of exome-based, co-dominant single nucleotide polymorphism markers in hexaploid wheat (*Triticum aestivum* L.). Plant Biotechnol J 11:279–295
85. Rufo R, Alvaro F, Royo C et al (2019) From landraces to improved cultivars: assessment of genetic diversity and population structure of Mediterranean wheat using SNP markers. PLoS One 14:e0219867
86. Ravel C, Praud S, Murigneux A et al (2006) Single-nucleotide polymorphism frequency in a set of selected lines of bread wheat (*Triticum aestivum* L.). Genome 49:1131–1139
87. Edwards D, Forster JW, Chagné D et al (2007) What are SNPs? In: Oraguzie NC, Rikkerink EHA, Gardiner SE et al (eds) Association mapping in plants. Springer, New York, NY, pp 41–52
88. Montenegro JD, Golicz AA, Bayer PE et al (2017) The pangenome of hexaploid bread wheat. Plant J 90:1007–1013
89. Shavrukov Y (2016) Comparison of SNP and CAPS markers application in genetic research in wheat and barley. BMC Plant Biol 16:11
90. Rimbert H, Darrier B, Navarro J et al (2018) High-throughput SNP discovery and genotyping in hexaploid wheat. PLoS One 13: e0186329
91. Dobosy JR, Rose SD, Beltz KR et al (2011) RNase H-dependent PCR (rhPCR): improved specificity and single nucleotide polymorphism detection using blocked cleavable primers. BMC Biotechnol 11:80
92. Long YM, Chao WS, Ma GJ et al (2017) An innovative SNP genotyping method adapting to multiple platforms and throughputs. Theor Appl Genet 130:597–607
93. Jatayev S, Kurishbayev A, Zotova L et al (2017) Advantages of Amplifluor-like SNP markers over KASP in plant genotyping. BMC Plant Biol 17:254
94. Fang T, Lei L, Li G et al (2020) Development and deployment of KASP markers for multiple alleles of Lr34 in wheat. Theor Appl Genet 133:2183–2195
95. Zhu J, Pearce S, Burke A et al (2014) Copy number and haplotype variation at the *VRN-A1* and central *FR-A2* loci are associated with frost tolerance in hexaploid wheat. Theor Appl Genet 127:1183–1197
96. Abyzov A, Urban AE, Snyder M et al (2011) CNVnator: an approach to discover, genotype, and characterize typical and atypical CNVs from family and population genome sequencing. Genome Res 21:974–984
97. Würschum T, Langer SM, Longin CFH et al (2018) A three-component system incorporating *Ppd-D1*, copy number variation at *Ppd-B1*, and numerous small-effect quantitative trait loci facilitates adaptation of heading time in winter wheat cultivars of worldwide origin. Plant Cell Environ 41:1407–1416
98. Würschum T, Longin CFH, Hahn V et al (2017) Copy number variations of *CBF*

genes at the *Fr-A2* locus are essential components of winter hardiness in wheat. Plant J 89:764–773

99. Berard A, Paslier MCL, Dardevet M et al (2009) High-throughput single nucleotide polymorphism genotyping in wheat (*Triticum* spp.). Plant Biotechnol J 7:364–374
100. Bernardo A, Wang S, Amand PS et al (2015) Using next generation sequencing for multiplexed trait-linked markers in wheat. PLoS One 10:e0143890
101. Masouleh AK, Waters DLE, Reinke RF et al (2009) A high-throughput assay for rapid and simultaneous analysis of perfect markers for important quality and agronomic traits in rice using multiplexed MALDI-TOF mass spectrometry. Plant Biotechnol J 7:355–363
102. Varshney RK, Singh VK, Hickey JM et al (2016) Analytical and decision support tools for genomics-assisted breeding. Trends Plant Sci 21:354–363
103. Semagn K, Babu R, Hearne S et al (2014) Single nucleotide polymorphism genotyping using Kompetitive allele specific PCR (KASP): overview of the technology and its application in crop improvement. Mol Breed 33:1–14
104. Gabriel S, Ziaugra L, Tabbaa D (2009) SNP genotyping using the Sequenom MassARRAY iPLEX platform. Curr Protoc Hum Genet 60:2.12.11–12.12.18
105. Tobler A, Andersen M, Paner T et al (2005) The SNPlex genotyping system: a flexible and scalable platform for SNP genotyping. J Biomol Tech 16:398–406
106. Jing H-C, Bayon C, Kanyuka K et al (2009) DArT markers: diversity analyses, genomes comparison, mapping and integration with SSR markers in *Triticum monococcum*. BMC Genomics 10:458
107. Edet OU, Gorafi YSA, Nasuda S et al (2018) DArTseq-based analysis of genomic relationships among species of tribe *Triticeae*. Sci Rep 8:16397
108. Sohail Q, Shehzad T, Kilian A et al (2012) Development of diversity array technology (DArT) markers for assessment of population structure and diversity in *Aegilops tauschii*. Breed Sci 62:38–45
109. Akhunov E, Nicolet C, Dvorak J (2009) Single nucleotide polymorphism genotyping in polyploid wheat with the Illumina GoldenGate assay. Theor Appl Genet 119:507–517
110. Trebbi D, Maccaferri M, De Heer P et al (2011) High-throughput SNP discovery and genotyping in durum wheat (*Triticum durum* Desf.). Theor Appl Genet 123:555–569
111. Ganal MW, Polley A, Graner EM et al (2012) Large SNP arrays for genotyping in crop plants. J Biosci 37:821–828
112. Thomson MJ (2014) High-throughput SNP genotyping to accelerate crop improvement. Plant Breed Biotechnol 2:195–212
113. Cavanagh CR, Chao S, Wang S et al (2013) Genome-wide comparative diversity uncovers multiple targets of selection for improvement in hexaploid wheat landraces and cultivars. Proc Natl Acad Sci U S A 110:8057–8062
114. Wang S, Wong D, Forrest K et al (2014) Characterization of polyploid wheat genomic diversity using a high-density 90,000 single nucleotide polymorphism array. Plant Biotechnol J 12:787–796
115. Winfield MO, Allen AM, Burridge AJ et al (2016) High-density SNP genotyping array for hexaploid wheat and its secondary and tertiary gene pool. Plant Biotechnol J 14:1195–1206
116. Allen AM, Winfield MO, Burridge AJ et al (2017) Characterization of a wheat Breeders' Array suitable for high-throughput SNP genotyping of global accessions of hexaploid bread wheat (*Triticum aestivum* L.). Plant Biotechnol J 15:390–401
117. Cui F, Zhang N, Fan XL et al (2017) Utilization of a wheat 660K SNP array-derived high-density genetic map for high-resolution mapping of a major QTL for kernel number. Sci Rep 7:3788
118. Boeven PH, Longin CF, Leiser WL et al (2016) Genetic architecture of male floral traits required for hybrid wheat breeding. Theor Appl Genet 129:2343–2357
119. Norman A, Taylor J, Tanaka E et al (2017) Increased genomic prediction accuracy in wheat breeding using a large Australian panel. Theor Appl Genet 130:2543–2555
120. Elshire RJ, Glaubitz JC, Sun Q et al (2011) A robust, simple genotyping-by-sequencing (GBS) approach for high diversity species. PLoS One 6:e19379
121. Poland JA, Brown PJ, Sorrells ME et al (2012) Development of high-density genetic maps for barley and wheat using a novel two-enzyme genotyping-by-sequencing approach. PLoS One 7:e32253
122. Li H, Vikram P, Singh RP et al (2015) A high density GBS map of bread wheat and its application for dissecting complex disease resistance traits. BMC Genomics 16:216
123. Bernhardt N, Brassac J, Dong X et al (2020) Genome-wide sequence information reveals recurrent hybridization among diploid wheat wild relatives. Plant J 102:493–506

124. Kolmer JA, Herman A, Ordonez ME et al (2020) Endemic and panglobal genetic groups, and divergence of host-associated forms in worldwide collections of the wheat leaf rust fungus *Puccinia triticina* as determined by genotyping by sequencing. Heredity (Edinb) 124:397–409
125. Bhatta M, Morgounov A, Belamkar V et al (2018) Unlocking the novel genetic diversity and population structure of synthetic hexaploid wheat. BMC Genomics 19:591
126. Rahimi Y, Bihamta MR, Taleei A et al (2019) Genome-wide association study of agronomic traits in bread wheat reveals novel putative alleles for future breeding programs. BMC Plant Biol 19:541
127. Liu Y, Salsman E, Fiedler JD et al (2019) Genetic mapping and prediction analysis of FHB resistance in a hard red spring wheat breeding population. Front Plant Sci 10:1007
128. Gutierrez-Gonzalez JJ, Mascher M, Poland J et al (2019) Dense genotyping-by-sequencing linkage maps of two synthetic W7984×Opata reference populations provide insights into wheat structural diversity. Sci Rep 9:1–15
129. Li D, Zhang J, Liu H et al (2019) Characterization of a wheat–tetraploid *Thinopyrum elongatum* 1E(1D) substitution line K17–841-1 by cytological and phenotypic analysis and developed molecular markers. BMC Genomics 20:963
130. Lozada DN, Mason RE, Sarinelli JM et al (2019) Accuracy of genomic selection for grain yield and agronomic traits in soft red winter wheat. BMC Genet:20
131. Ollier M, Talle V, Brisset A-L et al (2020) QTL mapping and successful introgression of the spring wheat-derived QTL *Fhb1* for fusarium head blight resistance in three European triticale populations. Theor Appl Genet 133:457–477
132. Yuan C, Singh RP, Liu D et al (2020) Genome-wide mapping of adult plant resistance to leaf rust and stripe rust in CIMMYT wheat line Arableu# 1. Plant Dis 104:1455–1464
133. Hussain W, Baenziger PS, Belamkar V et al (2017) Genotyping-by-sequencing derived high-density linkage map and its application to QTL mapping of flag leaf traits in bread wheat. Sci Rep 7:1–15
134. Ma J, Lin Y, Tang S et al (2020) A genome-wide association study of coleoptile length in different Chinese wheat landraces. Front Plant Sci 11:677
135. Jamil M, Ali A, Gul A et al (2019) Genome-wide association studies of seven agronomic traits under two sowing conditions in bread wheat. BMC Plant Biol 19:149
136. Brasier K, Ward B, Smith J et al (2020) Identification of quantitative trait loci associated with nitrogen-use efficiency in winter wheat. PLoS One 15:e0228775
137. Pradhan S, Babar MA, Bai G et al (2020) Genetic dissection of heat-responsive physiological traits to improve adaptation and increase yield potential in soft winter wheat. BMC Genomics 21:1–15
138. Zhou Y, Conway B, Miller D et al (2017) Quantitative trait loci mapping for spike characteristics in hexaploid wheat. Plant Genome 10:1–15
139. Sehgal D, Rosyara U, Mondal S et al (2020) Incorporating genome-wide association mapping results into genomic prediction models for grain yield and yield stability in CIMMYT spring bread wheat. Front Plant Sci 11:197
140. Elbasyoni IS, Lorenz AJ, Guttieri M et al (2018) A comparison between genotyping-by-sequencing and array-based scoring of SNPs for genomic prediction accuracy in winter wheat. Plant Sci 270:123–130
141. Golicz AA, Bayer PE, Edwards D (2014) Skim-based genotyping by sequencing. In: Batley J (ed) Plant genotyping. Methods in molecular biology (methods and protocols). Humana Press, New York, NY, pp 257–269
142. Bayer PE, Ruperao P, Mason AS et al (2015) High-resolution skim genotyping by sequencing reveals the distribution of crossovers and gene conversions in *Cicer arietinum* and *Brassica napus*. Theor Appl Genet 128:1039–1047
143. Gardiner L-J, Wingen LU, Bailey P et al (2019) Analysis of the recombination landscape of hexaploid bread wheat reveals genes controlling recombination and gene conversion frequency. Genome Biol 20
144. Xu Y, Li P, Yang Z et al (2017) Genetic mapping of quantitative trait loci in crops. Crop J 5:175–184
145. Cui F, Fan X, Zhao C et al (2014) A novel genetic map of wheat: utility for mapping QTL for yield under different nitrogen treatments. BMC Genet 15
146. Nadeem MA, Nawaz MA, Shahid MQ et al (2017) DNA molecular markers in plant breeding: current status and recent advancements in genomic selection and genome editing. Biotechnol Biotechnol Equip 32:261–285
147. Cabral AL, Jordan MC, Larson G et al (2018) Relationship between QTL for grain shape, grain weight, test weight, milling yield, and

plant height in the spring wheat cross RL4452/'AC Domain'. PLoS One 13: e0190681
148. Zhao M, Wang G, Leng Y et al (2018) Molecular mapping of fusarium head blight resistance in the spring wheat line ND2710. Phytopathology 108:972–979
149. Gupta PK, Kulwal PL, Jaiswal V (2019) Association mapping in plants in the post-GWAS genomics era. Adv Genet 104:75–154
150. Morrell PL, Buckler ES, Ross-Ibarra J (2011) Crop genomics: advances and applications. Nat Rev Genet 13:85–96
151. Kearsey MJ, Farquhar AGL (1998) QTL analysis in plants; where are we now. Heredity 80:137–142
152. Ma J, Wingen LU, Orford S et al (2015) Using the UK reference population Avalon x cadenza as a platform to compare breeding strategies in elite Western European bread wheat. Mol Breed 35:70
153. Zhang Z, Ersoz E, Lai CQ et al (2010) Mixed linear model approach adapted for genome-wide association studies. Nat Genet 42:355–360
154. Hu W, Gao D, Wu H et al (2020) Genome-wide association mapping revealed syntenic loci *QFhb-4AL* and *QFhb-5DL* for fusarium head blight resistance in common wheat (*Triticum aestivum* L.). BMC Plant Biol 20:29
155. Wu L, Zhang Y, He Y et al (2019) Genome-wide association mapping of resistance to fusarium head blight spread and deoxynivalenol accumulation in Chinese elite wheat germplasm. Phytopathology 109:1208–1216
156. Gao L, Turner MK, Chao S et al (2016) Genome-wide association study of seedling and adult plant leaf rust resistance in elite spring wheat breeding lines. PLoS One 11: e0148671
157. Sapkota S, Hao Y, Johnson J et al (2019) Genome-wide association study of a world-wide collection of wheat genotypes reveals novel quantitative trait loci for leaf rust resistance. Plant Genome 12:1–14
158. Juliana P, Singh RP, Singh PK et al (2018) Genome-wide association mapping for resistance to leaf rust, stripe rust and tan spot in wheat reveals potential candidate genes. Theor Appl Genet 131:1405–1422
159. Kang Y, Barry K, Cao F et al (2020) Genome-wide association mapping for adult resistance to powdery mildew in common wheat. Mol Biol Rep 47:1241–1256
160. Lv G, Dong Z, Wang Y et al (2020) Identification of genetic loci of black point in Chinese common wheat by genome-wide association study and linkage mapping. Plant Dis. PDIS-12-19-2733-RE
161. Safdar LB, Andleeb T, Latif S et al (2020) Genome-wide association study and QTL meta-analysis identified novel genomic loci controlling potassium-use efficiency and agronomic traits in bread wheat. Front Plant Sci 11:70
162. Shokat S, Sehgal D, Vikram P et al (2020) Molecular markers associated with agro-physiological traits under terminal drought conditions in bread wheat. Int J Mol Sci 21:3156
163. Martinez SA, Godoy J, Huang M et al (2018) Genome-wide association mapping for tolerance to preharvest sprouting and low falling numbers in wheat. Front Plant Sci 9:141
164. Chen J, Zhang F, Zhao C et al (2019) Genome-wide association study of six quality traits reveals the association of the *TaRPP13L1* gene with flour colour in Chinese bread wheat. Plant Biotechnol J 17:2106–2122
165. Liu J, Feng B, Xu Z et al (2018) A genome-wide association study of wheat yield and quality-related traits in Southwest China. Mol Breed 38:1
166. Johnson M, Kumar A, Oladzadabbasabadi A et al (2019) Association mapping for 24 traits related to protein content, gluten strength, color, cooking and milling quality using balanced and unbalanced data in durum wheat [*Triticum turgidum* L. var. Durum (Desf.)]. Front Genet 10:717
167. Jernigan KL, Godoy JV, Huang M et al (2018) Genetic dissection of end-use quality traits in adapted soft white winter wheat. Front Plant Sci 9:271
168. Wang S-X, Zhu Y-L, Zhang D-X et al (2017) Genome-wide association study for grain yield and related traits in elite wheat varieties and advanced lines using SNP markers. PLoS One 12:e0188662
169. Alqudah AM, Haile JK, Alomari DZ et al (2020) Genome-wide and SNP network analyses reveal genetic control of spikelet sterility and yield-related traits in wheat. Sci Rep 10:2098
170. Li X, Xu X, Liu W et al (2020) Dissection of superior alleles for yield-related traits and their distribution in important cultivars of wheat by association mapping. Front Plant Sci 11:175
171. Liu Y, Salsman E, Wang R et al (2020) Meta-QTL analysis of tan spot resistance in wheat. Theor Appl Genet 133:2363

172. Steiner B, Buerstmayr M, Wagner C et al (2019) Fine-mapping of the fusarium head blight resistance QTL *Qfhs.ifa-5A* identifies two resistance QTL associated with anther extrusion. Theor Appl Genet 132:2039–2053
173. Randhawa HS, Mutti JS, Kidwell K et al (2009) Rapid and targeted introgression of genes into popular wheat cultivars using marker-assisted background selection. PLoS One 4:e5752
174. Bernardo R (2008) Molecular markers and selection for complex traits in plants: learning from the last 20 years. Crop Sci 48:1649–1664
175. Brar GS, Brûlé-Babel AL, Ruan Y et al (2019) Genetic factors affecting fusarium head blight resistance improvement from introgression of exotic Sumai 3 alleles (including *Fhb1*, *Fhb2*, and *Fhb5*) in hard red spring wheat. BMC Plant Biol 19:179
176. Platten JD, Cobb JN, Zantua RE (2019) Criteria for evaluating molecular markers: comprehensive quality metrics to improve marker-assisted selection. PLoS One 14:e0210529
177. Randhawa HS, Asif M, Pozniak C et al (2013) Application of molecular markers to wheat breeding in Canada. Plant Breed 132:458–471
178. Randhawa MS, Singh RP, Dreisigacker S et al (2018) Identification and validation of a common stem rust resistance locus in two bi-parental populations. Front Plant Sci 9:1788
179. Guzy-Wrobelska J, Szarekjo I (2003) Molecular and agronomic evaluation of wheat DH lines obtained through maize pollination and anther culture methods. Plant Breed 122:305–313
180. Tadesse W, Sanchez-Garcia M, Tawkaz S et al (2019) Doubled haploid production in wheat. In: Ordon F, Friedt W (eds) Advances in breeding techniques for cereal crops. Burleigh Dodds Science Publishing, Cambridge, UK, pp 93–116
181. Fox SL, Wise IL, Smith MH et al (2013) Shaw hard red spring wheat. Can J Plant Sci 93:299–305
182. Depauw RM, Knox RE, Mccaig TN et al (2011) Carberry hard red spring wheat. Can J Plant Sci 91:529–534
183. Graf RJ, Hucl P, Orshinsky BR et al (2003) McKenzie hard red spring wheat. Can J Plant Sci 83:565–569
184. Singh AK, Clarke JM, Knox RE et al (2012) Transcend durum wheat. Can J Plant Sci 92:809–813
185. Humphreys DG, Townley-Smith TF, Lukow O et al (2010) Burnside extra strong hard red spring wheat. Can J Plant Sci 90:79–84
186. Depauw RM, Townley-Smith TF, Humphreys G et al (2005) Lillian hard red spring wheat. Can J Plant Sci 85:397–401
187. Kumar S, Fox SL, Mitchell Fetch J et al (2019) AAC magnet Canada Western red spring wheat. Can J Plant Sci 99:988–996
188. Bokore FE, Cuthbert RD, Knox RE et al (2019) Mapping quantitative trait loci associated with common bunt resistance in a spring wheat (*Triticum aestivum* L.) variety Lillian. Theor Appl Genet 132:3023–3033
189. Kassa MT, Haas S, Schliephake E et al (2016) A saturated SNP linkage map for the orange wheat blossom midge resistance gene *Sm1*. Theor Appl Genet 129:1507–1517
190. Berraies S, Knox RE, Depauw RM et al (2020) Effectiveness of multigenerational transfer of Sumai-3 fusarium head blight resistance in hard red spring wheat breeding populations. Can J Plant Sci 100:156–174
191. Tshikunde NM, Mashilo J, Shimelis H et al (2019) Agronomic and physiological traits, and associated quantitative trait loci (QTL) affecting yield response in wheat (*Triticum aestivum* L.): a review. Front Plant Sci:10
192. Maeoka RE, Sadras VO, Ciampitti IA et al (2020) Changes in the phenotype of winter wheat varieties released between 1920 and 2016 in response to in-furrow fertilizer: biomass allocation, yield, and grain protein concentration. Front Plant Sci 10:1786
193. Peng J, Richards DE, Hartley NM et al (1995) 'Green revolution' genes encode mutant gibberellin response modulators. Nature 400:256–261
194. Evenson RE, Gollin D (2003) Assessing the impact of the green revolution, 1960 to 2000. Science 300:758–762
195. Guo Z, Song Y, Zhou R et al (2010) Discovery, evaluation and distribution of haplotypes of the wheat *Ppd-D1* gene. New Phytol 185:841–851
196. Wilhelm EP, Boulton MI, Al-Kaff N et al (2013) *Rht-1* and *Ppd-D1* associations with height, GA sensitivity, and days to heading in a worldwide bread wheat collection. Theor Appl Genet 126:2233–2243
197. Bernardo R (2016) Bandwagons I, too, have known. Theor Appl Genet 129:2323–2332
198. Beres BL, Hatfield JL, Kirkegaard JA et al (2020) Toward a better understanding of genotype × environment × management interactions—a global wheat initiative

agronomic research strategy. Front Plant Sci 11:828

199. Chen XM (2005) Epidemiology and control of stripe rust [*Puccinia striiformis* f. sp. *tritici*] on wheat. Can J Plant Pathol 27:314–337
200. Liu R, Lu J, Zhou M et al (2020) Developing stripe rust resistant wheat (*Triticum aestivum* L.) lines with gene pyramiding strategy and marker-assisted selection. Genet Resour Crop Evol 67:381–391
201. Kuchel H, Fox R, Reinheimer J et al (2007) The successful application of a marker-assisted wheat breeding strategy. Mol Breed 20:295–308
202. Chen H, Iqbal M, Yang R-C et al (2016) Effect of *Lr34/Yr18* on agronomic and quality traits in a spring wheat mapping population and implications for breeding. Mol Breed 36
203. Bai B, Du JY, Lu QL et al (2014) Effective resistance to wheat stripe rust in a region with high disease pressure. Plant Dis 98:891–897
204. Abhinandan K, Skori L, Stanic M et al (2018) Abiotic stress signaling in wheat—an inclusive overview of hormonal interactions during abiotic stress responses in wheat. Front Plant Sci 9:734
205. Gratani L (2014) Plant phenotypic plasticity in response to environmental factors. Adv Bot 2014:1–17
206. Wei B, Jing R, Wang C et al (2009) *Dreb1* genes in wheat (*Triticum aestivum* L.): development of functional markers and gene mapping based on SNPs. Mol Breed 23:13–22
207. Liu M, Wang Z, Xiao H et al (2018) Characterization of *TaDREB1* in wheat genotypes with different seed germination under osmotic stress. Hereditas 155:26
208. Zhang J, Xu Y, Chen W et al (2015) A wheat *1-FEH w3* variant underlies enzyme activity for stem WSC remobilization to grain under drought. New Phytol 205:293–305
209. Rasheed A, Mujeeb-Kazi A, Ogbonnaya FC et al (2018) Wheat genetic resources in the post-genomics era: promise and challenges. Ann Bot 121:603–616
210. Zhao C, Cui F, Wang X et al (2012) Effects of 1BL/1RS translocation in wheat on agronomic performance and quality characteristics. Field Crops Res 127:79–84
211. Singh AK, Depauw RM, Knox RE et al (2016) AAC Durafield durum wheat. Can J Plant Sci 96:719–725
212. Barakat MN, Saleh MS, Al-Doss AA et al (2015) Mapping of QTLs associated with abscisic acid and water stress in wheat. Biol Plant 59:291–297
213. Malik S, Mehboob UR, Malik TA (2015) Genetic mapping of potential QTLs associated with drought tolerance in wheat. J Anim Plant Sci 25:1032–1040
214. Christopher J, Christopher M, Jennings R et al (2013) QTL for root angle and number in a population developed from bread wheats (*Triticum aestivum*) with contrasting adaptation to water-limited environments. Theor Appl Genet 126:1563–1574
215. Bai C, Liang Y, Hawkesford MJ (2013) Identification of QTLs associated with seedling root traits and their correlation with plant height in wheat. J Exp Biol 64:1745–1753
216. Talukder SK, Babar MA, Vijayalakshmi K et al (2014) Mapping QTL for the traits associated with heat tolerance in wheat (*Triticum aestivum* L.). BMC Genet 15:97
217. Azam F, Chang X, Jing R (2015) Mapping QTL for chlorophyll fluorescence kinetics parameters at seedling stage as indicators of heat tolerance in wheat. Euphytica 202:245
218. Tiwari C, Wallwork H, Kumar U et al (2013) Molecular mapping of high temperature tolerance in bread wheat adapted to the eastern Gangetic plain region of India. Field Crops Res 154:201–210
219. Yu M, Chen G (2013) Conditional QTL mapping for waterlogging tolerance in two RILs populations of wheat. Springerplus 2:245
220. Ban Y, Ishikawa G, Ueda H et al (2020) Novel quantitative trait loci for low grain cadmium concentration in common wheat (*Triticum aestivum* L.). Breed Sci 70:331–341
221. Dai GB, Zhang D, Hong DL (2013) Validation of quantitative trait loci for aluminum tolerance in Chinese wheat landrace FSW. Euphytica 192:172–179
222. Wani SH, Choudhary M, Kumar P et al (2018) Marker-assisted breeding for abiotic stress tolerance in crop plants. In: Gosal S, Wani S (eds) Biotechnologies of crop improvement. Springer, Cham, pp 1–23
223. Souza EJ, Graybosch RA, Guttieri MJ (2002) Breeding wheat for improved milling and baking quality. J Crop Prod 5:39–74
224. Zhao LA, Zhang KP, Liu B et al (2010) A comparison of grain protein content QTLs and flour protein content QTLs across environments in cultivated wheat. Euphytica 174:325–335
225. Simmonds NW (1995) The relation between yield and protein in cereal grain. J Sci Food Agric 67:309–315
226. Mesfin A, Frohberg RC, Anderson JA (1999) RFLP markers associated with high grain protein from *Triticum turgidum* L. var.

*dicoccoides* introgressed into hard red spring wheat. Crop Sci 39:508–513

227. Distelfeld A, Uauy C, Fahima T et al (2006) Physical map of the wheat high-grain protein content gene *Gpc-B1* and development of a high-throughput molecular marker. New Phytol 169:753–763
228. Jin H, Wen W, Liu J et al (2016) Genome-wide QTL mapping for wheat processing quality parameters in a Gaocheng 8901/ Zhoumai 16 recombinant inbred line population. Front Plant Sci 7:1032
229. Sun DJ, He ZH, Xia XC et al (2005) A novel STS marker for polyphenol oxidase activity in bread wheat. Mol Breed 16:209–218
230. Geng H, Xia X, Zhang L et al (2012) Development of functional markers for a lipoxygenase gene *TaLox-B1* on chromosome 4BS in common wheat. Crop Sci 52:568–576
231. He X, Zhang Y, He Z et al (2008) Characterization of *phytoene synthase 1* gene (*Psy1*) located on common wheat chromosome 7A and development of a functional marker. Theor Appl Genet 116:213–221
232. Zhang C, Dong C, He X et al (2011) Allelic variants at the *TaZds-D1* locus on wheat chromosome 2DL and their association with yellow pigment content. Crop Sci 151:1580–1590
233. Wei J, Geng H, Zhang Y et al (2015) Mapping quantitative trait loci for peroxidase activity and developing gene-specific markers for *TaPod-A1* on wheat chromosome 3AL. Theor Appl Genet 128:2067–2076

# Chapter 4

# Genome-Wide Association Studies (GWAS) in Cereals

**Braulio J. Soto-Cerda, Akshaya Vasudevan, André Laroche, and Raja Ragupathy**

## Abstract

The advent of high-throughput next-generation sequencing technologies in the last decade, coupled with its substantial decrease in cost in the recent years and development of complementary array-based genotyping platforms, has revolutionized the generation of genome-wide markers and propelled several statistical methods for unearthing marker-phenotype association. This chapter provides an outline of the conceptual principles and steps of methods widely used for genome-wide association studies (GWAS) in cereals. Specifically, we focussed on presenting practical steps, starting from assembling populations suitable for phenotyping, genotyping platforms, estimation of population structure from genome-wide markers, estimation of linkage disequilibrium (LD), and methods of GWAS. We also highlighted the available sources of cereal genome assemblies and major software packages used for GWAS, namely, TASSEL-5.0, PLINK, and GAPIT-3.0.

**Key words** Marker-trait association, Kinship, Mapping panel, Causal allele, Population structure, Linkage disequilibrium, Common genetic variants, Fine mapping, High-throughput genotyping, SNP effect, Single-locus GWAS

## 1 Introduction

Phenotypes are manifestations of genes and their interaction with the environment. Identifying genomic regions underlying natural phenotypic variation is fundamental to crop improvement, and achieved by forward genetic approaches such as QTL-mapping and genome-wide association studies (GWAS). Employing conventional genetic mapping strategies involves utilizing a biparental mapping population like $F_2$, doubled haploid (DH), and recombinant inbred lines (RIL). However, occurrence of only one or two crossover per chromosome per meiosis and involvement of only a few generations of advancements in obtaining these mapping populations demand a larger population for fine mapping a genetic region.

Andriy Bilichak and John D. Laurie (eds.), *Accelerated Breeding of Cereal Crops*, Springer Protocols Handbooks, https://doi.org/10.1007/978-1-0716-1526-3_4, © Springer Science+Business Media, LLC, part of Springer Nature 2022

GWAS is a statistical-genetic method for detecting significant "genotype-phenotype" associations. It utilizes a panel of diverse accessions encompassing historical recombination events that have accumulated over thousands of years, in contrast to classical QTL-mapping strategies employing RIL or DH mapping populations, relying on crossover events over a few generations. This facilitates increased QTL resolution in GWAS in comparison with the conventional QTL-mapping strategies, along with the opportunity to assess multiple alleles at the same time. In addition to diversity panel, nested association mapping (NAM) and multiparent advanced generation intercross (MAGIC) populations are also used in GWAS. NAM panel involves crossing a single reference genotype to multiple parents followed by crossing the resultant biparental cross families, and can harness the benefits of linkage and association mapping strategies [1]. MAGIC populations are created by several generations of intercrossing among multiple founder lines [2]. These populations are advantageous from the point of view of increasing rare allele frequency and balancing the overall allele frequencies, thus permitting the capture of favorable QTL alleles of low frequency.

Firstly, GWAS includes assembling a panel of accessions, which is abundant in genetic diversity and has a weak population structure and relatedness and a rapid LD decay. A weak population structure implies that the individuals in a population are not equally related to each other, as strong genetic relatedness can lead to a large number of false positives. Linkage disequilibrium refers to the non-random association between alleles or the correlation between alleles in a population.

Statistical evaluation in association mapping is based on linear model analysis, i.e., regression of a phenotype on a genomic variant. The P-value from the study signifies the probability of obtaining a "phenotype-genotype" association by mere chance (*see* **Note 1**).

With the advancement in next-generation and third-generation sequencing technologies and associated cost reduction in its adoption, several cereals have been sequenced, and their gold-standard reference genomes are available (Table 1). This simplifies the process of identifying genome-wide markers – single nucleotide polymorphisms (SNPs) from whole genome resequencing or genome complexity reduction-based strategies. In addition, a strategy involving splitting the raw sequence data into k-mers, determining their association with a trait of interest, and then localizing these significant k-mers onto the reference has been suggested, in order to account for insertion-deletion (InDel) and structural variations [3] (*see* **Note 2**).

**Table 1**
**Cereals with reference genome sequences, with an application in GWAS**

| Sl. no. | Crop | Scientific name | Chromosome number | Genome size | Reference | GWAS using the reference sequence |
|---|---|---|---|---|---|---|
| 1 | Hexaploid wheat | *Triticum aestivum* | $2n = 6x = 42$ | ~17 Gb | [21] | Agronomic traits under drought and heat stress conditions [22] |
| 2 | Tetraploid wheat | *Triticum turgidum* ssp. *durum* | $2n = 4x = 28$ | ~12 Gb | [23] | Fusarium head blight resistance [24] |
| 3 | Rice | *Oryza sativa* ssp. *japonica* | $2n = 24$ | 420 Mb | [25] | Root morphology traits [26] |
| 4 | Maize | *Zea mays* | $2n = 20$ | ~2.4 Gb | [27] | Early development, maturity, and other agronomic traits [28] |
| 5 | Barley | *Hordeum vulgare* | $2n = 2x = 14$ | ~5.3 Gb | [29] | Heading time, days to maturity, grain yield along with four other agronomic traits [30] |
| 6 | Rye | *Secale cereale* | $2n = 2x = 14$ | 8.1 Gb | [31] | – |
| 7 | Sorghum | *Sorghum bicolor* | $2n = 20$ | ~730 Mb | [32] | Inflorescence architecture [33] |

## 2 Methodology

### 2.1 Plant Material and Phenotyping

Initially, a panel of accessions with large genetic diversity has to be assembled. Population structure and the relatedness among the panel of accessions are to be determined to find a panel with weak population structure and rapid linkage disequilibrium decay. In association studies, pre-breeding material derived from elite cereal cultivars [4], lines constituting advanced yield trials in cereal improvement programs [5], progenitors, and wild relatives can be utilized (*see* **Note 3**). The accessions should be seeded in an experimental design like the completely randomized design, alpha lattice design, or a modified augmented design, based on seed availability, over multiple years and/or multiple locations for generating high-quality phenotyping datasets. The phenotypic data for the traits of interest is recorded in every field season for all the traits of interest. The phenotypic data can be analyzed using a mixed model, and best linear unbiased prediction (BLUP) values can be estimated to account for the random effects.

### 2.2 High-Throughput Genotyping

Single nucleotide polymorphism data can be generated using any of the NGS strategies, including genotyping-by-sequencing (GBS), exome capture and sequencing, and whole genome resequencing. The genotyping data can also be called using genotyping arrays, which are capable of capturing diversity across gene pools [6]. The sequencing reads should be aligned to the crop-kind's reference genome (Table 1) using read-alignment tools like BWA [7] and Bowtie2 [8]. The sequence alignment map (SAM) file format should be converted to binary alignment map format using the SAMtools [9] utility. The multi-sample variant calling can be performed using different tools including the Genome Analysis Toolkit (GATK) [10] and mpileup program in SAMtools. Furthermore, bioinformatics pipeline like Fast-GBS v2.0 [11] and array genotyping-specific software are available to process the GBS and SNP-array data, respectively. The identified SNPs need to be filtered for quality control, to avoid spurious associations. SNPs are filtered based on the following criteria: (a) a call rate of less than 0.09, i.e., they are called in minimum of 10% of individuals in the population; (b) if they are monomorphic; (c) if the minimum allele frequency (MAF) is less than 0.05; and (d) a minimum of three reads must be aligned to the SNP.

### 2.3 Estimation of Population Structure

Population structure is studied using genome-wide SNPs in STR UCTURE v.2.3.4 [12]. A predefined number of genetic cluster (K) from 1 to 7 and 50,000 burn-in iterations followed by 100,000 Markov chain Monte Carlo (MCMC) replications across 5 independent runs for each K value can be employed. The number of clusters (K) can be calculated with the Evanno method [13], in the R-package POPHELPER or using DeltaK statistic estimated with STRUCTURE HARVESTER, a web-based platform for executing Evanno method. The genetic structure can be visualized by constructing a dendrogram using the neighbor-joining (NJ) algorithm implemented in PAST 3.0 [14] or TASSEL v.5.2.31 [15]. Also, principal component (PC) plot analysis can be conducted to decipher the genetic structure in PAST 3.0 [14], TASSEL v.5.2.31 [15], or GAPIT-3.0 [16].

### 2.4 Linkage Disequilibrium

The LD is measured as the squared correlation coefficient ($r^2$) estimated using the allele frequency. The LD can be calculated using the filtered set of SNPs on TASSEL. The GAPIT version 3.0, a R package, can also be employed for LD analysis, in which the LD decay is displayed as a plot with the distance between markers along the x-axis and the $r^2$ values along the y-axis. In case of polyploid genomes, LD is estimated for individual sub-genomes, involving the respective set of SNPs.

### 2.5 Genome-Wide Association Analysis

The statistical analysis for GWAS can be carried using single-locus and multi-locus models. In single-locus models, each SNP is assessed individually for its association with the phenotype.

**Table 2**
**Tools for GWAS analysis in cereals**

| Sl. no. | Tool | Common statistical models | Availability | Reference |
|---|---|---|---|---|
| 1 | Tassel | General linear model (GLM), mixed linear model (MLM) | https://www.maizegenetics.net/tassel | [15] |
| 2 | GAPIT 3 | GLM, MLM, compressed MLM (CMLM), enriched CMLM (ECMLM), multiple locus mixed linear model (MLMM), settlement of MLM under progressively exclusive relationship (SUPER), fixed and random model circulating probability unification (FarmCPU), and Bayesian-information and linkage-disequilibrium iteratively nested keyway (BLINK) | http://zzlab.net/GAPIT/GAPIT.library.R http://zzlab.net/GAPIT/gapit_functions.txt | [16] |
| 3 | PLINK | MLM | https://www.cog-genomics.org/plink/1.9/assoc | [17] |
| 4 | mrMLM | Multi-locus random-SNP-effect mixed linear model (mrMLM), FASTmrMLM, multi-locus random-SNP-effect efficient mixed model association (FASTmrEMMA), iterative modified-sure independence screening empirical-Bayesian LASSO (ISIS EM-BLASSO), polygene-background-control-based least angle regression plus empirical Bayes (pLARmEB), and Kruskal-Wallis test with empirical Bayes under polygenic background control (pKWmEB) | http://cran.r-project.org/web/packages/mrMLM/index.html | [18] |

However, this strategy doesn't fit well for quantitative traits, reducing the power of the analysis. Further, this statistical approach necessitates the application of multiple test correction, increasing the probability of type II error (false negatives) with stringent criteria. In case of multi-locus models, initially the SNPs are screened using a single-locus model, and later the set of filtered-SNPs is evaluated using a multi-locus approach.

Single-locus models: The general linear model-Q (GLM-Q) and GLM-PCA involve the population structure and principal components, respectively. The mixed linear model (MLM) under PLINK takes into account the kinship matrix (K), in addition to the population structure (Q) [17]. The estimation of Q, PCA, and K, followed by association analysis with single-locus models, can be implemented in TASSEL [15]. However, the covariates can be obtained from other programs like the Q-matrix from STRUC TURE [12].

Multi-locus models: Association analysis using six multi-locus models, multi-locus random-SNP-effect mixed linear model (mrMLM), FASTmrMLM, FASTmrEMMA, ISIS EM-BLASSO, pLARmEB, and pKWmEB, can be performed with the R-package mrMLM tools for genome-wide association study [18]. The Q-matrix from STRUCTURE and K-matrix from TASSEL are taken into account using the fileKin and filePS arguments in the package. The critical log of odds score can be set to greater than 3.0, for detecting quantitative trait nucleotides significantly associated with the trait of interest.

The links to TASSEL, GAPIT, PLINK, and mrMLM are given in Table 2.

In summary, GWAS considers the following steps:

1. Assembly of the association panel: It is advisable to use all sources of information available for a target species such as passport data, historical phenotypic data, and molecular data to define an association panel. The use of core collections from germplasm, which have been assembled with the purpose of removing genetic redundancy to provide a manageable number of genetically diverse genotypes, is a good starting point for attaining successful GWAS.
2. Genetic structure analysis: Once the association panel has been genotyped with genome-wide markers with sufficient density and distribution, genetic structure can be elucidated. Create a genotypic file in Excel with genotype names in columns and marker data in rows (Fig. 1). To set genotypic matrix to STR UCTURE format, replace SNP alleles (A, T, C, G) (Fig. 1a) with numerical coding (1, 3, 2, 4), and transpose columns and rows (Fig. 1b). Missing genotypic data (NA) must be coded as −9. Save the data file as tab delimited with a suitable filename, i.e., <Genotypicdata.txt>. Download STRUCTURE and its manual (https://web.stanford.edu/group/pritchardlab/structure_software/release_versions/v2.3.4/html/structure.html), and follow the instructions to run the software. To determine the number of clusters or subpopulations ($K$) using the Evanno method, compile results in a compressed file (Fig. 1c), and upload it into STRUCTURE HARVESTER (http://taylor0.biology.ucla.edu/structureHarvester/). Once the optimum number of K has been determined, construct the Q-matrix in the format needed for TASSEL organizing population membership values (Q values) in columns and genotype names in rows in Excel as indicated in Fig. 1. Save the file as tab delimited with an appropriate name.
3. Kinship analysis: GWAS using mixed linear model (MLM) requires the estimation of kinship relationships among genotypes to minimize false positives. Upload the genotypic data in

**A**

| | Geno1 | Geno2 | Geno3 | Geno4 | Geno5 | Geno6 | Geno7 | Geno8 | Geno9 | Geno10 |
|---|---|---|---|---|---|---|---|---|---|---|
| SNP1 | T | T | T | T | T | T | T | T | T | T |
| SNP2 | T | T | T | T | T | T | T | T | T | T |
| SNP3 | T | T | T | T | T | T | T | T | T | T |
| SNP4 | T | T | T | T | T | T | T | T | T | T |
| SNP5 | C | C | C | C | C | C | C | C | C | C |
| SNP6 | G | G | G | G | G | A | G | G | A | G |
| SNP7 | T | T | T | T | T | A | T | T | A | T |
| SNP8 | T | T | C | T | T | T | T | NA | NA | T |
| SNP9 | G | G | A | G | G | G | G | G | G | G |
| SNP10 | A | A | G | A | A | A | A | A | A | A |
| SNP11 | G | G | A | G | G | G | G | G | A | G |
| SNP12 | G | G | A | G | G | G | G | G | A | G |

**C**

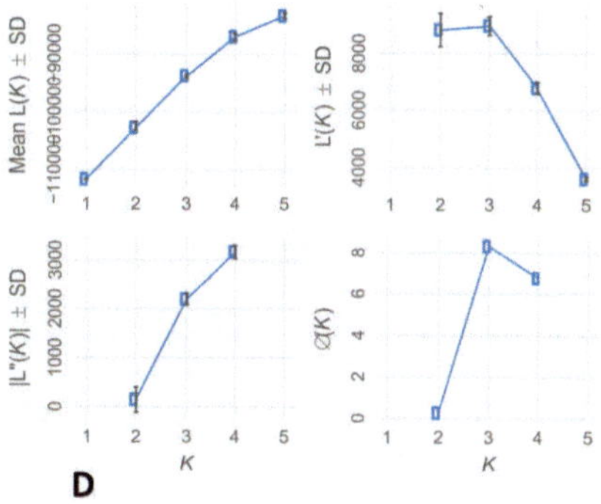


**B**

| | SNP1 | SNP2 | SNP3 | SNP4 | SNP5 | SNP6 | SNP7 | SNP8 | SNP9 | SNP10 | SNP11 | SNP12 |
|---|---|---|---|---|---|---|---|---|---|---|---|---|
| Geno1 | 3 | 3 | 3 | 3 | 2 | 4 | 3 | 3 | 4 | 1 | 4 | 4 |
| Geno2 | 3 | 3 | 3 | 3 | 2 | 4 | 3 | 3 | 4 | 1 | 4 | 4 |
| Geno3 | 3 | 3 | 3 | 3 | 2 | 4 | 3 | 2 | 1 | 4 | 1 | 1 |
| Geno4 | 3 | 3 | 3 | 3 | 2 | 4 | 3 | 3 | 4 | 1 | 4 | 4 |
| Geno5 | 3 | 3 | 3 | 3 | 2 | 4 | 3 | 3 | 4 | 1 | 4 | 4 |
| Geno6 | 3 | 3 | 3 | 3 | 2 | 1 | 1 | 3 | 4 | 1 | 4 | 4 |
| Geno7 | 3 | 3 | 3 | 3 | 2 | 4 | 3 | 3 | 4 | 1 | 4 | 4 |
| Geno8 | 3 | 3 | 3 | 3 | 2 | 4 | 3 | -9 | 4 | 1 | 4 | 4 |
| Geno9 | 3 | 3 | 3 | 3 | 2 | 1 | 1 | -9 | 4 | 1 | 1 | 1 |
| Geno10 | 3 | 3 | 3 | 3 | 2 | 4 | 3 | 3 | 4 | 1 | 4 | 4 |

**D**

| <Covariate> | | | |
|---|---|---|---|
| <Trait> | Q1 | Q2 | Q3 |
| Geno1 | 1 | 0 | 0 |
| Geno2 | 0.909 | 0.001 | 0.09 |
| Geno3 | 0.999 | 0 | 0 |
| Geno4 | 1 | 0 | 0 |
| Geno5 | 0.999 | 0 | 0 |
| Geno6 | 0.689 | 0.055 | 0.255 |
| Geno7 | 0.962 | 0.001 | 0.037 |
| Geno8 | 0.812 | 0 | 0.188 |
| Geno9 | 0.762 | 0 | 0.237 |
| Geno10 | 0.565 | 0 | 0.435 |

**Fig. 1** Genetic structure analysis using STRUCTURE. (**a**) Matrix of SNP markers data in its original format. (**b**) Transformed SNP markers data into STRUCTURE format. (**c**) Graphic output of Evanno analysis indicating the optimum number of clusters (K). (**d**) Q-matrix constructed in TASSEL format

TASSEL, select "Analysis," and "Relatedness" and then select "Kinship" (Fig. 2). Save the kinship matrix as tab delimited with an appropriate name.

4. GWAS analysis using GLM: Upload into TASSEL the genotypic, Q-matrix, and phenotypic data files. Then, select the three files and go to "Data" option, and select "Intersection" to combine the three files (Fig. 3). After that, select the "Analysis" option and "Association" and select GLM, and click "Ok" and run the analysis. After a few seconds, the analysis will be completed and results can be visualized.
5. GWAS analysis using MLM: In addition to the three previous files, you must upload the kinship matrix into TASSEL. The procedure is similar to the GLM, but after combining the genotypic, Q-matrix and phenotypic data files into a single concatenated file, you need to select the concatenated file plus the kinship matrix simultaneously (Fig. 4). After that, select the "Analysis" option, "Association," and select MLM and click "Run" to carry out the analysis. After a few minutes, the analysis will be completed and results can be visualized.
6. Quality control of GWAS results: After GWAS analysis has been completed, you need to check if the selected model either GLM or MLM fitted well to control *P*-value inflation and, as

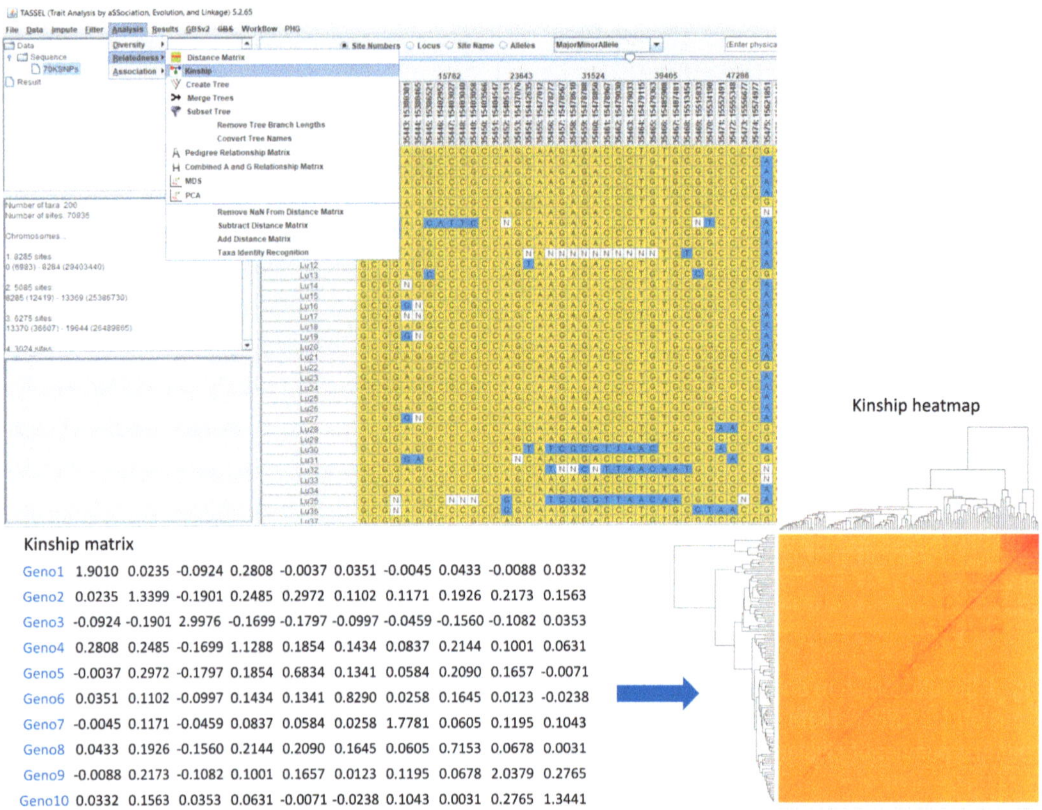


**Fig. 2** Kinship relationship analysis in TASSEL

a result, the occurrence of false-positive associations. To examine the rate of false-positive associations, quantile-quantile (Q-Q) plots are useful. Q-Q plots allow comparing the distribution between the observed P-values generated by the GWAS model and the expected P-values. When a substantial deviation occurs from the very beginning of the distribution, it indicates that the model failed in controlling type I error (false positives), and a large proportion of significant marker-trait associations are false. To produce a Q-Q plot in TASSEL, after running the analysis, you must select "Results" and then "QQ Plot," to visualize the distribution of expected and observed P-values, representing statistical significance of associations (Fig. 5).

7. Visualization of GWAS results: Since next-generation sequencing has permitted the identification of thousands of genetic markers (mostly SNPs), the graphical display of observed P-values helps to better understand which chromosome regions play pivotal roles in determining the QTL architecture of a target trait. An easy way to achieve that is using Manhattan plots (Fig. 6).

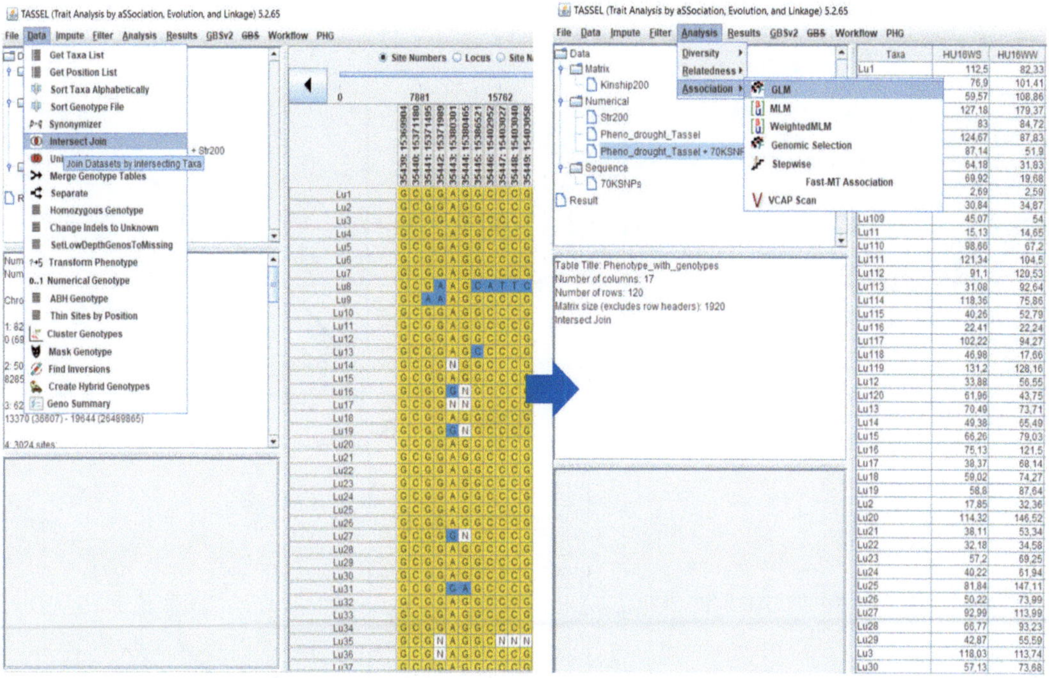


**Fig. 3** Procedure to conduct GWAS using GLM in TASSEL

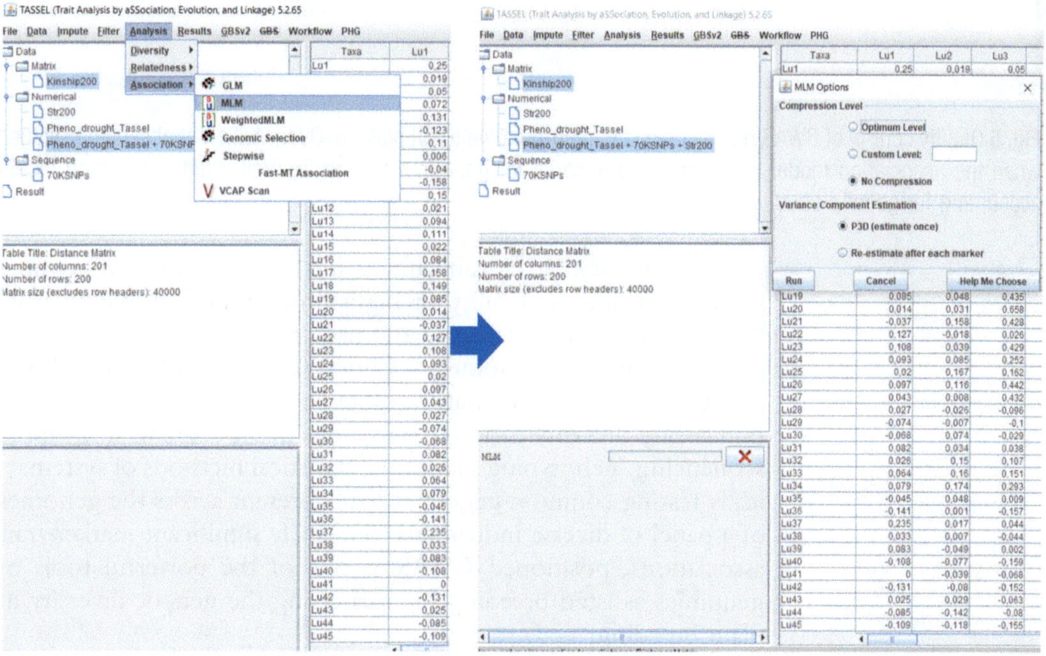


**Fig. 4** Procedure to conduct GWAS using MLM in TASSEL

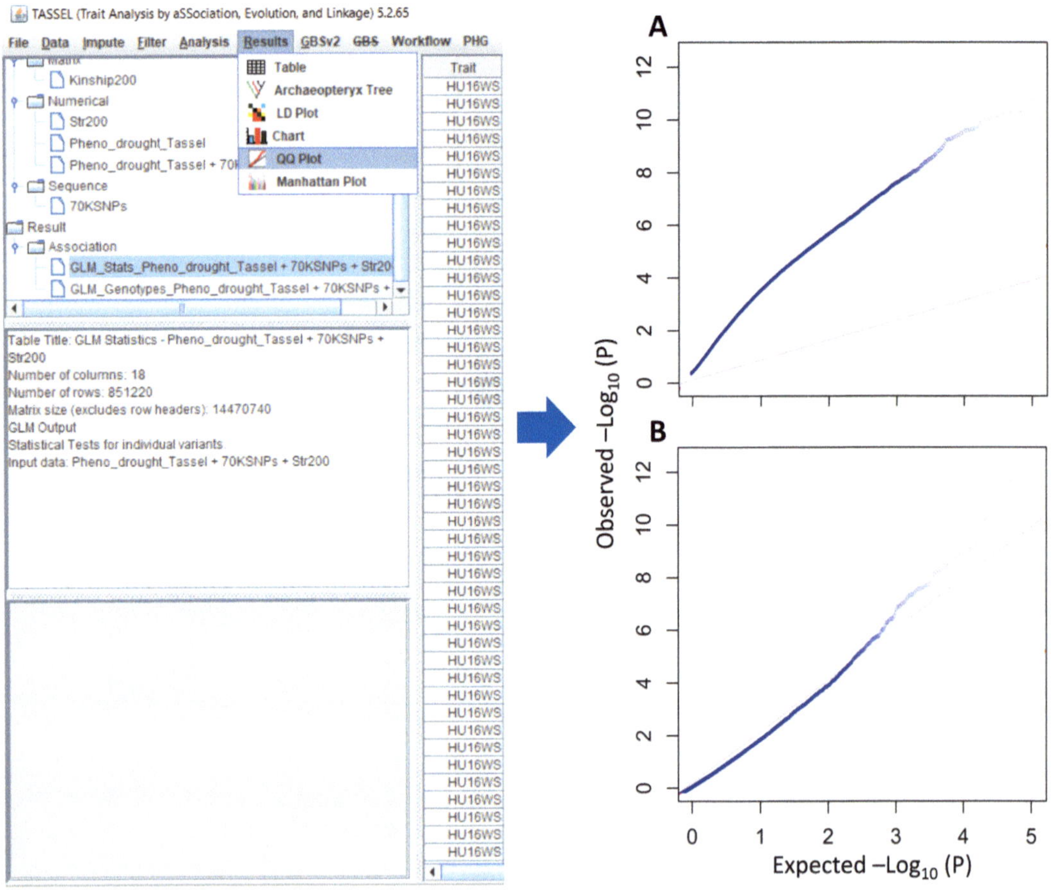


**Fig. 5** Quality control of GWAS results using quantile-quantile (Q-Q) plots (**a**) Overinflation of observed P-values when the association model lacks power to account for genetic structure (**b**) Well-fitted association model accounting for genetic structure

In summary, the advancements in the past decade in genome sequencing and availability of high-throughput genotyping platforms have greatly facilitated the use of well-curated germplasm panels in genome-phenome association studies. GWAS unravels the genetic architecture of natural phenotypic variation of complex traits using diversity panels. Continuous improvements in genome sequencing/genotyping platforms, statistical methods of systematically testing common genetic variants present across the genomes of a panel of diverse individuals to identify significant marker-trait associations, positioned GWAS as one of the powerful tools of genomics-assisted breeding for harnessing the genetic diversity in plant breeding.

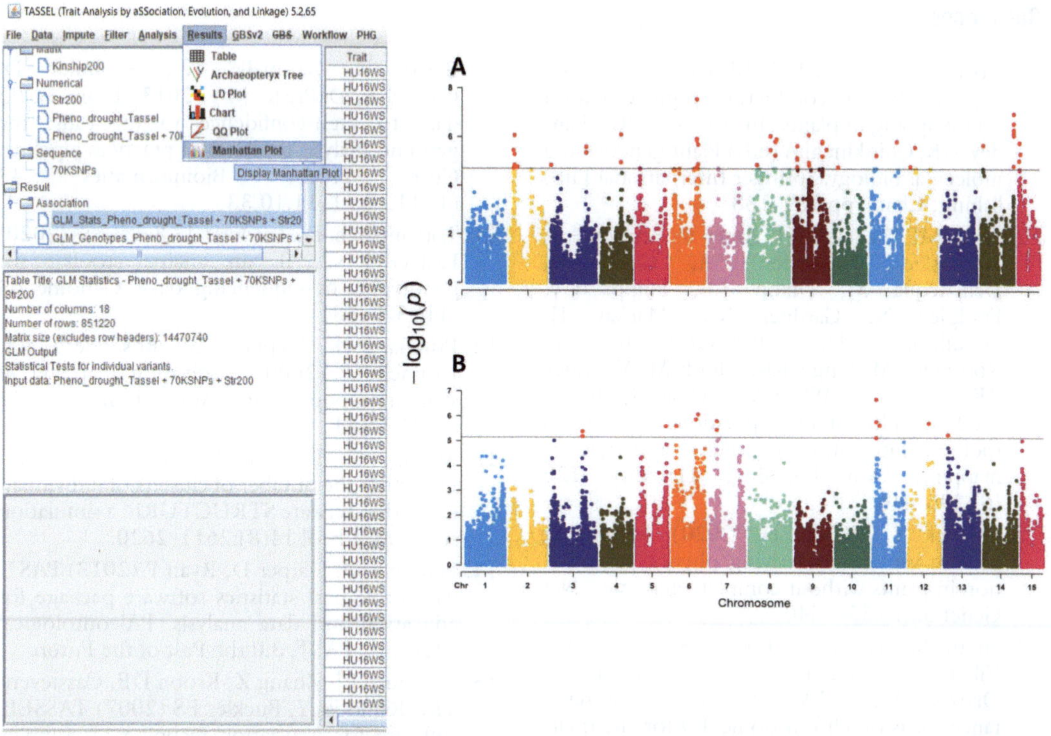


**Fig. 6** Visualization of GWAS results using Manhattan plots. (**a**) Manhattan plot exhibiting an excessive number of significant marker-trait associations as an effect of genetic structure. (**b**) Manhattan plot showing adequate control of false-positive associations. Red dots represent significant associations

## 3 Notes

1. For a summary on the statistical methods involved in GWAS, please refer [19].
2. Recently, an integrated GWAS pipeline was published, namely, HAPPI GWAS [20], with a simple easy-to-follow manual, which will be useful for the researchers for implementing GWAS in their programs.
3. A multitude of factors such as species (self-fertilizing, etc.), sample size, genetic heterogeneity, kinship/population structure, SNP-effect size, and allele frequency contribute to the accuracy of dissecting the genetic architecture of natural phenotypic variance of complex traits by finding statistically significant association of common genetic variants with traits of agronomic interests. In practical terms, one of the prime bottlenecks is quality and speed of generation of accurate phenotypic datasets, and hence, wider adoption of high-throughput field phenotyping methods will greatly help geneticists and plant breeders in fine mapping and introgressing diverse alleles from germplasm resources.

## References


1. Cockram J, Mackay I (2018) Genetic mapping populations for conducting high-resolution trait mapping in plants. In: Varshney RK, Pandey MK, Chitikineni A (eds) Plant genetics and molecular biology. Springer International Publishing, Cham, pp 109–138
2. Scott MF, Ladejobi O, Amer S, Bentley AR, Biernaskie J, Boden SA, Clark M, Dell'Acqua M, Dixon LE, Filippi CV, Fradgley N, Gardner KA, Mackay IJ, O'Sullivan D, Percival-Alwyn L, Roorkiwal M, Singh RK, Thudi M, Varshney RK, Venturini L, Whan A, Cockram J, Mott R (2020) Multi-parent populations in crops: a toolbox integrating genomics and genetic mapping with breeding. Heredity 125 (6):396–416
3. Voichek Y, Weigel D (2020) Identifying genetic variants underlying phenotypic variation in plants without complete genomes. Nat Genet 52(5):534–540
4. Singh S, Sehgal D, Kumar S, Arif MAR, Vikram P, Sansaloni CP, Fuentes-Dávila G, Ortiz C (2020) GWAS revealed a novel resistance locus on chromosome 4D for the quarantine disease Karnal bunt in diverse wheat pre-breeding germplasm. Sci Rep 10(1):5999
5. Sehgal D, Rosyara U, Mondal S, Singh R, Poland J, Dreisigacker S (2020) Incorporating genome-wide association mapping results into genomic prediction models for grain yield and yield stability in CIMMYT spring bread wheat. Front Plant Sci 11:197
6. Winfield MO, Allen AM, Burridge AJ, Barker GLA, Benbow HR, Wilkinson PA, Coghill J, Waterfall C, Davassi A, Scopes G, Pirani A, Webster T, Brew F, Bloor C, King J, West C, Griffiths S, King I, Bentley AR, Edwards KJ (2016) High-density SNP genotyping array for hexaploid wheat and its secondary and tertiary gene pool. Plant Biotechnol J 14 (5):1195–1206
7. Li H, Durbin R (2009) Fast and accurate short read alignment with burrows-wheeler transform. Bioinformatics 25(14):1754–1760
8. Langmead B, Salzberg S (2013) Bowtie2. Nat Methods 9:357–359
9. Li H, Handsaker B, Wysoker A, Fennell T, Ruan J, Homer N, Marth G, Abecasis G, Durbin R (2009) The sequence alignment/map format and SAMtools. Bioinformatics 25 (16):2078–2079
10. Van der Auwera GA, Carneiro MO, Hartl C, Poplin R, del Angel G, Levy-Moonshine A, Jordan T, Shakir K, Roazen D, Thibault J, Banks E, Garimella KV, Altshuler D, Gabriel S, DePristo MA (2013) From FastQ data to high-confidence variant calls: the genome analysis toolkit best practices pipeline. Curr Protoc Bioinformatics 43 (1):11.10.11–11.10.33
11. Torkamaneh D, Laroche J, Belzile F (2020) Fast-GBS v2. 0: an analysis toolkit for genotyping-by-sequencing data. Genome 63 (11):577–581
12. Pritchard JK, Stephens M, Rosenberg NA, Donnelly P (2000) Association mapping in structured populations. Am J Hum Genet 67 (1):170–181
13. Evanno G, Regnaut S, Goudet J (2005) Detecting the number of clusters of individuals using the software STRUCTURE: a simulation study. Mol Ecol 14(8):2611–2620
14. Hammer Ø, Harper D, Ryan P (2013) PAST: paleontological statistics software package for education and data analysis. PAleontological STatistics, PAST, 3.0 the Past of the Future
15. Bradbury PJ, Zhang Z, Kroon DE, Casstevens TM, Ramdoss Y, Buckler ES (2007) TASSEL: software for association mapping of complex traits in diverse samples. Bioinformatics 23 (19):2633–2635
16. Tang Y, Liu X, Wang J, Li M, Wang Q, Tian F, Su Z, Pan Y, Liu D, Lipka AE, Buckler ES, Zhang Z (2016) GAPIT version 2: an enhanced integrated tool for genomic association and prediction. Plant Genome 9:2. plantgenome2015.2011.0120
17. Purcell S, Neale B, Todd-Brown K, Thomas L, Ferreira MAR, Bender D, Maller J, Sklar P, de Bakker PIW, Daly MJ, Sham PC (2007) PLINK: a tool set for whole-genome association and population-based linkage analyses. Am J Hum Genet 81(3):559–575
18. Zhang Y-W, Tamba CL, Wen Y-J, Li P, Ren W-L, Ni Y-L, Gao J, Zhang Y-M (2020) mrMLM v4.0: an R platform for multi-locus genome-wide association studies. bioRxiv. 2020.2003.2004.976464
19. Hayes B (2013) Overview of statistical methods for genome-wide association studies (GWAS). In: Gondro C, van der Werf J, Hayes B (eds) Genome-wide association studies and genomic prediction. Humana Press, Totowa, NJ, pp 149–169
20. Slaten ML, Chan YO, Shrestha V, Lipka AE, Angelovici R (2020) HAPPI GWAS: holistic analysis with pre- and post-integration GWAS. Bioinformatics 36(17):4655–4657

21. IWGSC, Appels R, Eversole K, Stein N, Feuillet C, Keller B, Rogers J, Pozniak CJ, Choulet F, Distelfeld A, Poland J, Ronen G, Sharpe AG, Barad O, Baruch K, Keeble-Gagnère G, Mascher M, Ben-Zvi G, Josselin A-A, Himmelbach A, Balfourier F, Gutierrez-Gonzalez J, Hayden M, Koh C, Muehlbauer G, Pasam RK, Paux E, Rigault P, Tibbits J, Tiwari V, Spannagl M, Lang D, Gundlach H, Haberer G, Mayer KFX, Ormanbekova D, Prade V, Šimková H, Wicker T, Swarbreck D, Rimbert H, Felder M, Guilhot N, Kaithakottil G, Keilwagen J, Leroy P, Lux T, Twardziok S, Venturini L, Juhász A, Abrouk M, Fischer I, Uauy C, Borrill P, Ramirez-Gonzalez RH, Arnaud D, Chalabi S, Chalhoub B, Cory A, Datla R, Davey MW, Jacobs J, Robinson SJ, Steuernagel B, van Ex F, BBH W, Benhamed M, Bendahmane A, Concia L, Latrasse D, Bartoš J, Bellec A, Berges H, Doležel J, Frenkel Z, Gill B, Korol A, Letellier T, Olsen O-A, Singh K, Valárik M, van der Vossen E, Vautrin S, Weining S, Fahima T, Glikson V, Raats D, Číhalíková J, Toegelová H, Vrána J, Sourdille P, Darrier B, Barabaschi D, Cattivelli L, Hernandez P, Galvez S, Budak H, JDG J, Witek K, Yu G, Small I, Melonek J, Zhou R, Belova T, Kanyuka K, King R, Nilsen K, Walkowiak S, Cuthbert R, Knox R, Wiebe K, Xiang D, Rohde A, Golds T, Čížková J, Akpinar BA, Biyiklioglu S, Gao L, N'Daiye A, Kubaláková M, Šafář J, Alfama F, Adam-Blondon A-F, Flores R, Guerche C, Loaec M, Quesneville H, Condie J, Ens J, Maclachlan R, Tan Y, Alberti A, Aury J-M, Barbe V, Couloux A, Cruaud C, Labadie K, Mangenot S, Wincker P, Kaur G, Luo M, Sehgal S, Chhuneja P, Gupta OP, Jindal S, Kaur P, Malik P, Sharma P, Yadav B, Singh NK, Khurana JP, Chaudhary C, Khurana P, Kumar V, Mahato A, Mathur S, Sevanthi A, Sharma N, Tomar RS, Holušová K, Plíhal O, Clark MD, Heavens D, Kettleborough G, Wright J, Balcárková B, Hu Y, Salina E, Ravin N, Skryabin K, Beletsky A, Kadnikov V, Mardanov A, Nesterov M, Rakitin A, Sergeeva E, Handa H, Kanamori H, Katagiri S, Kobayashi F, Nasuda S, Tanaka T, Wu J, Cattonaro F, Jiumeng M, Kugler K, Pfeifer M, Sandve S, Xun X, Zhan B, Batley J, Bayer PE, Edwards D, Hayashi S, Tulpová Z, Visendi P, Cui L, Du X, Feng K, Nie X, Tong W, Wang L (2018) Shifting the limits in wheat research and breeding using a fully annotated reference genome. Science 361(6403): eaar7191

22. Li L, Mao X, Wang J, Chang X, Reynolds M, Jing R (2019) Genetic dissection of drought and heat-responsive agronomic traits in wheat. Plant Cell Environ 42(9):2540–2553

23. Maccaferri M, Harris NS, Twardziok SO, Pasam RK, Gundlach H, Spannagl M, Ormanbekova D, Lux T, Prade VM, Milner SG, Himmelbach A, Mascher M, Bagnaresi P, Faccioli P, Cozzi P, Lauria M, Lazzari B, Stella A, Manconi A, Gnocchi M, Moscatelli M, Avni R, Deek J, Biyiklioglu S, Frascaroli E, Corneti S, Salvi S, Sonnante G, Desiderio F, Marè C, Crosatti C, Mica E, Özkan H, Kilian B, De Vita P, Marone D, Joukhadar R, Mazzucotelli E, Nigro D, Gadaleta A, Chao S, Faris JD, Melo ATO, Pumphrey M, Pecchioni N, Milanesi L, Wiebe K, Ens J, MacLachlan RP, Clarke JM, Sharpe AG, Koh CS, Liang KYH, Taylor GJ, Knox R, Budak H, Mastrangelo AM, Xu SS, Stein N, Hale I, Distelfeld A, Hayden MJ, Tuberosa R, Walkowiak S, Mayer KFX, Ceriotti A, Pozniak CJ, Cattivelli L (2019) Durum wheat genome highlights past domestication signatures and future improvement targets. Nat Genet 51(5):885–895

24. Sari E, Knox RE, Ruan Y, Henriquez MA, Kumar S, Burt AJ, Cuthbert RD, Konkin DJ, Walkowiak S, Campbell HL, Singh AK, Ross J, Lokuruge P, Hsueh E, Boyle K, Sidebottom C, Condie J, Yates S, Pozniak CJ, Fobert PR (2020) Historic recombination in a durum wheat breeding panel enables high-resolution mapping of fusarium head blight resistance quantitative trait loci. Sci Rep 10(1):7567

25. Kawahara Y, de la Bastide M, Hamilton JP, Kanamori H, McCombie WR, Ouyang S, Schwartz DC, Tanaka T, Wu J, Zhou S, Childs KL, Davidson RM, Lin H, Quesada-Ocampo L, Vaillancourt B, Sakai H, Lee SS, Kim J, Numa H, Itoh T, Buell CR, Matsumoto T (2013) Improvement of the *Oryza sativa* Nipponbare reference genome using next generation sequence and optical map data. Rice 6 (1):4

26. Kadam NN, Tamilselvan A, Lawas LMF, Quinones C, Bahuguna RN, Thomson MJ, Dingkuhn M, Muthurajan R, Struik PC, Yin X, Jagadish SVK (2017) Genetic control of plasticity in root morphology and anatomy of rice in response to water deficit. Plant Physiol 174(4):2302–2315

27. Jiao Y, Peluso P, Shi J, Liang T, Stitzer MC, Wang B, Campbell MS, Stein JC, Wei X, Chin C-S (2017) Improved maize reference genome with single-molecule technologies. Nature 546 (7659):524–527

28. Mayer M, Hölker AC, González-Segovia E, Bauer E, Presterl T, Ouzunova M, Melchinger AE, Schön C-C (2020) Discovery of beneficial

haplotypes for complex traits in maize landraces. Nat Commun 11(1):4954

29. Mascher M, Gundlach H, Himmelbach A, Beier S, Twardziok SO, Wicker T, Radchuk V, Dockter C, Hedley PE, Russell J, Bayer M, Ramsay L, Liu H, Haberer G, Zhang X-Q, Zhang Q, Barrero RA, Li L, Taudien S, Groth M, Felder M, Hastie A, Šimková H, Staňková H, Vrána J, Chan S, Muñoz-Amatriaín M, Ounit R, Wanamaker S, Bolser D, Colmsee C, Schmutzer T, Aliyeva-Schnorr L, Grasso S, Tanskanen J, Chailyan A, Sampath D, Heavens D, Clissold L, Cao S, Chapman B, Dai F, Han Y, Li H, Li X, Lin C, McCooke JK, Tan C, Wang P, Wang S, Yin S, Zhou G, Poland JA, Bellgard MI, Borisjuk L, Houben A, Doležel J, Ayling S, Lonardi S, Kersey P, Langridge P, Muehlbauer GJ, Clark MD, Caccamo M, Schulman AH, Mayer KFX, Platzer M, Close TJ, Scholz U, Hansson M, Zhang G, Braumann I, Spannagl M, Li C, Waugh R, Stein N (2017) A chromosome conformation capture ordered sequence of the barley genome. Nature 544(7651):427–433

30. Abed A, Belzile F (2019) Comparing single-SNP, multi-SNP, and haplotype-based approaches in association studies for major traits in barley. Plant Genome 12(3):190036

31. Rabanus-Wallace MT, Hackauf B, Mascher M, Lux T, Wicker T, Gundlach H, Baez M, Houben A, Mayer KFX, Guo L, Poland J, Pozniak CJ, Walkowiak S, Melonek J, Praz CR, Schreiber M, Budak H, Heuberger M, Steuernagel B, Wulff B, Börner A, Byrns B, Čížková J, Fowler DB, Fritz A, Himmelbach A, Kaithakottil G, Keilwagen J, Keller B, Konkin D, Larsen J, Li Q, Myśków B, Padmarasu S, Rawat N, Sesiz U, Biyiklioglu-Kaya S, Sharpe A, Šimková H, Small I, Swarbreck D, Toegelová H, Tsvetkova N, Voylokov AV, Vrána J, Bauer E, Bolibok-Bragoszewska H, Doležel J, Hall A, Jia J, Korzun V, Laroche A, Ma XF, Ordon F, Özkan H, Rakoczy-Trojanowska M, Scholz U, Schulman AH, Siekmann D, Stojałowski S, Tiwari VK, Spannagl M, Stein N (2021) Chromosome-scale genome assembly provides insights into rye biology, evolution and agronomic potential. Nat Genet 53(4):564–573

32. Paterson AH, Bowers JE, Bruggmann R, Dubchak I, Grimwood J, Gundlach H, Haberer G, Hellsten U, Mitros T, Poliakov A, Schmutz J, Spannagl M, Tang H, Wang X, Wicker T, Bharti AK, Chapman J, Feltus FA, Gowik U, Grigoriev IV, Lyons E, Maher CA, Martis M, Narechania A, Otillar RP, Penning BW, Salamov AA, Wang Y, Zhang L, Carpita NC, Freeling M, Gingle AR, Hash CT, Keller B, Klein P, Kresovich S, McCann MC, Ming R, Peterson DG, Mehboobur R, Ware D, Westhoff P, Mayer KFX, Messing J, Rokhsar DS (2009) The *Sorghum bicolor* genome and the diversification of grasses. Nature 457(7229):551–556

33. Zhou Y, Srinivasan S, Mirnezami SV, Kusmec A, Fu Q, Attigala L, Salas Fernandez MG, Ganapathysubramanian B, Schnable PS (2019) Semiautomated feature extraction from rgb images for sorghum panicle architecture GWAS. Plant Physiol 179(1):24–37

# Chapter 5

# CropGBM: An Ultra-Efficient Machine Learning Toolbox for Genomic Selection-Assisted Breeding in Crops

Yuetong Xu, John D. Laurie, and Xiangfeng Wang

## Abstract

Continued improvement and falling costs of DNA sequencing have accelerated the increase in genomic resources for crop plants. From these efforts, considerable genetic diversity has been found and is aiding in the identification of markers for breeding purposes. High-density molecular markers have allowed for marker-assisted selection of quantitative traits that are controlled by a small number of genes. Recently, whole genomic selection has been proposed where markers genome-wide are used to estimate the contribution of all loci to traits of interest. In this chapter we outline the steps needed to perform genomic selection using machine learning. We describe our method called Crop Genomic Breeding Machine (CropGBM) and demonstrate its use on diverse maize lines containing high-density markers.

**Key words** Marker-assisted selection, Genomic selection, Machine learning, Artificial intelligence

## 1 Introduction

Traditional breeding is primarily based on phenotype selection. Breeders select excellent offspring by observing crop phenotypes to achieve genetic improvement of target traits. Advances in molecular genetic technology have identified a wide range of genetic variations in crop genomes, from which a large number of molecular markers associated with traits are selected to accelerate the breeding process. Marker-assisted selection (MAS) [1] has been commonly utilized to assist breeding, and is usually only effective for qualitative traits controlled by single genes with major effects. For many important agronomic traits controlled by numerous quantitative trait loci (QTLs) in which each QTL contributes a minor effect to the phenotype, MAS is usually not effective [2–4]. First of all, markers adjacent to QTL sites with insignificant effect may be omitted. Secondly, due to the insufficient coverage

Software website: https://ibreeding.github.io/

Andriy Bilichak and John D. Laurie (eds.), *Accelerated Breeding of Cereal Crops*, Springer Protocols Handbooks, https://doi.org/10.1007/978-1-0716-1526-3_5, © Springer Science+Business Media, LLC, part of Springer Nature 2022

and low-density markers, the accumulative influence of the QTLs contributing to the traits cannot be accurately estimated. Therefore, MAS is usually unable to direct breeding for quantitative traits, such as yield and resistance traits. With the rapid advance of sequencing technology that enables high-density molecular markers, whole genomic selection (GS) technology has been proposed. GS utilizes genome-wide markers, which may avoid the omission of small-effect markers, to estimate the contribution of all loci in the genome to traits, based on the calculation of a genomic estimated breeding value (GEBV) derived from a GS predictive models with the genome-wide markers as input [5]. Assisted with the predicted phenotype or GEBV from the GS model, breeders thus can precisely select breeding material or design breeding schemes to greatly shorten the breeding cycle. For crops with heterosis, breeders evaluate inbred lines not only based on parental phenotypes but also the potential for creating superior hybrids. In the seed industry, GS usually uses 20% of the total samples as training population to construct a predictive model to predict the phenotype of the offspring of untested hybrid combinations, from which the general combining ability (GCA) is estimated to accelerate screening [6]. Also, GS can infer the genotype of the $F_1$ hybrid offspring based on their parental genotypes, which may greatly reduce the cost for genotyping [7].

Currently, the most commonly used GS algorithm is to apply regression analysis to predict phenotypes, which are mainly divided into two categories: the BLUP (best linear unbiased prediction) method represented by gBLUP (genomic BLUP) and rrBLUP (ridge regression BLUP) and the Bayesian method represented by Bayes-A and Bayes-B. The traditional BLUP-type method constructs a pedigree matrix inferred from the samples based on the breeding history, and then uses MLM (mixed linear model) to calculate the EBV (estimated breeding value) [8]. In contrast, gBLUP computes a pedigree matrix containing the genomic relationship between each pair of the samples based on the genotype of all the samples in one breeding population, and then uses this correlation matrix instead of the kinship matrix to genomically estimate EBV (referred to as GEBV) for each sample. Additionally, rrBLUP treats the labeling effect as a random effect, assuming it is a standard normal distribution, and then sums up the labeling effects to estimate GEBV [9]. The Bayesian method assumes that the labeling effect obeys a certain prior distribution, so problems such as hyper-parameter optimization for the prior distribution exist. However, if the mark set is too large, the performance of the Bayesian method is not comparable to the BLUP-based method [5]. Although gBLUP has been widely used in building GS models due to its high efficiency and accuracy, disadvantages still exist especially when given extra-large sample sets. First, processing of genotypic data is complicated, as gBLUP requires the conversion of genotypes from the character forms of A, T, G, and C to digital forms of 0, 1, and 2, according to the minor allele frequencies (MAFs) for each SNP site; when the population is changed, this conversion has to be redone which may cost significant amounts of computing time. Second, gBLUP struggles to capture the complex nonlinear relationships between genotype and phenotypes, as gBLUP is based on linear mixed-effect models. Third, the kinship matrix derived from one designated population is non-extensible, which means that if the training population is changed, the kinship matrix has to be reconstructed, especially when the genetic composition and population structure are complex; otherwise, problematic model fitting may occur that generates a great deal of false-positive results. Fourth, the linear model of gBLUP is efficient at solving regression problems; however, many agronomic traits are multi-classification problems which require a more appropriate method for solving rather than using a regression model.

In the recent years, high-throughput genotyping and phenotyping technologies have been rapidly advancing concomitantly with lowering costs. This has resulted in the rapid accumulation of genotypic and phenotypic data from breeding materials, inbred lines, and $F_1$ hybrids collected from actual breeding programs. Thus, the ever-accumulating large datasets are gradually forming a Big Data environment, which increases the feasibility of utilizing machine learning-based paradigms to perform data-driven decision-making for breeding. To overcome the innate shortcomings of traditional GS methods, adoption of machine learning (ML) theory and algorithms for genomic selection has been highly anticipated. Several types of ML algorithms including SVM (support-vector machine), RF (random forest), GB (gradient boost), LGB (light gradient boost), and other machine learning methods are gradually being applied to biological problems such as genomic selection [10–12].

ML methods do not require the distribution and variance of labeling effects, and can fully explore the nonlinear relationship between labels by continuously learning data and optimizing parameters. According to our unpublished research (Yan et al., unpublished), an ensemble learning paradigm – gradient boost (GB) in ML outperformed other types of ML models. One of its variants LightGBM (gradient boosting machine) exhibited great advantages over gBLUP, rrBLUP, and Bayesian methods, as well as other types of ML algorithms. A great advantage is its ultra-efficiency in terms of computing when training models. Performance testing on a large population containing 100,000 samples showed that, while rrBLUP may take almost 1 month to accomplish model training, LightGBM only used 9 min on a small-scale server. Thus, we implemented the LightGBM algorithm as a toolbox, called CropGBM (Crop Genomic Breeding Machine), to carry out genomic selection for breeding. In this chapter, we illustrate the usage of CropGBM in terms of utilizing machine learning for genomic selection-assisted breeding in crops, including streamline analysis of data preprocessing, population structure analysis, feature selection, and phenotype prediction. The CropGBM toolbox and example dataset are freely available for academic research proposes at https://ibreeding.github.io.

## 2 Example Dataset

The example dataset used here for demonstrating the functionality of CropGBM is composed of 6210 maize $F_1$ hybrids generated by crossing 207 maternal lines with 30 paternal lines, including genotypes of 4903 SNPs and phenotypes of flowering time (DTT, days to tasseling), plant stature (PH, plant height), and kernel yield (EW, ear weight). The 4903 SNPs were selected from a total of 14.8 million SNPs generated from whole-genome resequencing of 1458 inbred lines that were used for constructing the $F_1$ hybrid population. The example dataset are freely available at https://github.com/YuetongXU/CropGBM_Tutorial-data.

### 2.1 Hardware Configuration of the Server

There is no specific hardware requirement for running CropGBM programs. The choice of hardware is determined according to the sample size and number of SNPs. The configuration of the server used for developing CropGBM is as follows.

| | |
|---|---|
| **CPU** | Xeon CPU (E5–2665 2.40GHz, 8 Cores) × 2 |
| **Memory** | 128 Gb |
| **GPU** | NVIDIA GeForce GTX-P8 1080 × 4 |

### 2.2 Software Environment and Preinstalled Libraries

**Linux operating system**: Ubuntu 16.04.5 LTS, 18.04.2 LTS, is recommended.

**CropGBM**: A toolbox that is a one-stop solution for genomic selection-assisted breeding based on the gradient boosting algorithm with ensemble learning paradigm. It features a streamlined analytical pipeline with genotype processing, phenotype processing, population structure analysis, SNP screening, genotype-to-phenotype (G2P) prediction, data visualization, and other functional modules. Additionally, it includes t-SNE and OPTICS to analyze and visualize population structure. The kernel package used in CropGBM is the "LightGBM" package to perform regression and classification problem. Installation instruction of CropGBM may be accessed via https://ibreeding.github.io/.

**Plink1.9**: PLINK is required to perform a set of large-scale analyses in a computationally efficient manner, to process the genotype and phenotype data. The tool of Plink1.9 may be obtained from the download address http://www.cog-genomics.org/plink/1.9/.

**GPU**: CropGBM supports GPU acceleration to speed up the process of model training with three to five times increase. If your server is equipped with a GPU card, you may want to check the ability of the GPU in terms of scientific computing via https://developer.nvidia.com/cuda-gpus. This website may return the GPU type, and help the users select the appropriate compute unified device architecture (CUDA) toolkit to provide the necessary environment for GPU-accelerated computing. The CUDA download address is https://developer.nvidia.com/cuda-downloads.

## 3 Analytical Procedure of Running CropGBM

CropGBM supports two modes to configure parameters prior to running the program, using configuration files to perform streamlined analysis or direct command lines to add parameters. The configuration file documents the parameters and the corresponding values in one single file, so that it is convenient for users to uniformly manage and reuse a large number of parameters.

### 3.1 Preprocessing of Genotype and Phenotype Data

CropGBM calls the software Plink1.9 for preprocessing of genotypic data. To ensure a high quality of SNP set, the following steps need to be done, including screening of incomplete samples and SNP sites according to deletion rate and MAFs, imputation of missing genotypes based on high-frequency genotypes, and removal of redundant SNPs based on linkage. Most importantly, CropGBM converts the character-based genotypes (such as AA, AC, CC) to digit-based genotypes (0, 1, 2). Additionally, CropGBM also supports statistical analysis of genotype data such as missing rate, heterozygosity rate, and MAF distribution, so that the users may examine the quality of the genotype data. Moreover, system-level phenotype variation may exist to influence G2P prediction, which is caused by population stratification. Thus, for certain situations, preprocessing of phenotype data is also required. CropGBM utilizes the Z-score algorithm to remove systematic bias of phenotype across different populations. The detailed usage is shown below.

### 3.2 Genotype Data Preprocessing

#### 3.2.1 Filter and View Overall Data

```
$ cropgbm -c configfile.params -o cropgbm_result/ -pg all --file-prefix genofile --fileformat bed --snpmaxmiss 0.10 --samplemax-miss 0.10 --maf_max 0.05 --r2_cutoff 0.7
```

This command filters the data based on the SNP miss rate (*--snpmaxmiss 0.10*), the sample miss rate (*--samplemaxmiss 0.10*), the minimum allele frequency (*--maf_max 0.05*), and r ^ 2 (*–r2_cutoff 0.7*) to generate files in BED and PED formats and distribution histograms showing the overall situation of the data (Fig. 1).

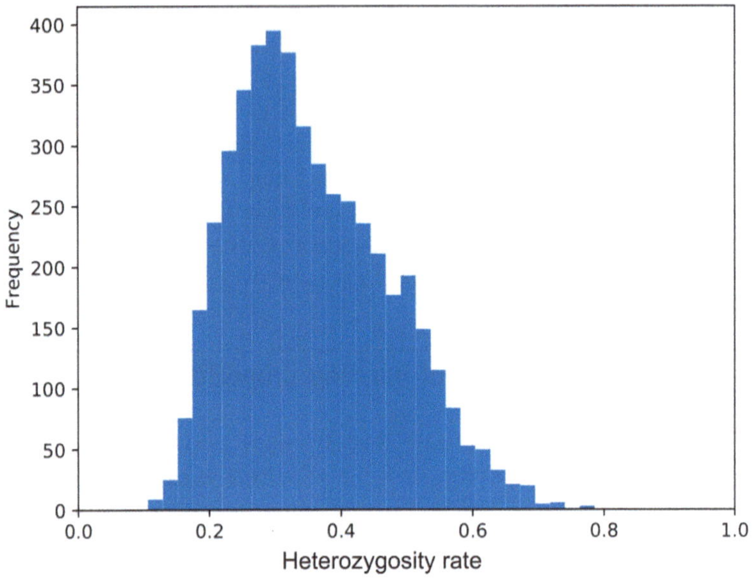


**Fig. 1** Distribution of heterozygosity rate

#### 3.2.2 Extract and Remove Specific ID of Samples and SNP Data

```
$ cropgbm -c configfile.params -o cropgbm_result/ -pg filter --fileprefix genofile --keep-sampleid-path ksampleid_file.txt --extract-snpid-path ksnpid_file.txt
$ cropgbm -c configfile.params -o cropgbm_result/ -pg filter --fileprefix genofile --remove-sampleid-path rsampleid_file.txt --exclude-snpid-path rsnpid_file.txt
```

This command extracts and removes the data of specific samples according to the sample ID by --keep-sampleid-path or --remove-sampleid-path, and the data of a specific SNP in all samples according to the SNP ID by --extract-snpid-path or --exclude-snpid-path.

### 3.3 Conversion of Character-Based Genotypes to Digits

```
$ cropgbm -c configfile.params -o cropgbm_result/ -pg filter --fileprefix genofile --recode --remove-sampleid-path rsampleid_file.txt --exclude-snpid-path rsnpid_file.txt
```

The parameter --recode indicates that the recoding operation is performed on the genotype data. The change rule is 00-> 0, 01-> 1, 10-> 1, 11-> 2. The genofile_filter.geno in the output file is converted genotype file.

### 3.4 Running the CropGBM with Command-Line Parameters as a Pipeline

```
$ cropgbm -c configfile.params -o cropgbm_result/ -pg all --fileprefix genofile --fileformat ped --snpmaxmiss 0.10 --snpmaxmiss 0.10 --maf_max 0.05 --r2_cutoff 0.7 --recode --keep-sampleid-path ksampleid_file.txt --extract-snpid-path ksnpid_file.txt
```

Except for the -c and –pg parameters, all of the other parameters can be set through the configuration file and omitted on the command line.

## 4 Phenotype Data Preprocessing

### 4.1 Extract and Visualize Phenotype Data

```
$ cropgbm -c configfile.params -o cropgbm_result/ -pp --phe-plot
--phefile-path phefile.txt --phefile-header --phefile-sep
```

The phenotype file contains at least two columns of data. By default, CropGBM treats the first column as the sample ID and the second column as the phenotype data which can be visualized as a histogram distribution (Fig. 2).

### 4.2 Extract and Visualize the Phenotype Data According to Sample ID

```
$ cropgbm -c configfile.params -o cropgbm_result/ -pp --phe-plot
--phefile-path phefile.txt --ppgroupfile-path phefile.txt --
ppgroupfile-sep ',' --ppgroupid-column 3 --ppgroupfile-header
```

This command generates a phefile_scatter.pdf file to facilitate examining whether population stratification of phenotype occurs in different groups of samples (Fig. 3).

### 4.3 Normalization of Phenotype Data with Z-Score

```
$ cropgbm -c configfile.params -o cropgbm_result/ -pp -phe-
plot -phefile-path phefile.txt -ppgroupfile-path phefile.txt -
ppgroupfile-sep ',' -ppgroupid-column 3 -ppgroupfile-header -
phe_norm -norm-mode z-score
```

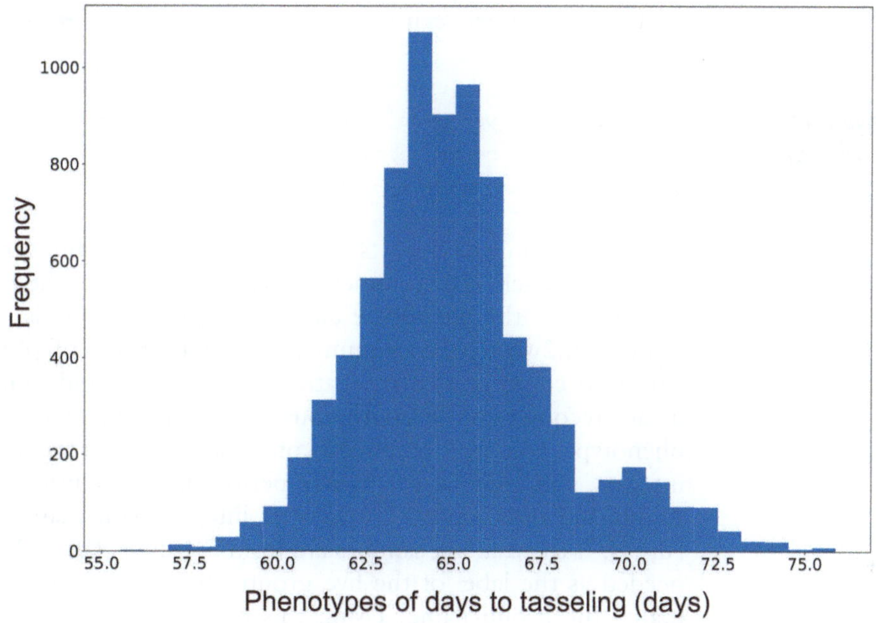


**Fig. 2** Phenotype distribution represented by histogram plot

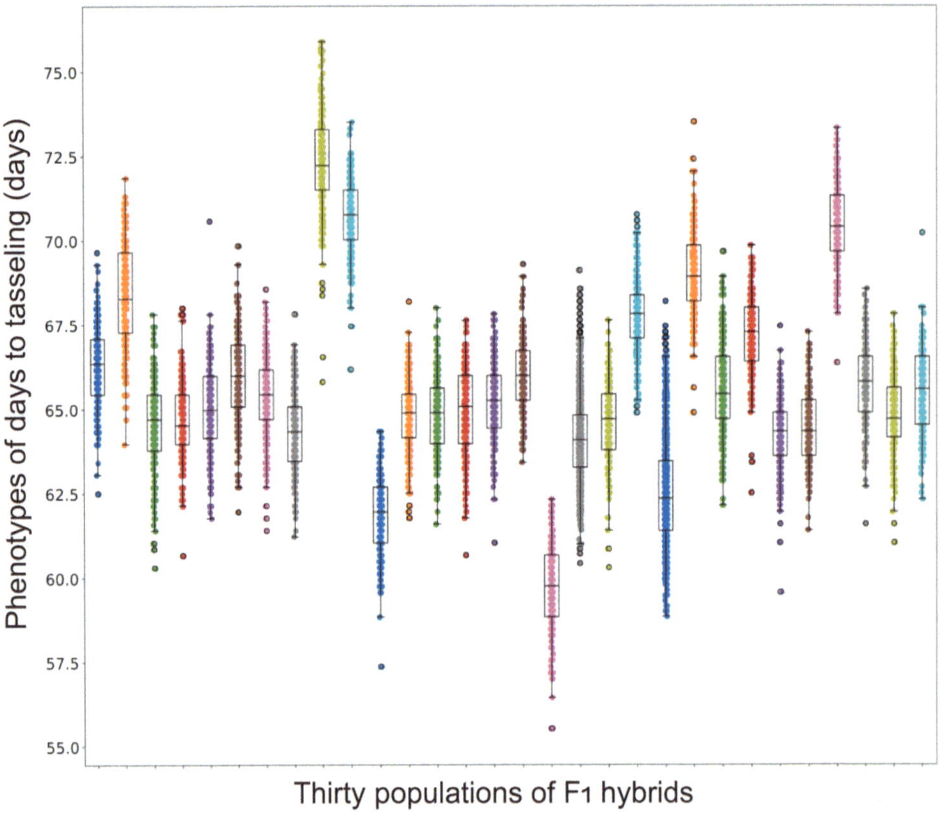


**Fig. 3** Phenotype distribution of 30 populations of $F_1$ hybrids before normalization

This command performs Z-score normalization on phenotype data to remove between-group stratification (*--norm-mode z-score*) (Fig. 4).

#### 4.4 Conversion of Phenotype Data

```
# recode phenotype data into continuous non-negative integer
$ cropgbm -c configfile.params -o cropgbm_result/ -pp -phe-recode word2num -phefile-path phefile.txt
```

This command converts phenotype data into continuous non-negative integers. The parameter *--phe-recode* specifies the recoding operation on the phenotype data. The optional value is [word2-num, num2word]. Word2num means conversion of phenotype data into continuous nonnegative integer form, and num2word means reconversion of continuous nonnegative integer form into phenotype data. This requires a conversion table corresponding to integers and phenotype. When performing classification tasks, LightGBM only accepts consecutive integers with example labels [0, N]. If the training samples are from 5 groups, [0, 1, 2, 3, 4] is needed as the label of the five groups, but this usually does not match the group name. Using this parameter, the program can

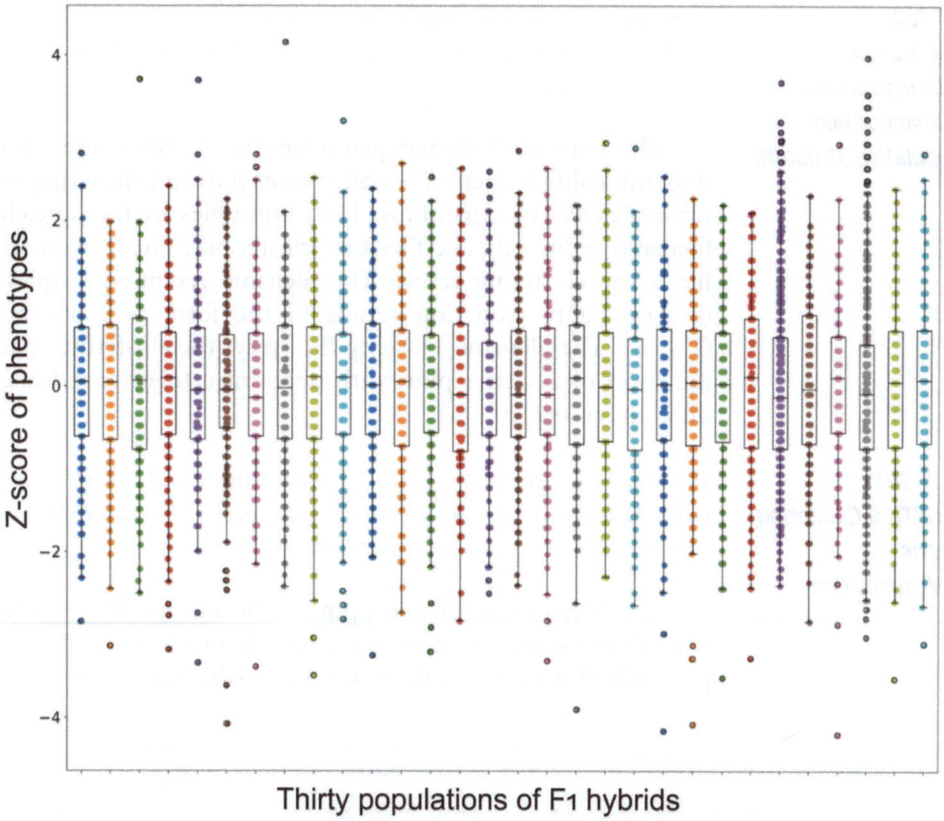


**Fig. 4** Phenotype distribution of 30 populations of $F_1$ hybrids after Z-score normalization

implement a reversible conversion between sample labels and [0, N] consecutive integers, providing compatible phenotype data for downstream classification tasks. The phefile.word2num in the output file is a correspondence file between phenotype data and continuous nonnegative integer.

## 5 Population Structure Analysis

As population stratification may cause model overfitting, it is important to understand population structure and the genetic composition of the samples before constructing the model using CropGBM. Then, the training and testing dataset may be appropriately partitioned to avoid overfitting. In addition to the commonly used PCA (principal component analysis) and K-means clustering algorithm, CropGBM also integrates nonlinear algorithms—t-SNE and OPTICS—to visualize the population structure.

### 5.1 PCA and K-Means Clustering Analysis and Visualization of Population Structure

```
$ cropgbm -c configfile.params -o cropgbm_result/ -s --genofile-path filename_filter.geno --structure_plot --redim-mode pca --cluster-mode kmeans --n-clusters 30
```

This command clusters genotype data based on user-specified dimensionality reduction (*--redim-mode pca*) and clustering (*--cluster-mode kmeans*) algorithms. The output files are filename.cluster, filename_redim.pdf, and filename_cluster.pdf. The filename.cluster file is the clustering result. The filename_redim.pdf displays the dimensionality reduction results in the form of a scatter plot (Fig. 5). The filename_cluster.pdf displays the clustering results in the form of a scatter plot, with different categories indicated by different colors (Fig. 6).

### 5.2 t-SNE and OPTICS Clustering Analysis and Visualization

```
$ cropgbm -c configfile.params -o cropgbm_result/ -s --genofile-path filename_filter.geno --structure_plot --redim-mode tsne --window-size 5 --cluster-mode optics
```

The filename_reachability.pdf in the output file displays the reachable distance between each sample in the form of a scatter plot, which is output only when the *--redim-mode* value is t-SNE

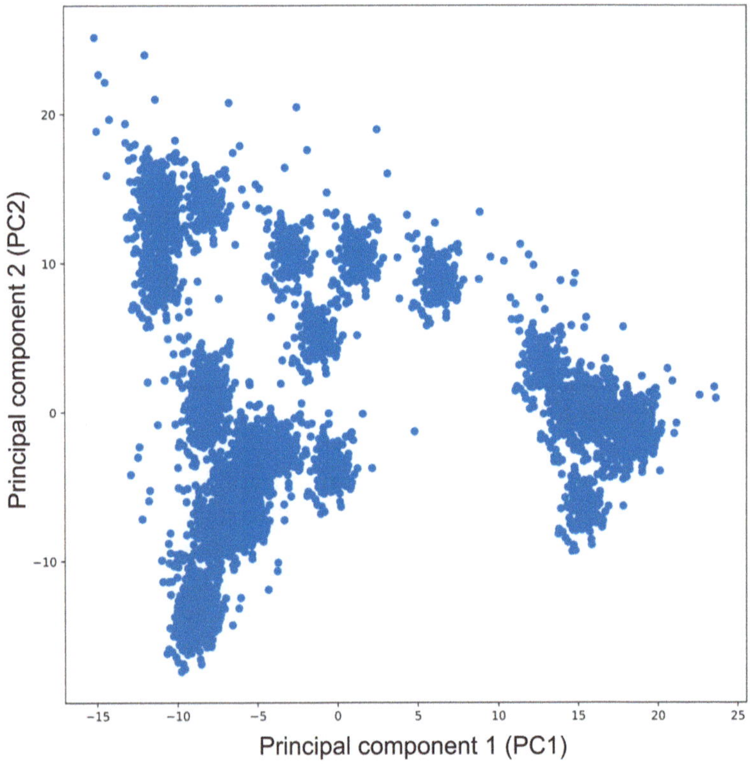


**Fig. 5** PCA representation of the 30 populations indicates strong population stratification

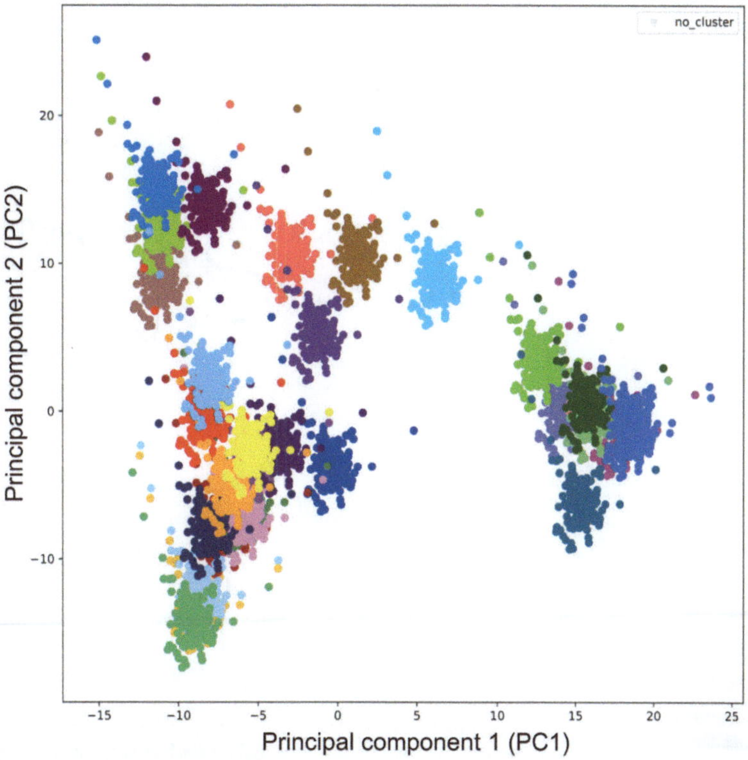


**Fig. 6** PCA representation of the 30 populations. Samples in the same population are colored by the same colors

(Fig. 7). Different categories are represented by different colors, and discrete points are represented by black dots. At the same time, a plot of population structure constructed by the t-SNE algorithm is also generated with different categories represented by different colors (Fig. 8).

## 6 Constructing Genomic Selection GS Model with Training Samples

Construction of the GS model is the core functionality of CropGBM. Similar to other ML algorithms, the quality of the parameters is essential to the precision and performance of the predictive model to ensure the robustness and precision of the model. To avoid problematic overfitting, multiple validation sets are highly recommended to derive the most optimal parameters with the cross-validation analysis.

### 6.1 Cross-Validation Analysis

```
$ cropgbm -c configfile.params -o cropgbm_result/ -e -cv --traingeno train.geno --trainphe train.phe --cv-nfold 5 --min-detal 0.5
```

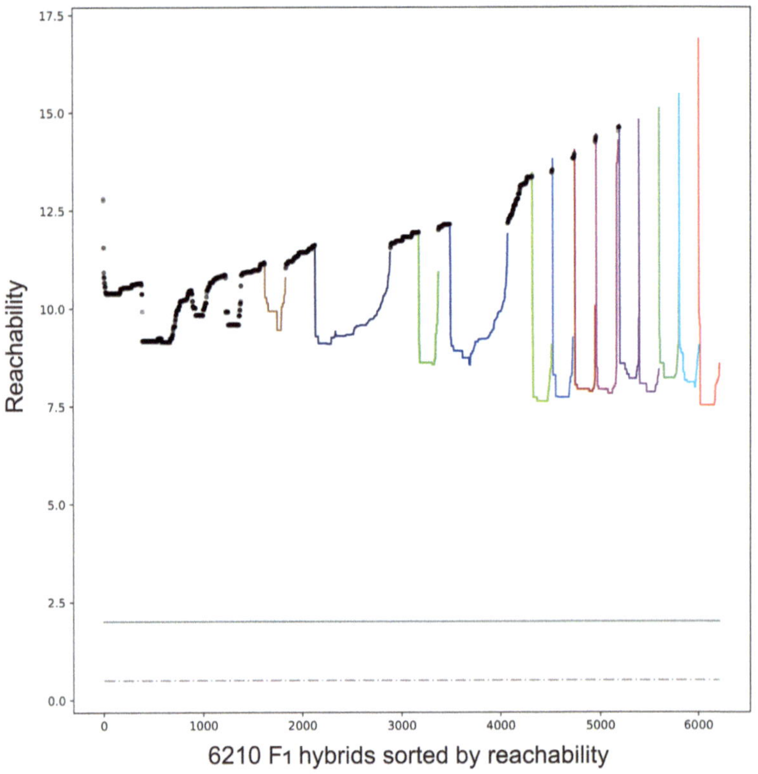


**Fig. 7** OPTICS plot of the 6210 $F_1$ hybrids to illustrate the genetic similarity among samples

This command will estimate the accuracy of the model under different iterations. The accuracy is measured by mean square error (MSE) between the prediction result and the actual phenotype value. It's worth noting that, when the difference between training and testing set is considerably large, model precision may not be objective and it is necessary to sacrifice precision to ensure model robustness.

### 6.2 Training Model

```
$ cropgbm -c configfile.params -o cropgbm_result/ -e -t --train-geno train.phe --trainphe train.phe --validgeno valid.geno --validphe valid.phe
```

This command will estimate the prediction accuracy of the model on the validation set at different iterations. The output model file is train.lgb_model, in which model structure and parameters are documented for the subsequent prediction of the phenotypes of testing samples. If a validation set is not provided, the *--validgeno* and *--validphe* parameters can be omitted. It is important to note that CropGBM only recognizes different SNPs by index, not by column name. If the SNP of the same column is different

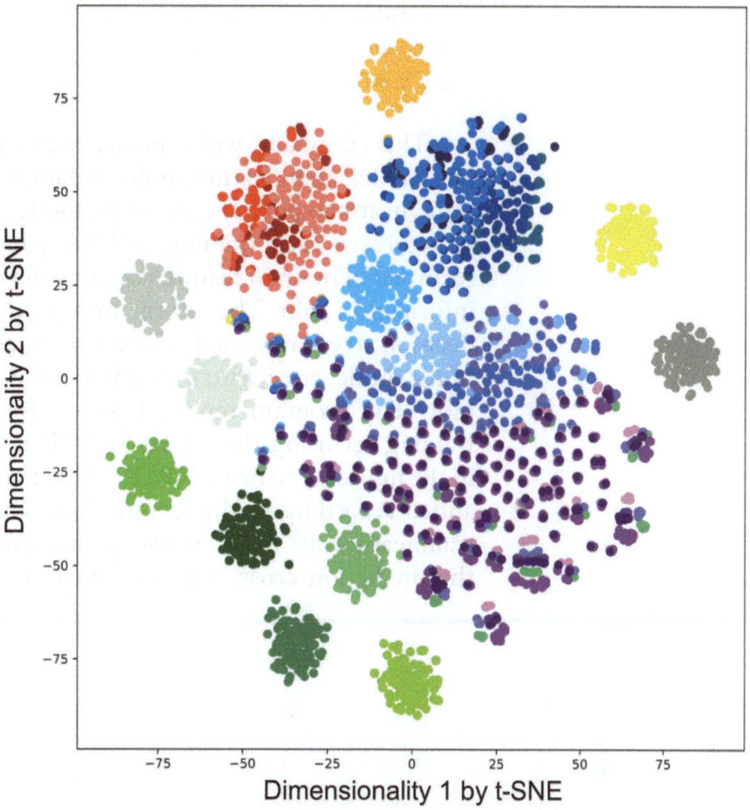


**Fig. 8** t-SNE plot to illustrate the population stratification of the 6210 $F_1$ hybrids

between the two datasets, the program cannot recognize it. Therefore, the training set, testing set, and validation set must be consistent in terms of using the same SNP index; otherwise, the prediction result is non-reference.

## 7 Feature Selection Functionality

Similar to other ML algorithms, CropGBM may perform feature selection analysis during the training process, so that features with high predictive effectiveness may be automatically identified. CropGBM derives a so-called parameter of information gain (IG) to rank the predictive effectiveness for each feature. As a matter of fact, the higher the IG, the higher the probability of the genotype being associated with the trait. Therefore, feature selection in CropGBM is similar to the GWAS analysis, which can be used to identify SNPs with significant association to a target trait. This means that a large number of nonrelevant SNPs may be removed and a highly condensed marker panel may be constructed according to the feature selection functionality in CropGBM.

```
$ cropgbm -c configfile.params -o cropgbm_result/ -e -t -sf
--bygain-boxplot --traingeno train.geno --trainphe train.phe
--gain-min 0.05 --colorbar-max 0.6 --cv-times 5
```

This command will generate five output files, including train.lgb_model, train.feature, train_bygain.pdf, train_random.pdf, and train_heatmap.pdf. The train.lgb_model is the model file. The tree structure of each training and the gain value of each node are recorded. The train.feature contains the selected feature with high information gain. The information gain value of each SNP in each decision tree is recorded. The train_bygain.pdf shows the variation of the model error with the addition of SNP in the form of a scatter plot. The program repeated the fivefold cross-validation on the training set using the added SNP (Fig. 9). The x-axis coordinate in the figure is the new SNP ID added in the model, and the order is added according to the featureGain_sum value of the SNP in the train.feature file from largest to smallest. The y-axis coordinate is the prediction error. The train_random.pdf shows the change of

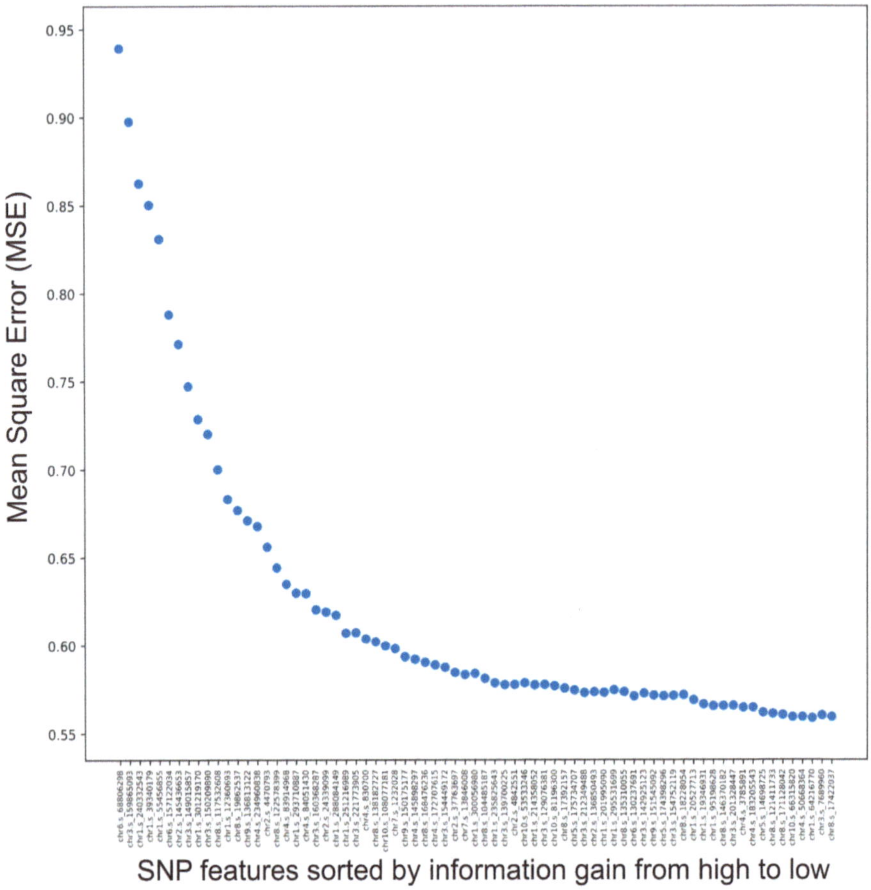


**Fig. 9** SNP features sorted by feature importance based on the value of information gain inferred by CropGBM

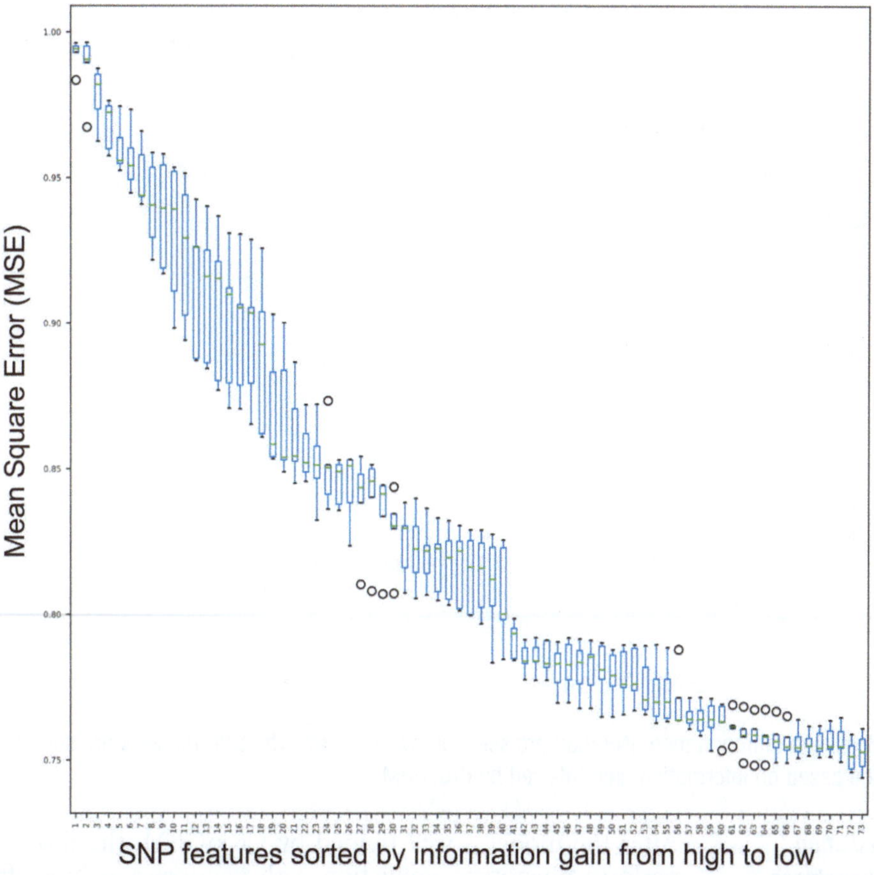


**Fig. 10** Randomized SNP features sorted by feature importance based on the value of information gain inferred by CropGBM

model error with the addition of SNP in the form of boxplots (Fig. 10). The x-axis in the figure is the number of SNPs used by the model. Since the SNP used in each cross-validation is randomly extracted from all SNP, there is no SNP ID. The y-axis is the prediction error. The train_heatmap.pdf files shows the gain value and change regularity of each SNP in different decision tree in the form of a heat map, that is, the information in the train.feature file (Fig. 11). The x-axis coordinate in the figure is SNP ID, which is arranged in descending order from the featureGain_sum value of the SNP in the train.feature file; the y-axis coordinate is the index of the decision tree.

Additionally, CropGBM may use different sets of SNPs during the modeling process, especially when the SNP set is excessively redundant. Thus, SNPs significantly associated with phenotypes may not be accurately identified. Therefore, it is recommended that users select no more than 10,000 SNPs for modeling to ensure the accuracy of the feature selection functionality.

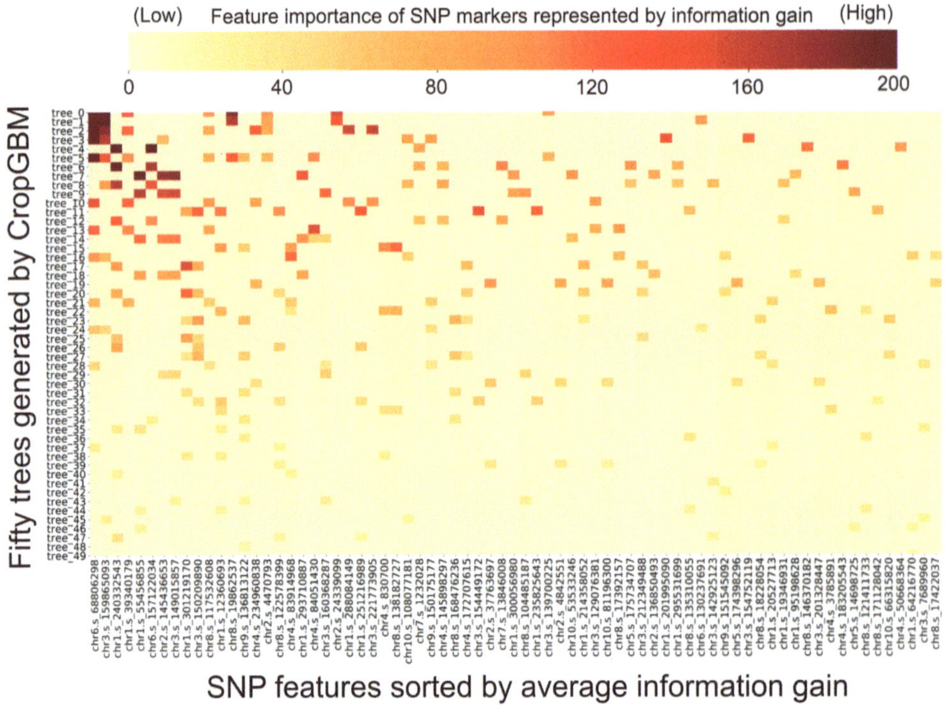


**Fig. 11** Heat map of importance SNP markers selected by CropGBM. The SNP markers are sorted by feature importance based on information gain inferred by CropGBM

### *7.1 Prediction of the Phenotypes of Testing Samples*

After the model is well trained by CropGBM, the final step is to perform phenotype prediction with genotypes as input from the testing population. As long as the prediction step is accomplished, the entire pipeline of CropGBM is finished. The trained model may be repeatedly used for prediction, and the result file is recorded in train.predict:

```
$ cropgbm -c configfile.params -o cropgbm_result/ -e -p --
testgeno test.geno --modelfile-path train.lgb_model
```

## 8 Concluding Remarks

In this chapter, we demonstrate the basic functionality and utility of CropGBM with an example dataset generated from a maize breeding program to illustrate the power of machine learning to perform genomic selection-assisted breeding. According to our benchmark testing, the most advantageous merit of CropGBM is ultra-high efficiency in terms of model training compared to rrBLUP and gBLUP. For example, rrBLUP took over 17 hours and 116 Gb memory to finish model training. In comparison, CropGBM only took 8 minutes and 20 Gb memory on the same server. When the sample size increased to 100,000 samples, rrBLUP failed to train the model. On the same dataset, CropGBM only used 15 minutes and 40 Gb memory with only CPU computing used. If GPU acceleration is enabled, this procedure may be reduced to only 4 min. With the rapid advance of genotyping technology and drone-carried machine vision systems to automatically collect crop traits, phenotypes will not be the limit to traditional agronomic traits, but expansion to include physiological traits captured by hyperspectral cameras will be possible [13]. Therefore, it is foreseeable that breeding data may explode in the near future forming a Big Data environment for the seed industry. Thus, the ultra-high efficient CropGBM toolbox will be an important, one-stop solution to construct data-driven decision-making models for crop breeding.

In addition, gBLUP or rrBLUP derives the linear, additive effect of alleles in a binary fashion (biallelic), namely, presence or absence corresponding to 0 or 1. They also require converting the genotypes into 0, 1, and 2 for regression analysis. However, when the number of alleles exceeds two, the biallelic effect may not be properly inferred by a linear model when the SNP is in a non-biallelic status. When the crop species features a polyploidy genome such as wheat, the situation becomes even more complicated. If the prediction is only carried on biallelic SNPs, a large amount of genetic information may be lost. Thus, one critical feature of CropGBM is that it may use one-hot coding scheme to represent genotypes rather than only using 0, 1, 2 to represent the frequency of bi-alleles in the population. This is likely another important reason why the precision of CropGBM is slightly higher than rrBLUP for a certain circumstance.

Moreover, the feature selection functionality is another advantageous merit of CropGBM compared to rrBLUP, which may be used to identify trait-associated SNPs. By this means, the initial SNP set containing tens of thousands of markers can be significantly condensed to a small marker panel containing less than 100 markers. Therefore, this genotyping platform with the ability for multiplexing samples may be used to further reduce genotyping cost. Our analysis proved that the SNPs selected by CropGBM with high predictive effectiveness are mostly located within the QTL of functionally known genes previously identified by GWAS or QTL mapping. Thus, CropGBM offers an avenue for gene discovery and identification of important genetic variation in breeding programs. In summary, the successful application of CropGBM in genomic selection indicated that machine learning facilitated with artificial intelligence is a promising technology to facilitate Big Data-driven breeding.

## Acknowledgments

This work was supported by the National Science Foundation of China (31871706): Utilization of Machine Learning Strategy to Predict Maize Heterosis.

### References

1. Lande R, Thompson R (1990) Efficiency of marker-assisted selection in the improvement of quantitative traits. Genetics 124 (3):743–756
2. Bernardo R (2008) Molecular markers and selection for complex traits in plants: learning from the last 20 years. Crop Sci 48:1649–1664. https://doi.org/10.2135/cropsci2008.03.0131
3. Xu Y, Crouch J (2008) Marker-assisted selection in plant breeding: from publications to practice. Crop Sci 48:391–407. https://doi.org/10.2135/cropsci2007.04.0191
4. Heffner E, Sorrells M, Jannink J-L (2009) Genomic selection for crop improvement. Crop Sci 49:1–12. https://doi.org/10.2135/cropsci2008.08.0512
5. Meuwissen THE, Hayes BJB, Goddard MEM (2001) Prediction of total genetic value using genome-wide dense marker maps. Genetics 157:1819–1829
6. Xu S, Zhu D, Zhang Q (2014) Predicting hybrid performance in rice using genomic best linear unbiased prediction. Proc Natl Acad Sci U S A 111:12456–12461. https://doi.org/10.1073/pnas.1413750111
7. Kadam DC, Potts SM, Bohn MO, Lipka AE, Lorenz AJ (2016) Genomic prediction of single crosses in the early stages of a maize hybrid breeding pipeline. G3 (Bethesda, Md) 6 (11):3443–3453. https://doi.org/10.1534/g3.116.031286
8. Henderson CR (1984) Applications of linear models in animal breeding. University of Guelph, Guelph
9. Bernardo R, Yu J (2007) Prospects for genomewide selection for quantitative traits in maize. Crop Sci 47:1082–1090. https://doi.org/10.2135/cropsci2006.11.0690
10. Maros M, Capper D, Jones D, Hovestadt V, Deimling A, Pfister S, Benner A, Zucknick M, Sill M (2020) Machine learning workflows to estimate class probabilities for precision cancer diagnostics on DNA methylation microarray data. Nat Protoc 15:479–512. https://doi.org/10.1038/s41596-019-0251-6
11. Dorman SN, Baranova K, Knoll JHM, Urquhart Brad L, Mariani G, Carcangiu ML, Rogan Peter K (2016) Genomic signatures for paclitaxel and gemcitabine resistance in breast cancer derived by machine learning. Mol Oncol 10 (1):85–100. https://doi.org/10.1016/j.molonc.2015.07.006
12. Qiu Z, Cheng Q, Song J, Tang Y, Ma C (2016) Application of machine learning-based classification to genomic selection and performance improvement, vol 9771. doi:https://doi.org/10.1007/978-3-319-42291-6_41
13. Shakoor N, Lee S, Mockler T (2017) High throughput phenotyping to accelerate crop breeding and monitoring of diseases in the field. Curr Opin Plant Biol 38:184–192. https://doi.org/10.1016/j.pbi.2017.05.006

# Chapter 6

# Doubled Haploid Production through Microspore Culture

**Theia Jensen, Kyle Bodell, Fengying Jiang, and John D. Laurie**

## Abstract

Isolated microspore culture (IMC) for the production of doubled haploid (DH) plants is a method used by plant breeders to rapidly obtain homozygous breeding lines. Breeders cut many years off the time it takes to create commercial lines by recovering plants from chromosome doubling of haploids. Although there are several DH production methods, IMC has great potential and is highly desirable because large numbers of cells can be isolated and cultured with relative ease. Even with this advantage, however, several factors inhibit universal adoption of microspore culture, especially in wheat. Problematic areas are the rapid death of microspores in culture, the failure of microspore to undergo embryogenesis, and the high percentage of albinos that can be seen in resulting DH plants of certain genotypes. This chapter highlights these challenges and reviews research being done toward finding solutions for IMC improvement with a focus on wheat.

**Key words** Isolated microspore culture, Androgenesis, Doubled haploid, Albinism, Epigenetic

## 1 Introduction

The process of obtaining plants from microspores is achieved through either isolated microspore culture (IMC) or the culturing of whole anthers. Microspore embryogenesis is referred to as androgenesis, and although the microspores are haploid, spontaneous chromosome doubling often occurs to result in doubled haploid (DH) plants. Methods for DH production in wheat include not only anther culture and IMC but also wheat-maize interspecific pollination. Wheat-maize hybridization is an example of gynogenesis, in which emasculated wheat flowers are pollinated with maize pollen. The maize genome is eliminated naturally following fertilization resulting in the development of haploid embryos containing only the genetic material of the wheat mother plant [1]. Embryo rescue is required, and unlike the spontaneous doubling that can occur through microspore culture, plants derived from maize hybridization require colchicine treatment for obtaining DH plants.

Andriy Bilichak and John D. Laurie (eds.), *Accelerated Breeding of Cereal Crops*, Springer Protocols Handbooks,
https://doi.org/10.1007/978-1-0716-1526-3_6, © Springer Science+Business Media, LLC, part of Springer Nature 2022

The choice of DH production method varies by laboratory and by species. Although maize pollination is the preferred method used by wheat breeders because of consistency in regeneration from diverse genotypes, IMC and anther culture have the highest potential to produce the largest numbers of DH plants. This is because there are considerably more microspores per floret than egg cells and when regeneration is high, large numbers of DH plants can be obtained. Theoretically and practically IMC is the most desirable of the microspore-related methods. It is less time-consuming than anther culture, provides better nutrient availability for cells, allows for better observation of embryogenesis events, and avoids growth of diploid structures that can originate from anther walls [2]. In a side-by-side comparison, IMC in barley resulted in significantly more green plants compared to anther culture [3].

DH plants contain two identical copies of each chromosome and are incredibly useful for breeding programs. DHs allow for selection and fixation of genotypes of interest from F1 hybrids and reduce the time required for making homozygous individuals. Extensive genotypic variation can be captured because each microspore represents unique recombination events. DH plants can be used alongside marker-assisted selection for introduction of specific traits into elite lines, as has been performed for stripe-rust resistance in barley [4, 5]. Furthermore, since the IMC method starts from individual cells, there are many useful applications beyond production of doubled haploid plants, including potential for genetic transformation, genome editing, and the study of recombination [6]. Transformation of microspores before cell division offers the potential for avoidance of genetic chimeras, as has been attempted recently using genome editing [7, 8]. In addition, DH crops are useful for creating chromosome maps for different species [4].

Much of the early work in microspore DH production focused on empirical optimization of culture conditions and stress treatments for androgenesis induction. Powerful genomics tools are now being applied for measuring gene expression in microspores as they transition through androgenesis [9–11]. Studies looking at epigenetic factors are also aiding in the further understanding of normal microspore to pollen development and the microspore to globular proembryo and the proembryo to embryo/plantlet transitions [12–20]. Even with this gain in our understanding of androgenesis, however additional obstacles remain that hinder the complete adoption of this technology, including rapid microspore death in culture and albinism. Here we discuss these major challenges and highlight the important steps in cereal IMC (Fig. 1), focusing on wheat.

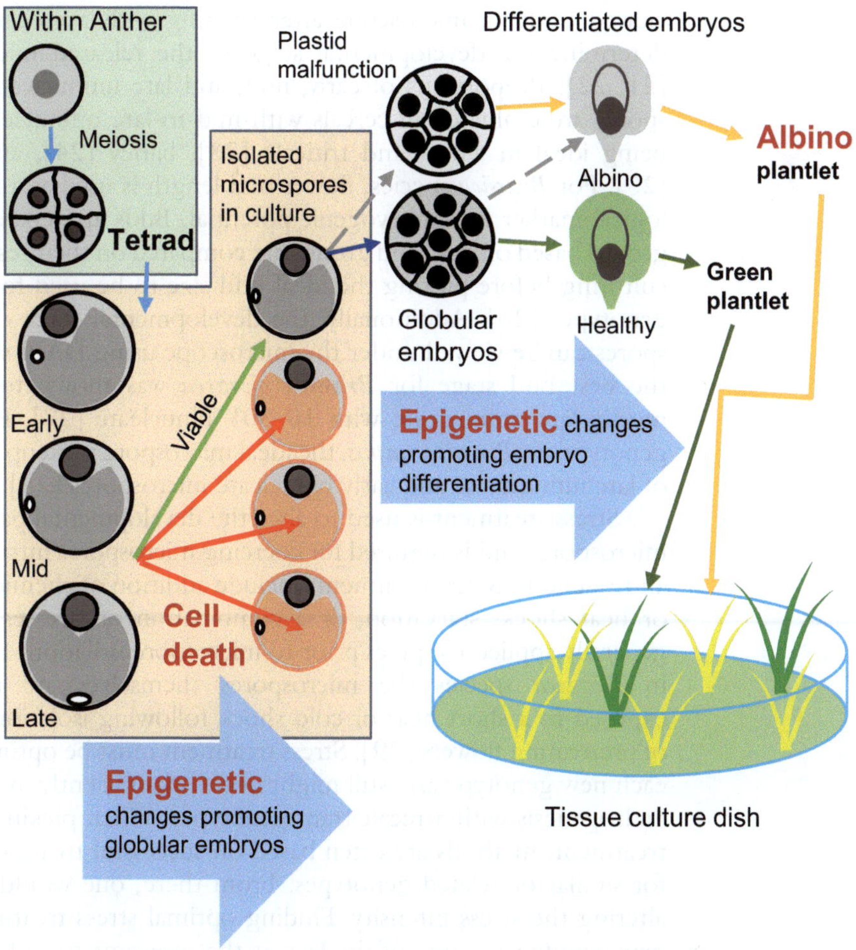


**Fig. 1** A schematic overview of IMC starting from a tetrad of microspores after meiosis in anthers (top left) to plantlets in culture (bottom right). Producing doubled haploid plants from microspores faces various challenges, including rapid cell death in culture, epigenetic barriers to androgenesis, and albino plant formation

## 2 Androgenesis Induction through IMC

Cereal IMC requires stress to promote competence for androgenesis, microspore isolation, culturing to induce embryos, and steps to recover green plants for transfer to soil. Protocols are similar between cultivars and species with slight alterations optimized for each specific case. Here we outline the overall process with an emphasis on wheat.

### 2.1 Handling of Spikes for IMC

Plant morphological markers are important indicators used to ensure collection of microspores at the ideal developmental stage. In cereals, the positional relationship of the spike to the end of the boot or internodal distance is often utilized. In wheat and barley, anthers are collected from mid-spike florets, then crushed, and viewed under a microscope after staining with acetocarmine to determine the developmental stage of the released microspores [21, 22]. Proportions of early, mid, and late uninucleate microspores are evaluated in cereals with mid-to-late uninucleate stage being ideal in wheat and triticale [23], barley [24], and maize [25]. For *Brassica* species, flower bud length is used as a morphological marker for embryogenic potential. Buds are separated into groups based on size, and groups are compared on their response to culturing before picking the ideal bud size to be used for specific genotypes [26]. Additionally, the developmental stage of microspores can be viewed under the microscope using DAPI stain, and the best bud stage for *Brassica oleracea* was shown to contain mostly late uninucleate with 10–30% binucleate [27]. For some genotypes in *Brassica napus*, the ideal microspore makeup is mostly of late uninucleate and early binucleate microspores [28].

Stress treatment is used to alter the developmental pathway of microspores and is required for coercing microspores into entering androgenesis. Stress treatments include addition of chemicals, cold or heat shock, starvation, or a combination of stresses and are generally applied to spikes prior to microspore isolation. However, in *Brassica* species the microspores themselves are generally exposed to a short heat or cold shock following isolation instead of pretreating flowers [29]. Stress treatment must be optimized for each new genotype and still might not work efficiently in inducing androgenesis within recalcitrant genotypes. When picking a stress treatment, methods are often based on successful treatments used for similar or related genotypes. From there, one would start by altering the stress intensity. Finding optimal stress treatments for new genotypes is one of the factors that prevents the adoption of androgenesis for many breeders.

### 2.2 Isolation of Microspores

To prevent contamination of cell cultures, all plant material is surface sterilized at the start of IMC procedures. Surface sterilization is routinely performed using aqueous solutions of diluted bleach or ethanol where plant material is submerged and agitated for a set amount of time that varies between lab, species, and cultivar. The sterilization is followed by several washes with sterile water to remove excess bleach or ethanol. All media used for isolation and culture are filter sterilized or autoclaved prior to use.

Methods for extracting microspores from anthers vary by species and lab but generally involve crushing or blending of flowers (Fig. 2a). In wheat and barley, florets are blended in a sterile blender with media [21, 22, 30]. In maize either the anthers are removed and blended in a sterile blender with media or spikelets are blended [25]. Flowers are removed from the inflorescence in *Brassica* species and suspended in media for crushing using a sterile piston or rod [28, 29].

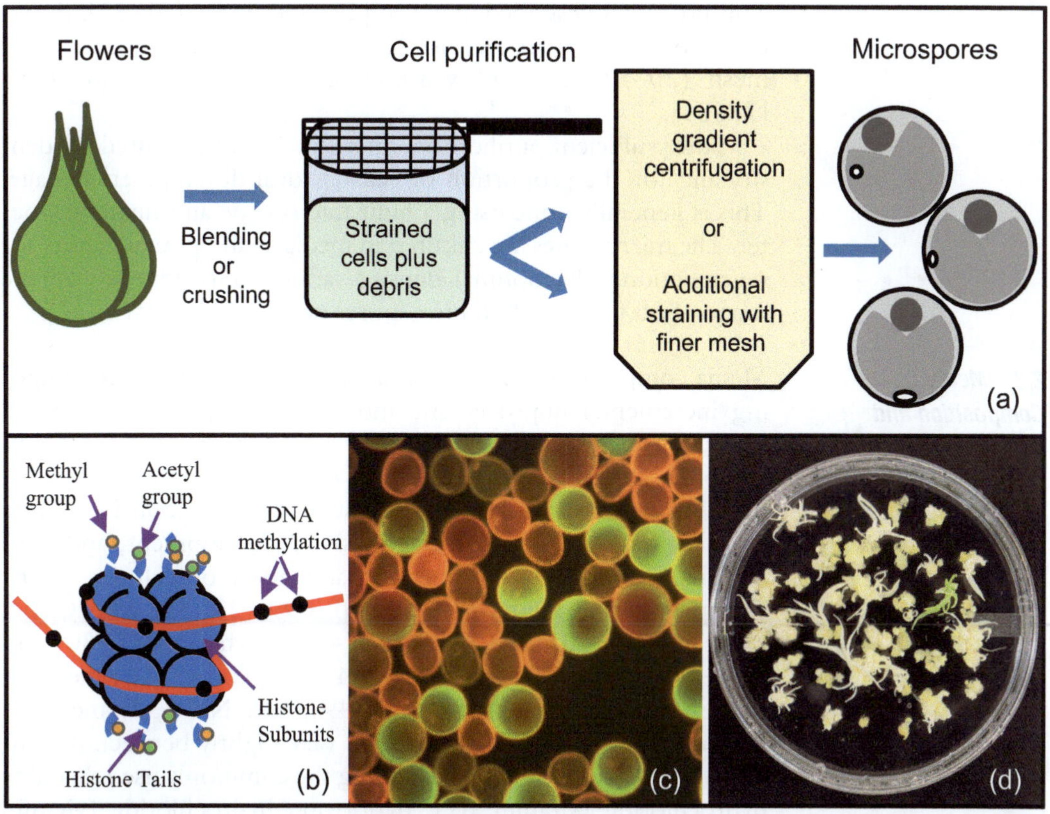


**Fig. 2** Steps for successful IMC. (**a**) Obtaining purified microspores is important for IMC. Harvested flowers are surface sterilized, blended, and strained to remove debris. Microspores are then further purified for culture by density gradient centrifugation or finer mesh straining. (**b**) Epigenetic fine-tuning of gene expression can include DNA methylation and histone tail modifications. Treatments that facilitate changing epigenetic programming help androgenesis and improve IMC. (**c**) Many microspores die shortly after isolation. Here live microspores are able to take up and process fluorescein diacetate making them green in color. Dead microspores appear red by uptake of propidium iodide. Being able to suppress cell death is important for improving IMC. (**d**) For certain genotypes recovering green plants can be problematic. Here only one green plant is visible among several albinos

Following initial extraction in liquid media, microspores must be further purified from debris before culturing (Fig. 2a). Passing the slurry through sieves, generally 100 μm in size, removes the large debris, and following with several washing steps removes much of the smaller debris. Microspores may, however, require further purification in many cultivars. Purification is generally done using density gradient centrifugation in which cells are separated based on their density. Cells are suspended in maltose with a small amount (approximately 1 mL) of media layered above the cell suspension. Centrifugation then results in a band of viable cells that forms at the interface which is subsequently collected for culturing.

This process is described in wheat [21] and barley [31–33]. Purification of microspores has also been accomplished using a smaller mesh (30–40 μm) and washing steps within *Brassica napus* [28, 34].

After sufficient purification, microspores are evaluated for density and for the proportion of cells at ideal developmental stage. This is generally done using a light microscope and hemocytometer. The microspores are cultured in media at an optimal density for regeneration. The optimal density varies but is generally in the range of 30,000–50,000 microspores per mL [21, 30, 34–37].

### 2.3 Media Composition and Tissue Culture

Media composition for microspore culture has experienced ongoing incremental improvement and has been the topic of much study. There are many different types of media, with different components for optimizing green plant production. Generally, there are nutrients for growth and survival of the cells in culture, hormones to induce embryogenesis and development, and antibacterial additives to prevent contamination of cultures. Some examples of the media used are NPB-99 [23, 38, 39], CIMC [21], MMS3 [40], and CHB [41, 42] in wheat, IMI [30] and FHG [24, 43] in barley, YP [25, 44] and N6 [45, 46] media in maize, and NLN [47, 48], MS [49], and NN [50] media in *Brassica*. Macro- and micronutrients vary slightly between the different media, as do vitamins including the commonly used thiamine hydrochloride (vitamin B1), pyridoxine hydrochloride (vitamin B6), and nicotinic acid (vitamin B3). Other organic compounds are commonly added such as myo-inositol and glutamine and glycine. Maltose is the commonly used sugar in media for IMC with some media having additional sugar sources as well. Hormones commonly added are kinetin, PAA (phenylacetic acid) in wheat, and BAP (6-benzylaminopurine) in barley. CIMC media [21] use antimicrobial agent cefotaxime, which was also shown to increase formation of embryo-like structures [23].

In addition to optimized media composition, one significant improvement to IMC can be the co-culture of microspores with ovaries. In wheat it was discovered that plating ovaries with microspores improves the amount of embryos but has little effect on the albino proportion [40]. Spikes from donor plants of the same variety are sterilized with bleach and rinsed several times with water before ovary extraction using forceps. The ovaries that are selected should be large with fluffy stigma, but not fully developed or pollinated. The use of ovary co-culture within wheat has become commonplace due to the significant improvement in embryos recovered with live ovary co-culture and ovary-conditioned media [51].

## 3 Epigenetic Factors to Consider for Androgenesis

Epigenetic modifications to DNA and associated histones are important for chromatin organization and gene expression and are considered important for androgenesis and recovery of DH plants [19, 52–54]. In plants, DNA methylation occurs in three contexts: CG, CHG, and CHH [55]. Methylation in the CG context is found throughout the genome, while CHG and CHH are found in regions marked for transcriptional silencing. In addition to this complex orchestration of DNA methylation at different regions of the genome, histone tail posttranslational modifications account for much of the epigenetic tuning of gene expression [56]. Histone-modifying enzymes that acetylate, deacetylate, and methylate are therefore active throughout plant development and response to external stimuli. In general, acetylation of histones is thought to relax chromatin and facilitate gene expression. Deacetylation and methylation are thought to have the reverse effect, especially at H3K9. Classical repressive marks are H3K9 dimethylation and H3K27 trimethylation. Coordination of these marks helps determine cellular response to stimuli and development. Because epigenetics plays such an important role in cell function, it is not surprising that it has been found to vary with cell type [57]. Recent work from several labs has demonstrated an association of epigenetic changes with androgenesis and for the recovery of plants in culture [52, 53, 58] (Fig. 2b).

Inhibition of histone deacetylases (HDACs) through the use of Trichostatin A (TSA) has been shown to improve IMC protocols by increasing the number of embryos and the proportion of green plants regenerated [52]. In wheat, TSA was shown to improve microspore cultures in a dose-dependent manner by increasing the number of embryos and the number of green plants regenerated [21]. In *B. rapa* HDAC inhibition improved plant regeneration and inhibited callus formation within culture [58]. TSA's mechanism of action involves inhibiting the HDACs, which allows histones to stay acetylated, and the chromatin to remain less tightly packed. Presumably this allows for more genes to partake in transcription and thus promotes cell totipotency [52]. In maize TSA was shown to increase histone 4 acetylation, which also had effects on global DNA methylation [53]. Additionally, it was theorized that the stress treatment meant to initiate the shift to embryogenesis works in a similar fashion, by interrupting the HDAC pathway and promoting increased totipotency in the cells by allowing more genes available for transcription [52].

Histone methylation can be manipulated by the molecule BIX-01294, which inhibits methylation of histone H3K9 [54]. This molecule was shown to increase the amount of proembryos recovered from microspore culture. BIX-01294 promotes embryogenesis induction in barley by reducing histone methylation as well as total DNA methylation [54]. The use of BIX-01294 can also halt the embryogenesis pathway at the proembryo stage if the cells are continuously exposed; therefore, it is recommended for use only briefly at the start of culture [54]. This molecule works by binding to H3K9 methyltransferase rendering them inactive, as well as inhibiting H3K9 Jumonji demethylase [59].

DNA methylation has also been implicated as having a significant impact on somatic embryogenesis, and the use of the 5-cytosine analog, 5-azacytidine (AzaC), has significant impacts on IMC [19, 60]. In barley microspore culture, a short treatment of AzaC in the culture medium up to the proembryo stage significantly increased the proportion of proembryos, as well as significantly reduced DNA methylation levels [19]. Interestingly the result of treatment with AzaC was remarkably similar to that of TSA treatment. When directly compared to TSA, both TSA and AzaC decreased H3K9 methylation, increased H4 acetylation, and decreased global methylation [53]. It is theorized that TSA primarily increases histone acetylation which then results in decreased histone methylation and DNA methylation, whereas AzaC decreases DNA methylation which increases histone acetylation and reduces histone methylation [53]. Collectively, these studies suggest that DNA methylation, histone acetylation, and histone methylation are all interconnected and that manipulation of this epigenetic system can promote cell competency and improve the switch to androgenesis and subsequent DH plant recovery.

## 4 Cell Death Considerations

A considerable obstacle for IMC is the death of microspores in culture (Fig. 2c). In wheat cultivar AC Andrew, 60–80% of isolated microspores are dead by the third day in culture [22]. Plant cells can die via one of the two pathways: vacuolar cell death and necrosis [61]. Vacuolar cell death resembles autophagy and is characterized by the increase in volume of lytic vacuoles that break down the components of cytoplasm. Necrosis, however, is a less defined cell death mechanism, characterized by early rupture of the plasma membrane and loss of cellular contents [61]. Much of the work improving IMC has been to optimize conditions to ensure maximum yield of multicellular structures (MCS) and resulting embryo-like structures (ELS) through decreasing cell death with media additives.

The death of microspores in culture can be partially attributed to the stress treatment required to induce the switch to the embryogenic pathway, with autophagy playing a role in inducing cell death [62]. Inhibition of cathepsin protease activity has been shown to diminish cell death and increase ELS formation in barley

[62]. Stress treatments likely generate reactive oxygen species (ROS), and the addition of the ROS scavenger dimethyl-tyrosine can increase the number of viable microspores as well as regenerated plants in triticale and wheat [22]. Antioxidants reduce oxidative stress and have been shown to prevent cell death caused by excess ROS [23]. Glutathione was shown to be the best antioxidant, increasing the number of ELS recovered as well as the percentage of green plants [23].

Anaerobic conditions in culture also promote cell death. The addition of Ficoll 400 at 300 g/L increased the amount of ELS and regenerated plants in two different durum wheat by preventing microspore settling in culture media [63]. Ficoll addition to triticale IMC increased the amount of callus tissue and derived plants, but its impact was genotype dependent [38]. Arabinogalactan proteins (AGPs) or arabinogalactans can help prevent programmed cell death (PCD) and increase the amount of MCS and ELS from microspore cultures [37]. AGPs are cell surface proteoglycans and have been shown to be essential for female gametogenesis in *Arabidopsis* [64]. AGP Larcoll suppresses cell death, and gum arabic is thought to mimic the effect of ovary co-culture as less cell death and increases in viable plants were obtained in the presence of gum arabic in the absence of ovaries [37].

Although much work has gone into mitigating the problem of microspore cell death in the early stages of IMC, studies have shown that PCD plays an integral role in the release of the ELS from the exine wall [65, 66]. Cells in the MCS in the area of the exine wall undergo PCD in order to release the cellular mass from the confines of the microspore wall [65, 66]. ELS could only be produced from MCS that had PCD occurring near the exine wall, a process thought to be important for progression of androgenesis [65]. So, in order to make further improvements to IMC, cell death needs to be suppressed in early days of culture, but not prevented at later stages when MCS need to break free for continued growth.

## 5 Problem with Albino Plant Regeneration

Often a large proportion of plants regenerated from IMC are albinos (Fig. 2d), and this can be partially attributed to genotype and external conditions as seen in Table 1. Although the causes of albinism in IMC are still not entirely understood, reviews have been dedicated to the topic [67, 68]. Albinism occurs when plant cells have lost their chlorophyll pigments or have incomplete differentiation of chloroplast membranes [68]. One of the early indications that genotype influences albinism comes from Larsen et al. [69], who demonstrated that crossing high green plant-yielding barley varieties resulted in progeny with similar green plant-yielding capabilities. When the progeny of low green plant-yielding varieties

**Table 1**
**Percentage of albino regenerants from various procedures, treatments, and genotypes**

| Species and cultivar | Treatment | Albino (%) | References |
|---|---|---|---|
| Wheat, Chris | No ovaries (control) | 13.3 | [40] |
| Wheat, Chris | Co-culture with 20 ovaries in 1 mL liquid media | 13.8 | [40] |
| Wheat, WED 202–16-2 | No media added to pretreatment (control) | 30 | [78] |
| Wheat, WED 202–16-3 | NPB98 medium added to pretreatment | 11 | [78] |
| Barley, Igri | Normal culture conditions (anther culture) | 12 | [73] |
| Barley, Cork | Normal culture conditions (anther culture) | 97 | [73] |
| Barley, Cork | No pretreatment (control) (anther culture) | 98 | [77] |
| Barley, Cork | Copper sulfate pretreatment (anther culture) | 80 | [77] |
| Triticale, Wanad | 190-2 M medium (control) | 75 | [80] |
| Triticale, Wanad | Starvation-inducing medium | 50 | [80] |
| Triticale, AC Alta | CHB3 medium | 41 | [38] |
| Triticale, AC Alta | NBP99 medium | 60 | [38] |
| Triticale, AC Alta | NBP99 + Ficoll | 55 | [38] |
| Triticale, AC Certa | CHB3 medium | 17 | [38] |
| Triticale, AC Certa | NBP99 medium | 11 | [38] |
| Triticale, AC Certa | NBP99 + Ficoll | 8 | [38] |

was crossed, offspring displayed similar low green plant results and genotype accounted for 76% of total variation in green plant proportion [69]. There has since been significant exploration into the effects of genetics in determining albino regeneration frequency [70–72].

In addition to genotype affecting albinism, numerous works have shown the effect of plastids on albino regeneration, and genes involved in plastid development have been associated with albino proportion [73–75]. Early amyloplast formation and the transcription of starch synthase genes Dpe2, GBSSI, and Sbe1 are all good indicators of high albino regeneration with higher rate of expression at the mid-to-late uninucleate stage in a high albino cultivar [33]. In this study, a high green plant-producing cultivar had significantly more plastids and higher proportion of proplastids compared to a high albino-producing cultivar. The researchers demonstrated that embryogenic response and regeneration of green plants in barley could be improved by avoiding early amyloplast development and high starch synthase gene expression through harvesting the microspores at the early-mid uninucleate stage. Studies like this highlight the importance of microspore developmental staging in IMC protocol optimization.

Although stress promotes the induction of androgenesis, it can also affect the proportion of albino plants recovered. Treating barley anthers in mannitol at 26 °C for 3 days was able to shift the green plant to albino ratio, with all mannitol pretreatments outperforming cold treatments at reducing albinos [76].

Many media additives have been shown to improve green plant regeneration and are likely to have varying effects based on the genotype of the crop. Several different factors allow for the increase in green plants: plant hormones, reducing cell death from ROS, and reducing amounts of starch in cells. These different avenues provide paths for researchers to improve their IMC protocols for maximizing green plant production. Some additives have been shown to mitigate or reduce the effect of amyloplasts and starch on the microspores and increase the amount of green plants within anther culture [77]. Copper sulfate in the pretreatment medium was shown to decrease starch in plastids and increase internal membranes and could increase the number of green plants regenerated in a concentration- and genotype-dependent manner [77]. The addition of NPB98 to the spike pretreatment solution increased green plants regenerated in a difficult genotype [78]. The addition of gum arabic to culture media with ovary co-culture increased the amount of green plants regenerated from recalcitrant wheat line "Chris" resulting in 180 green plants per 100,000 microspores [37]. The use of the AGP Larcoll similarly allowed for high green plant recovery [37].

The addition of plant hormones in regeneration and induction media has also been investigated as a means to reduce albinos. Addition of the cytokinin thidiazuron (TDZ) and the auxin dicamba can dramatically increase total plants regenerated and green plant proportions in barley, with a fivefold increase in green plants compared to the use of 6-benzylaminopurine [79].

## 6 Summary

Isolated microspore culture is a valuable method for plant breeders to rapidly obtain homozygous lines. In this chapter we have briefly outlined the microspore isolation procedure for wheat with reference to other cereals and oilseeds. We point out how stress is important for making microspores competent for culture and that stress treatments require optimization for different species and genotypes. Once in culture, microspore viability sharply declines with only a small portion able to take the steps toward becoming embryos. We are beginning to realize the importance of epigenetics in developmental transitions, and treatments that affect DNA methylation and histone acetylation are helping to improve IMC. For widespread adoption, IMC must be further improved and become genotype independent. This will require improvement or alternatives to stress treatments, suppression of cell death pathways, epigenetic and transcriptional reprogramming to facilitate embryogenesis, and finally avoidance of starch-forming amyloplasts and other pathways leading to albino plants. Through application of modern genomic tools and exploration of new culture media and methods, rapid improvement in IMC is anticipated in the coming years.

## Acknowledgments

This work was supported by the Alberta Funding Consortium grant (2018F142R) entitled "A Comparative Genomics Approach to Improve Doubled Haploid Breeding for Common and Durum Wheat." Generous contributors to these funds were the Alberta Agriculture and Forestry, Alberta Wheat Commission, and Saskatchewan Wheat Development Commission.

### References

1. O'Donoughue LS, Bennett MD (1994) Durum wheat haploid production using maize wide-crossing. Theor Appl Genet 89 (5):559–566. https://doi.org/10.1007/BF00222448
2. Ferrie AMR, Caswell KL (2011) Isolated microspore culture techniques and recent progress for haploid and doubled haploid plant production. Plant Cell Tiss Org 104 (3):301–309. https://doi.org/10.1007/s11240-010-9800-y
3. Li H, Devaux P (2005) Isolated microspore culture overperforms anther culture for green plant regeneration in barley (hordeum vulgare l.). Acta Physiol Plant 27(4):611–619. https://doi.org/10.1007/s11738-005-0065-8
4. Forster BP, Heberle-Bors E, Kasha KJ, Touraev A (2007) The resurgence of haploids in higher plants. Trends Plant Sci 12(8):368–375. https://doi.org/10.1016/j.tplants.2007.06.007
5. Toojinda T, Baird E, Booth A, Broers L, Hayes P, Powell W, Thomas W, Vivar H, Young G (1998) Introgression of quantitative trait loci (qtls) determining stripe rust resistance in barley: an example of marker-assisted line development. Theor Appl Genet 96 (1):123–131. https://doi.org/10.1007/s001220050718
6. Devaux P, Kilian A, Kleinhofs A (1995) Comparative mapping of the barley genome with male and female recombination-derived, doubled haploid populations. Mol Gen Genet 249 (6):600–608. https://doi.org/10.1007/BF00418029
7. Bilichak A, Sastry-Dent L, Sriram S, Simpson M, Samuel P, Webb S, Jiang F, Eudes F (2020) Genome editing in wheat microspores and haploid embryos mediated by delivery of zfn proteins and cell-penetrating peptide complexes. Plant Biotechnol J 18 (5):1307–1316. https://doi.org/10.1111/pbi.13296
8. Bhowmik P, Ellison E, Polley B, Bollina V, Kulkarni M, Ghanbarnia K, Song H, Gao C, Voytas DF, Kagale S (2018) Targeted mutagenesis in wheat microspores using crispr/cas9. Sci Rep 8(1):6502. https://doi.org/10.1038/s41598-018-24690-8
9. Zhang Y, Gao S-y, Liu H-h, Zhang X-l, Zeng A-s, Wang J-j, Hou X-l, Li Y (2020) Cdna-aflp analysis of differentially expressed genes during microspore embryogenesis in non-heading Chinese cabbage. In Vitro Cell Dev Biol Plant 56(1):18–28. https://doi.org/10.1007/s11627-019-10036-0
10. Bian J, Deng P, Zhan H, Wu X, Nishantha MD, Yan Z, Du X, Nie X, Song W (2019) Transcriptional dynamics of grain development in barley (hordeum vulgare l.). Int J Mol Sci 20(4):962
11. Belanger S, Marchand S, Jacques P-E, Meyers B, Belzile F (2018) Differential expression profiling of microspores during the early stages of isolated microspore culture using the responsive barley cultivar gobernadora.

G3-Genes Genom Genet 8(5):1603–1614. https://doi.org/10.1534/g3.118.200208

12. Borg M, Berger F (2015) Chromatin remodelling during male gametophyte development. Plant J 83(1):177–188. https://doi.org/10.1111/tpj.12856
13. Borg M, Jacob Y, Susaki D, LeBlanc C, Buendía D, Axelsson E, Kawashima T, Voigt P, Boavida L, Becker J, Higashiyama T, Martienssen R, Berger F (2020) Targeted reprogramming of h3k27me3 resets epigenetic memory in plant paternal chromatin. Nat Cell Biol. https://doi.org/10.1038/s41556-020-0515-y
14. Calarco JP, Borges F, Donoghue MTA, Van Ex F, Jullien PE, Lopes T, Gardner R, Berger F, Feijó JA, Becker JD, Martienssen RA (2012) Reprogramming of DNA methylation in pollen guides epigenetic inheritance via small rna. Cell 151(1):194–205. https://doi.org/10.1016/j.cell.2012.09.001
15. El-Tantawy AA, Solis MT, Risueno MC, Testillano PS (2014) Changes in DNA methylation levels and nuclear distribution patterns after microspore reprogramming to embryogenesis in barley. Cytogenet Genome Res 143 (1–3):200–208. https://doi.org/10.1159/000365232
16. Machczyńska J, Orłowska R, Mańkowski DR, Zimny J, Bednarek PT (2014) DNA methylation changes in triticale due to in vitro culture plant regeneration and consecutive reproduction. Plant Cell Tissue Org 119(2):289–299. https://doi.org/10.1007/s11240-014-0533-1
17. Machczyńska J, Zimny J, Bednarek PT (2015) Tissue culture-induced genetic and epigenetic variation in triticale (× triticosecale spp. Wittmack ex a. Camus 1927) regenerants. Plant Mol Biol 89(3):279–292. https://doi.org/10.1007/s11103-015-0368-0
18. Pandey P, Daghma DS, Houben A, Kumlehn J, Melzer M, Rutten T (2017) Dynamics of post-translationally modified histones during barley pollen embryogenesis in the presence or absence of the epi-drug trichostatin a. Plant Reprod 30(2):95–105. https://doi.org/10.1007/s00497-017-0302-5
19. Solís M-T, El-Tantawy A-A, Cano V, Risueño MC, Testillano PS (2015) 5-azacytidine promotes microspore embryogenesis initiation by decreasing global DNA methylation, but prevents subsequent embryo development in rapeseed and barley. Front Plant Sci 6:472. https://doi.org/10.3389/fpls.2015.00472
20. Wang F-X, Shang G-D, Wu L-Y, Xu Z-G, Zhao X-Y, Wang J-W (2020) Chromatin accessibility dynamics and a hierarchical transcriptional regulatory network structure for plant somatic embryogenesis. Dev Cell. https://doi.org/10.1016/j.devcel.2020.07.003
21. Jiang F, Ryabova D, Diedhiou J, Hucl P, Randhawa H, Marillia EF, Foroud NA, Eudes F, Kathiria P (2017) Trichostatin a increases embryo and green plant regeneration in wheat. Plant Cell Rep 36(11):1701–1706. https://doi.org/10.1007/s00299-017-2183-3
22. Sinha RK, Eudes F (2015) Dimethyl tyrosine conjugated peptide prevents oxidative damage and death of triticale and wheat microspores. Plant Cell Tissue Org 122(1):227–237. https://doi.org/10.1007/s11240-015-0763-x
23. Asif M, Eudes F, Goyal A, Amundsen E, Randhawa H, Spaner D (2013) Organelle antioxidants improve microspore embryogenesis in wheat and triticale. In Vitro Cell Dev Biol Plant 49(5):489–497. https://doi.org/10.1007/s11627-013-9514-z
24. Kasha KJ, Simion E, Oro R, Yao QA, Hu TC, Carlson AR (2001) An improved in vitro technique for isolated microspore culture of barley. Euphytica 120(3):379–385. https://doi.org/10.1023/A:1017564100823
25. Obert B, Szabó L, Mitykó J, Preťová A, Barnabás B (2005) Morphological events in cultures of mechanically isolated maize microspores. In Vitro Cell Dev Biol - Plant 41(6):775–782. https://doi.org/10.1079/IVP2005701
26. Seguí-Simarro JM, Rivas-Sendra A, Calabuig-Serna A, Camacho-Fernández C, Corral-Martínez P, Porcel R (2019) Embryogenic competence of microspores is associated with their ability to form a callosic, osmoprotective subintinal layer. J Exp Bot 70(4):1267–1281. https://doi.org/10.1093/jxb/ery458
27. da Silva Dias J (2003) *Protocol for broccoli microspore culture*, in *Doubled haploid production in crop plants*. Springer, pp 195–204
28. Rivas-Sendra A, Calabuig-Serna A, Seguí-Simarro JM (2017) Dynamics of calcium during in vitro microspore embryogenesis and in vivo microspore development in brassica napus and solanum melongena. Front Plant Sci 8:1177–1177. https://doi.org/10.3389/fpls.2017.01177
29. Gu H, Sheng X, Zhao Z, Yu H, Wang J (2014) Initiation and development of microspore embryogenesis and plant regeneration of brassica nigra. In Vitro Cell Dev Biol Plant 50 (5):534–540. https://doi.org/10.1007/s11627-014-9612-6
30. Li H, Devaux P (2003) High frequency regeneration of barley doubled haploid plants from isolated microspore culture. Plant Sci J 164 (3):379–386. https://doi.org/10.1016/S0168-9452(02)00424-7

31. Rodríguez-Serrano M, Bárány I, Prem D, Coronado MJ, Risueño MC, Testillano PS (2012) No, ros, and cell death associated with caspase-like activity increase in stress-induced microspore embryogenesis of barley. J Exp Bot 63 (5):2007–2024. https://doi.org/10.1093/jxb/err400
32. Esteves P, Belzile F (2014) Improving the efficiency of isolated microspore culture in six-row spring barley: I-optimization of key physical factors. Plant Cell Rep 33(6):993–1001. https://doi.org/10.1007/s00299-014-1583-x
33. Gajecka M, Marzec M, Chmielewska B, Jelonek J, Zbieszczyk J, Szarejko I (2020) Plastid differentiation during microgametogenesis determines green plant regeneration in barley microspore culture. Plant Sci J 291:110321. https://doi.org/10.1016/j.plantsci.2019.110321
34. Corral-Martínez P, Driouich A, Seguí-Simarro JM (2019) Dynamic changes in arabinogalactan-protein, pectin, xyloglucan and xylan composition of the cell wall during microspore embryogenesis in brassica napus. Front Plant Sci 10:332. https://doi.org/10.3389/fpls.2019.00332
35. Labbani Z, de Buyser J, Picard E (2007) Effect of mannitol pretreatment to improve green plant regeneration on isolated microspore culture in triticum turgidum ssp. Durum cv. 'Jennah khetifa'. Plant Breed 126 (6):565–568. https://doi.org/10.1111/j.1439-0523.2007.01399.x
36. Lantos C, Bóna L, Nagy É, Békés F, Pauk J (2018) Induction of in vitro androgenesis in anther and isolated microspore culture of different spelt wheat (triticum spelta l.) genotypes. Plant Cell Tissue Org 133(3):385–393. https://doi.org/10.1007/s11240-018-1391-z
37. Letarte J, Simion E, Miner M, Kasha KJ (2006) Arabinogalactans and arabinogalactan-proteins induce embryogenesis in wheat (triticum aestivum l.) microspore culture. Plant Cell Rep 24 (12):691–698. https://doi.org/10.1007/s00299-005-0013-5
38. Eudes F, Amundsen E (2005) Isolated microspore culture of Canadian 6× triticale cultivars. Plant Cell Tissue Org 82(3):233–241. https://doi.org/10.1007/s11240-005-0867-9
39. Zheng M, Liu W, Weng Y, Polle E, Konzak C (2001) Culture of freshly isolated wheat (triticum aestivum l.) microspores treated with inducer chemicals. Plant Cell Rep 20 (8):685–690. https://doi.org/10.1007/s00299-001-0393-0
40. Hu T, Kasha KJ (1997) Improvement of isolated microspore culture of wheat (triticum aestivum l.) through ovary co-culture. Plant Cell Rep 16(8):520–525. https://doi.org/10.1007/BF01142316
41. Patel M, Darvey NL, Marshall DR, Berry JO (2004) Optimization of culture conditions for improved plant regeneration efficiency from wheat microspore culture. Euphytica 140 (3):197–204. https://doi.org/10.1007/s10681-004-3036-z
42. Chu CC, Hill RD, Brule-Babel L (1990) High frequency of pollen embryoid formation and plant regeneration in triticum aestivum l. on monosaccharide containing media. Plant Sci J 66(2):255–262. https://doi.org/10.1016/0168-9452(90)90211-6
43. Kasha K, Simion E, Oro R, Shim Y (2003) *Barley isolated microspore culture protocol.* In: *Doubled haploid production in crop plants.* Springer, pp 43–47
44. Genovesi AD, Collins G (1982) In vitro production of haploid plants of corn via anther culture 1. Crop Sci 22(6):1137–1144
45. Pescitelli S, Johnson C, Petolino J (1990) Isolated microspore culture of maize: effects of isolation technique, reduced temperature, and sucrose level. Plant Cell Rep 8 (10):628–631
46. Nägeli M, Schmid J, Stamp P, Büter B (1999) Improved formation of regenerable callus in isolated microspore culture of maize: impact of carbohydrates, plating density and time of transfer. Plant Cell Rep 19(2):177–184
47. Ferrie A, Epp D, Keller W (1995) Evaluation of brassica rapa l. genotypes for microspore culture response and identification of a highly embryogenic line. Plant Cell Rep 14 (9):580–584
48. Lichter R (1981) Anther culture of brassica napus in a liquid culture medium. Z Pflanzenphysiol 103(3):229–237
49. Sato T, Nishio T, Hirai M (1989) Plant regeneration from isolated microspore cultures of Chinese cabbage (brassica campestris spp. Pekinensis). Plant Cell Rep 8(8):486–488
50. Nitsch C, Nitsch J (1967) The induction of flowering in vitro in stem segments of plumbago indica l. Planta 72(4):355–370
51. Zheng M, Weng Y, Liu W, Konzak C (2002) The effect of ovary-conditioned medium on microspore embryogenesis in common wheat (triticum aestivum l.). Plant Cell Rep 20 (9):802–807
52. Li H, Soriano M, Cordewener J, Muiño JM, Riksen T, Fukuoka H, Angenent GC, Boutilier K (2014) The histone deacetylase inhibitor trichostatin a promotes totipotency in the male gametophyte. Plant Cell 26(1):195–209

53. Yang F, Zhang L, Li J, Huang J, Wen R, Ma L, Zhou D, Li L (2010) Trichostatin a and 5-azacytidine both cause an increase in global histone h4 acetylation and a decrease in global DNA and h3k9 methylation during mitosis in maize. BMC Plant Biol 10(1):178. https://doi.org/10.1186/1471-2229-10-178
54. Berenguer E, Bárány I, Solís M-T, Pérez-Pérez Y, Risueño MC, Testillano PS (2017) Inhibition of histone h3k9 methylation by bix-01294 promotes stress-induced microspore totipotency and enhances embryogenesis initiation. Front Plant Sci 8:1161. https://doi.org/10.3389/fpls.2017.01161
55. Law JA, Jacobsen SE (2010) Establishing, maintaining and modifying DNA methylation patterns in plants and animals. Nat Rev Genet 11(3):204–220. https://doi.org/10.1038/nrg2719
56. Ueda M, Seki M (2020) Histone modifications form epigenetic regulatory networks to regulate abiotic stress response. Plant Physiol 182:15
57. Liu C, Lu F, Cui X, Cao X (2010) Histone methylation in higher plants. Annu Rev Plant Biol 61(1):395–420. https://doi.org/10.1146/annurev.arplant.043008.091939
58. Zhang L, Zhang Y, Gao Y, Jiang X, Zhang M, Wu H, Liu Z, Feng H (2016) Effects of histone deacetylase inhibitors on microspore embryogenesis and plant regeneration in pakchoi ( brassica rapa ssp. Chinensis l.). Sci Hortic 209:61–66. https://doi.org/10.1016/j.scienta.2016.05.001
59. Upadhyay AK, Rotili D, Han JW, Hu R, Chang Y, Labella D, Zhang X, Yoon Y-S, Mai A, Cheng X (2012) An analog of bix-01294 selectively inhibits a family of histone h3 lysine 9 jumonji demethylases. J Mol Biol 416(3):319–327. https://doi.org/10.1016/j.jmb.2011.12.036
60. Fraga HPF, Vieira LN, Caprestano CA, Steinmacher DA, Micke GA, Spudeit DA, Pescador R, Guerra MP (2012) 5-azacytidine combined with 2,4-d improves somatic embryogenesis of acca sellowiana (o. Berg) burret by means of changes in global DNA methylation levels. Plant Cell Rep 31(12):2165–2176. https://doi.org/10.1007/s00299-012-1327-8
61. van Doorn WG, Beers EP, Dangl JL, Franklin-Tong VE, Gallois P, Hara-Nishimura I, Jones AM, Kawai-Yamada M, Lam E, Mundy J, Mur LAJ, Petersen M, Smertenko A, Taliansky M, Van Breusegem F, Wolpert T, Woltering E, Zhivotovsky B, Bozhkov PV (2011) Morphological classification of plant cell deaths. Cell Death Differ 18(8):1241–1246. https://doi.org/10.1038/cdd.2011.36
62. Bárány I, Berenguer E, Solís M-T, Pérez-Pérez Y, Santamaría ME, Crespo JL, Risueño MC, Díaz I, Testillano PS (2018) Autophagy is activated and involved in cell death with participation of cathepsins during stress-induced microspore embryogenesis in barley. J Exp Bot 69(6):1387–1402. https://doi.org/10.1093/jxb/erx455
63. Cistué L, Romagosa I, Batlle F, Echávarri B (2009) Improvements in the production of doubled haploids in durum wheat (triticum turgidum l.) through isolated microspore culture. Plant Cell Rep 28(5):727–735. https://doi.org/10.1007/s00299-009-0690-6
64. Acosta-García G, Vielle-Calzada J-P (2004) A classical arabinogalactan protein is essential for the initiation of female gametogenesis in arabidopsis. Plant Cell 16(10):2614–2628. https://doi.org/10.1105/tpc.104.024588
65. Maraschin SF, Vennik M, Lamers GEM, Spaink HP, Wang M (2005) Time-lapse tracking of barley androgenesis reveals position-determined cell death within pro-embryos. Planta 220(4):531–540. https://doi.org/10.1007/s00425-004-1371-x
66. Maraschin SF, Gaussand G, Pulido A, Olmedilla A, Lamers GEM, Korthout H, Spaink HP, Wang M (2005) Programmed cell death during the transition from multicellular structures to globular embryos in barley androgenesis. Planta 221(4):459–470. https://doi.org/10.1007/s00425-004-1460-x
67. Makowska K, Oleszczuk S (2014) Albinism in barley androgenesis. Plant Cell Rep 33(3):385–392. https://doi.org/10.1007/s00299-013-1543-x
68. Kumari M, Clarke HJ, Small I, Siddique KHM (2009) Albinism in plants: a major bottleneck in wide hybridization, androgenesis and doubled haploid culture. Crit Rev Plant Sci 28(6):393–409. https://doi.org/10.1080/07352680903133252
69. Larsen ET, Tuvesson IKD, Andersen SB (1991) Nuclear genes affecting percentage of green plants in barley (hordeum vulgare l.) anther culture. Theor Appl Genet 82(4):417–420. https://doi.org/10.1007/BF00588593
70. Krzewska M, Czyczyło-Mysza I, Dubas E, Gołębiowska-Pikania G, Żur I (2015) Identification of qtls associated with albino plant formation and some new facts concerning green versus albino ratio determinants in triticale (×triticosecale wittm.) anther culture. Euphytica 206(1):263–278. https://doi.org/10.1007/s10681-015-1509-x
71. Torp AM, Hansen AL, Andersen SB (2001) Chromosomal regions associated with green plant regeneration in wheat (triticum aestivum

l.) anther culture. Euphytica 119(3):377. https://doi.org/10.1023/A:1017554129904

72. Muñoz-Amatriaín M, Castillo AM, Chen XW, Cistué L, Vallés MP (2008) Identification and validation of qtls for green plant percentage in barley (hordeum vulgare l.) anther culture. Mol Breed 22(1):119–129. https://doi.org/10.1007/s11032-008-9161-y
73. Caredda S, Devaux P, Sangwan RS, Proult I, Clément C (2004) Plastid ultrastructure and DNA related to albinism in androgenetic embryos of various barley (hordeum vulgare l.) cultivars. Plant Cell Tissue Org 76 (1):35–43. https://doi.org/10.1023/A:1025812621775
74. Munoz-Amatriain M, Svensson JT, Castillo AM, Close TJ, Valles MP (2009) Microspore embryogenesis: assignment of genes to embryo formation and green vs. albino plant production. Funct Integr Genomics 9(3):311–323. https://doi.org/10.1007/s10142-009-0113-3
75. Caredda S, Doncoeur C, Devaux P, Sangwan RS, Clément C (2000) Plastid differentiation during androgenesis in albino and non-albino producing cultivars of barley (hordeum vulgare l.). Sex Plant Reprod 13(2):95–104. https://doi.org/10.1007/s004970000043
76. Oleszczuk S, Sowa S, Zimny J (2006) Androgenic response to preculture stress in microspore cultures of barley. Protoplasma 228 (1):95–100. https://doi.org/10.1007/s00709-006-0179-x
77. Jacquard C, Nolin F, Hécart C, Grauda D, Rashal I, Dhondt-Cordelier S, Sangwan RS, Devaux P, Mazeyrat-Gourbeyre F, Clément C (2009) Microspore embryogenesis and programmed cell death in barley: effects of copper on albinism in recalcitrant cultivars. Plant Cell Rep 28(9):1329–1339. https://doi.org/10.1007/s00299-009-0733-z
78. Liu W, Zheng M, Konzak C (2002) Improving green plant production via isolated microspore culture in bread wheat (triticum aestivum l.). Plant Cell Rep 20(9):821–824. https://doi.org/10.1007/s00299-001-0408-x
79. Esteves P, Clermont I, Marchand S, Belzile F (2014) Improving the efficiency of isolated microspore culture in six-row spring barley: Ii-exploring novel growth regulators to maximize embryogenesis and reduce albinism. Plant Cell Rep 33(6):871–879. https://doi.org/10.1007/s00299-014-1563-1
80. Żur I, Dubas E, Golemiec E, Szechyńska-Hebda M, Gołębiowska G, Wędzony M (2009) Stress-related variation in antioxidative enzymes activity and cell metabolism efficiency associated with embryogenesis induction in isolated microspore culture of triticale (x triticosecale wittm.). Plant Cell Rep 28 (8):1279–1287. https://doi.org/10.1007/s00299-009-0730-2

# Chapter 7

# Chemical and Physical Mutagenesis Approaches for Identification of Herbicide and Drought Tolerance Traits in Wheat

**Neha Vaid, Marcus A. Samuel, Sateesh Kagale, and Raju Soolanayakanahally**

## Abstract

Since their inception, plant species have gathered numerous genetic polymorphisms which became fodder for evolutionary processes to shape the modern-day plants, agriculture and our food habits. However, while the current agricultural practices have substantially narrowed down this gene pool variability, the demand for novel and improved traits is ever persistent. Therefore, over decades, various techniques for artificial induction of mutations have been optimized. Indeed, induced mutagenesis has become central to the process of generating novel traits in crop cultivars and in understanding of gene function, genetic pathways and evolutionary processes.

In this chapter, using the complex polyploid crop species wheat as a case study, we have enlisted a step-by-step procedure for induction of polymorphisms by ethyl methanesulphonate and gamma radiation as representatives for chemical and physical mutagens, respectively. We explain the mutant screening protocols in detail using herbicide resistance and drought stress tolerance as traits of interest. We also discuss the state-of-the-art genomic techniques and pipelines that can be followed for identification of the causal mutations in polyploid crops, such as wheat.

**Key words** Mutagenesis, Ethyl methanesulphonate (EMS), Gamma radiation, Mutant screening, Wheat, Polyploid crops

## 1 Introduction

Natural and artificial selection of agronomic traits depends on the existing genetic diversity in a plant population. Trait-based selection has led to successful domestication of wild cultivars and to the continuous improvement of crop varieties in modern agriculture. For example, selection of spontaneous natural mutations that produced dwarf wheat and rice varieties resulted in 'Green Revolution' between 1965 and 1980. Genetic diversity in natural accessions and relative species has also helped in identifying gene function and understanding evolutionary processes [1–3]. However, several

Andriy Bilichak and John D. Laurie (eds.), *Accelerated Breeding of Cereal Crops*, Springer Protocols Handbooks,
https://doi.org/10.1007/978-1-0716-1526-3_7, © Springer Science+Business Media, LLC, part of Springer Nature 2022

studies have noted a decline in plant genetic diversity over the last few decades [4–7]. Modern agricultural practices have contributed to this phenomenon by recurrent use of a narrow gene pool in breeding programs and by global introduction and widespread cultivation of the same varieties which has led to decline and extinction of locally adapted germplasm.

The changing environmental conditions and rapidly evolving pathogen strains constantly threaten crop yields. There is also an immense demand for higher crop productivity, and introduction of value-adding traits, such as mineral fortification or gluten-free grain. However, for introduction of novel traits in crops, and to generate varieties fit for future climatic, edaphic and disease or pest pressures, plant breeders are constrained by the limited natural genetic diversity [8]. Additionally, crop improvement by introgression of trait-associated genetic loci from wild relatives to commercial cultivars is often accompanied with 'linkage drag' that introduces unwanted phenotypes from the genetic elements neighbouring the genes of interest [9]. In polyploid crop species, complex allelic interactions between the multiple genomes can further limit the success of breeding approaches. Similarly, traditional breeding is seldom successful in selecting traits associated with recessive alleles. Therefore, for several decades, plant breeders have complemented the traditional breeding approaches with mutational breeding which entails direct selection of desired traits in a population, circumventing the above-mentioned issues.

Mutational breeding involves artificial introduction of mutations and genetic variation in plants using physical, chemical and biological mutagens. The technique was first introduced in 1928 with the use of X-rays on maize seeds to identify plants with altered fertility and fitness [10]. Following this, several mutagens have been tested for their efficacies in inducing genetic and phenotypic variation in plants. Physical mutagens consist of high-energy ionizing particles and radiation, including alpha and beta particles, fast-moving neutrons, gamma rays, X-rays, ultraviolet radiation, etc. The physical mutagens act by breaking chemical bonds in DNA molecules resulting in nucleotide substitution, frameshift (insertion and/or deletion) mutations, large-scale chromosomal breaks and rearrangements [11]. Chemical mutagens, on the other hand, act by triggering base-pair transitions, deaminating nucleotides and interrupting DNA replication machinery causing strand breaks, mostly leading to single base conversions. The most commonly used chemical mutagens are alkylating agents, such as ethyl methanesulphonate (EMS) and methyl/ethyl nitrosourea; base analogues, such as 5 bromouracil; and intercalating agents such as ethidium bromide. Over the last six decades, 3332 mutagenized crops from 257 economically important species have been registered with the United Nations International Atomic Energy Agency and FAO (https://mvd.iaea.org/). Cereal crops, which are staple food for the majority of the world population, account for approximately 50% (1572) of these registered mutants. These include PNR series of rice varieties, initially generated as high-yielding rice cultivars, and later used as a parent to add traits such as aroma, short stature, early maturation and disease resistance to other commercial varieties [12, 13]. In wheat, mutant cultivars tolerant to aluminium toxicity were developed, which were eventually employed for trait transference [14]. This cultivar also aided in the identification of a molecular marker for aluminium toxicity tolerance trait [15, 16].

Several plant propagules, including seeds, pollen grains, calli, etc., have been found to be conducive to mutagenesis. The choice of the starting material depends on ease of use, plant species and the phenotype to be studied. For seed-propagated plants, due to their ease of use and scalability, seeds are the most commonly used starting material for mutagenesis. The resulting $M_1$ generation is composed of chimeric plants, mostly with somatic mutations [17]. The subsequent $M_2$ generation, however, filters out the mosaic somatic mutants, and can be screened for the desired phenotype [17]. Mutagenesis of pollen cells or a callus culture, on the other hand, removes the intermittent chimeric generation [18]. The mutagenized pollen cells are used for manual pollination resulting in hemizygous mutant population, which can be studied in $M_1$ (for dominant or semidominant alleles) or $M_2$ (for recessive alleles) generation [18, 19]. Calluses derived from haploid pollen or microspores, however, regenerate an independent homozygous mutant plant. Callus mutagenesis is the only available method for vegetatively propagated plants. However, for cereals with extremely large genomes, obtaining sufficient number of mutagenized plants to represent mutations in an entire genome can be quite laborious and time-consuming. Mutagenesis protocols for pollen and callus tissue have been described previously [19, 20]. In this chapter, due to broader interest, we will describe the protocol for induced mutagenesis using seeds as the starting material. For mutagenesis, fresh seed stock with high germination rate is preferable.

The choice of physical or chemical mutagen depends on factors, such as the type of mutations intended, frequency of mutations sought, equipment availability and the plant material to be used. As mentioned previously, physical mutagens result in higher frequency of insertion-deletion mutations and large fragment deletion, whereas, chemical mutagens mostly result in single base changes [11, 21]. In general, chemical agents show higher frequency of mutations per genome as compared to physical agents [21], and do not require any specialized equipment. For physical mutagens, the plant material to be mutagenized can also dictate the choice of mutagen. Ionizing particles (e.g. neutrons) have lower energy and penetration than X-rays or gamma rays making them suitable for starting material such as dry seeds [21, 22].

In this chapter, we provide a detailed protocol for widely used EMS mutagenesis and gamma irradiation for herbicide tolerance and drought tolerance traits in wheat, respectively. The chapter will cover a protocol for mutant screening and discuss methods for identification of phenotype-associated causal loci using state-of-the-art genomic techniques.

## 2 Materials

### 2.1 EMS Mutagenesis

*Sterilizing Solution*: 30% Chlorox/Sodium Hypochlorite, 70% Sterile Water, 0.05% Tween 20 70% ethanol (v/v) in water

100 mM phosphate buffer (pH 7.5) (freshly prepared, autoclaved to maintain sterile conditions)

0.5% ethyl methanesulphonate (EMS, also supplied as methanesulphonic acid ethyl ether, ethyl mesylate) solution in dimethyl sulphoxide (DMSO).

### 2.2 Gamma Irradiation

*Vacuum Desiccator*: 60% glycerol solution (v/v in water).

Air- and water-permeable paper bags (for seed treatments).

Gamma radiation source and equipment.

### 2.3 Screening and propagation of Mutants

*Screening of mutants on plates*: The screening of mutagenized plants has been summarized in Fig. 1.

In vitro growth medium—Dissolve 4.3 g Murashige and Skoog (MS) salts, 10 g sucrose (skip if photosynthesis or stress-related traits are analysed), 0.5 g MES and 6.8 g agar (or 8% agar, if the roots are to be phenotyped) in water. Add the supplement for selection of mutant seeds to the medium if it is not heat sensitive. If the supplement is heat sensitive, filter sterilize it and add to the MS medium before pouring in the Petri dishes. Adjust the pH of medium to 5.7 (with KOH or HCl) and total volume to 1 L with water. Autoclave and pour the medium to Petri dishes after it has cooled to approximately 50 °C.

*Screening of mutants in soil*: Soil.

Plant pots/trays.

Herbicide spray—Imazamox (2-[4,5-dihydro-4-methyl4-(1-methylethyl)-5-oxo-1-imidazol-2-yl]-5-methyl-3-pyridinecarboxylic acid) (40% w/v).

Pressurized spray container.

Growth chamber set at 22 °C/20 °C, 16 h/8 h photoperiod (day/night), with 150μM $m^{-2}$ $s^{-1}$ light intensity.

Scanner or camera to document the mutant phenotypes for analysis.

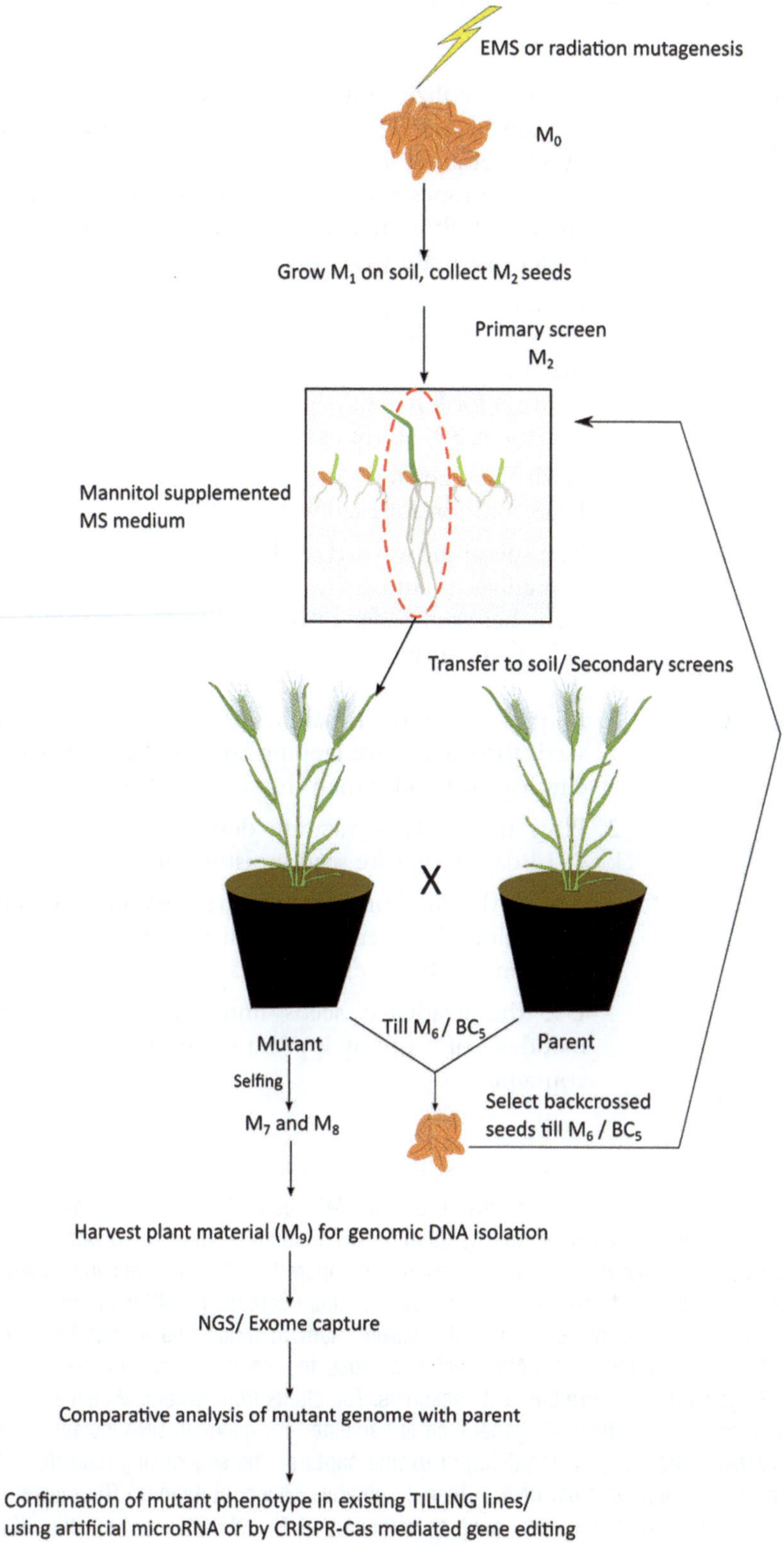


**Fig. 1** Pictorial representation of screening mutagenized wheat seeds for drought stress tolerance. The wheat seeds are mutagenized ($M_0$) using EMS or gamma radiations and the heterozygous $M_1$ plants are grown on

## 3 Methods

### 3.1 Mutagenesis

#### 3.1.1 EMS Mutagenesis

1. Surface sterilize 30,000 wheat seeds by incubating them in sodium hypochlorite sterilizing solution for 25 min followed by 70% ethanol for 1 min, under constant shaking. Wash the seeds five times with sterile water to remove hypochlorite and ethanol. Imbibe the seeds by soaking in 100 mM phosphate buffer overnight (*see* **Note 1**).
2. Decant the phosphate buffer and add freshly prepared 0.5% EMS solution in 100 mM phosphate buffer, in enough volume that the seeds are evenly soaked. Incubate the seeds in EMS solution for 8 h with constant shaking on an orbital shaker or a nutator at 30–50 rpm (*see* **Notes 2** and **3**).
3. Wash the seeds ten times with water (five times the volume of EMS solution for each wash).
4. Resuspend the washed seeds in the desired volume of water for immediate planting (recommended), or thoroughly dry the seeds between paper towels and store in airtight containers at 2–5 °C for future use.

#### 3.1.2 Gamma Irradiation

1. Aliquot the seeds in batches in paper envelopes. Ensure that seeds form a nonoverlapping layer in the envelope and that the envelope is smaller than the size of radiation chamber.
2. Place the seeds in vacuum desiccator over 60% glycerol for 7–14 days to ensure seed moisture of 10–14%.
3. Expose the equilibrated seeds to the gamma radiation source in irradiation chamber. The irradiation step must be performed and handled by an experienced user.
4. Use the irradiated seeds immediately for sowing (recommended) or store at low humidity at 2–5 °C in an airtight container.

**Fig. 1** (continued) soil, allowed to self-fertilize to produce M2 seeds. The primary screen for drought stress tolerance is conducted on $M_2$ seeds by growing them on MS medium supplemented with D-mannitol. The $M_2$ plants that show robust growth (marked with red oval) as compared to other mutants and controls are selected and can either be transferred to soil or subjected to secondary screenings. After the initial screenings, the resulting $M_2$ plants are backcrossed (BC) to the parent plant to reduce background mutations until five backcrosses ($BC_5$) or the $M_6$ generation. After each backcross, the resulting seeds are re-screened for the trait in either F1 or F2 generation, before the next backcross. The plants from $M_7$ and $M_8$ generations are allowed to self-fertilize. Plant material from $M_9$ generation is harvested for genomic DNA isolation, and sequencing analysis by next-generation sequencing (NGS) or exome capture. The sequencing data from the mutants is compared with the parental genomic DNA to identify putative causal mutations. The causal mutations can be further confirmed by various methods, such as screening of TILLING lines and specific RNA silencing by artificial microRNA, or by CRISPR/Cas9 gene editing method

### 3.2 Screening and Propagation of Mutants

#### 3.2.1 Screening Procedure

The $M_0$ seeds are propagated to $M_1$ generation, and the pooled $M_2$ seeds (*see* **Note 4**) are screened for phenotype of interest either on plates, hydroponics, or soil (*see* **Note 5**). It is essential that the controls are screened alongside the mutants (*see* **Note 6**). Here, we share screening protocols for two of our mutagenesis screens.

(a) On Petri plates: If sterile conditions have been maintained throughout the mutagenesis process, then the seeds can be directly sown on MS medium containing the additives for selection:

1. Under sterile conditions, pour MS medium containing additive for selection in sterile Petri dishes. For example, to identify EMS-mutagenized wheat plants with drought tolerance through ABA-dependent pathways, we have screened the $M_2$ generation for ABA sensitivity and $M_3$ generation for drought tolerance by using MS medium supplemented with 5–10μM ABA and 300 mM mannitol (supplemented MS medium), respectively (Vaid et al., unpublished data).
2. Using sterile forceps, place the seeds equidistant from each other, ensuring sufficient space for the growing roots and shoots. Untreated control seeds must be grown along with the mutants in each plate.
3. Incubate the plates at 22 °C/20 °C, 16 h/8 h photoperiod (day/night), with 150μM $m^{-2}$ $s^{-1}$ light intensity for 10 days. Change the position of plates within the growth chamber every day to minimize positional effects on the phenotype.
4. Scan/photograph each plate, and compare the mutant and control phenotype within each plate independently to choose mutants with phenotype of interest. In our screen for drought-tolerant wheat, ABA-hypersensitive lines in the $M_2$ generation were screened for shorter roots than control, and in the following $M_3$ generation, the same lines were screened in mannitol for longer root length than control plants.
5. Using forceps, gently remove selected seedlings from plates and transfer to soil. Cover the pots with transparent lids or place in a chamber with high humidity (50% RH) and reduced light intensity (80μM $m^{-2}$ $s^{-1}$) for 2 days or until the seedlings have established roots in the soil. Label each mutant and the generation carefully.
6. Collect $M_3$ seeds from each $M_2$ plant separately and label as an independent line. Store in a cool, dry place.

(b) On soil: Sow the seeds in soil and allow to germinate in a growth chamber. The density of seeding depends on the phenotype to be screened. For example, to select for tolerance phenotypes, such as herbicide tolerance, spread the seeds evenly in a large tray containing soil:

1. In case of physiological phenotype, such as leaf shape or photosynthetic traits, sow the seeds equidistant from each other to allow access to phenotyping at later growth stages.
2. To screen for herbicide tolerance trait, spray the plants with Imazamox solution at 3–5-leave growth stage.
3. The plants are screened 3 weeks post-spraying for survival and reduction in shoot biomass. Following the selection, transfer the surviving plants to pots and allow to grow until maturity.
4. Collect the seeds from each plant separately and label as an independent line. Store in a cool, dry place.

*3.2.2 Back Crosses to Reduce Background Mutations*

1. EMS- or irradiation-mediated mutagenesis is random and affects several genomic loci in addition to the loci of interest. To reduce non-consequential mutations from the genome, the selected mutants are backcrossed for at least three to four generations ($M_3$–$M_5$/$M_6$), followed by two generations ($M_7$–$M_8$) of selfing to obtain homozygous mutant plants. Each cycle of backcrossing is followed by screening of the mutant phenotype (*see* **Note** 7). Here speed breeding approach that reduces the generation time of wheat by approximately half can be followed [23].
2. If a large number of tolerant $M_2$ mutants were observed, then rank the mutants in order of potency for the phenotype sought using the scanned images (Sect. 3.2.1a, **step 4**). This allows for the mutants to be analysed sequentially in batches. In our screening for drought-tolerant wheat lines, the top 12 mutant lines in $M_2$ and $M_3$ generations that showed variable root length compared to controls in ABA and mannitol supplemented MS medium were further analysed preferentially (Vaid et al., unpublished).
3. Re-screen 15–30 $M_3$ seedlings per line along with parental control plants (as in Sect. 3.2.1) by growing them on supplemented MS medium for 10 days (*see* **Note 8**). In parallel, grow the parental genotype in non-supplemented MS medium as well.
4. After 10 days, note the phenotypes and select mutant lines with the phenotype of interest for propagation to the next generation (*see* **Note** 7).
5. Transfer 3–5 tolerant plants and parent genotype to soil. Following the speed breeding protocol, the plants are grown in 22 h/2 h photoperiod at 22 °C/17 °C (day/night) in

nutrient-rich soil. This photoperiod promotes fast vegetative growth leading to flowering at 30 days (as compared to 62 days after sowing in 16 h/8 h day/night photoperiod) and allows achieving up to six generations for wheat in 1 year. More details of the speed breeding principle and protocol can be found in [23].

6. At flowering stage, cross the mutant lines with the parental line, and collect the seeds generated upon maturity. Maintain each line independently (*see* **Note 9**).
7. Repeat **steps 2–5** from **4–5** independent lines from the previous generation until the $M_6$ generation. With each generation, 50% of the mutant genome should revert to the parental background, while the trait selection (in $F_1$ or $F_2$ generations, for dominant and recessive alleles, respectively (*see* **Note 7**)) would maintain the mutation associated with the desired phenotype. With four backcrosses ($M_3$–$M_6$), approximately 94% of the genome of the mutant line would revert to the parental genome, and most of the background mutations would be filtered out aiding in identification of the phenotype-associated mutation by genomic approach.
8. Re-screen $M_7$ and $M_8$ generations and transfer to soil as mentioned in **steps 2** and **4**. Allow both generations to self-fertilize and collect seeds from five independent lines separately. In the $M_8$ generation, 25% of the plants will be homozygous for the phenotype-causing mutation, and 100% of the seeds from these plants will exhibit the desired phenotype in the $M_9$ generation.

## 4 Identification and Characterization of Mutations

Physical and chemical mutagenesis result in several background mutations along with one or more causal mutations per plant. While backcrosses, as explained in Sect. 3.2.2, will remove a large proportion of the background mutations, other mutations might persist. The rate of mutations has been positively correlated with the increase in ploidy, implying higher chances of background mutations in a polyploid plant, such as wheat and oats [8, 24]. While the identification of the mutation associated with the phenotype is not essential for plant breeding purposes, it is pivotal for deciphering genetic and molecular pathways underlying the phenotype of interest. In the era of unprecedented advances in genomic techniques, whole genome sequencing (WGS) using next-generation sequencing (NGS) has become an efficient method for detection of mutations and polymorphisms in relatively smaller and less complex genomes, such as Arabidopsis and rice. However, NGS is still cost-prohibitive for re-sequencing of large polyploid genomes, such as wheat. A successful alternative to WGS is to selectively focus on sequencing of the coding region of the genome, a technique called exome capture [25–27]. Briefly, the sheared genomic DNA from an individual mutant line is tagged with line-specific barcode and hybridized to an array platform containing probes designed to represent all the expressed genes of the plant. After stringent washes, the genomic DNA is eluted and subjected to multiplexed paired-end sequencing reactions. For wheat, a NimbleGen 84 Mb exome capture array covering 286,799 exons from 82,511 transcripts [26] and a Roche NimbleGen 107 Mb exome array for hexaploid wheat [25] are available. For post-sequencing analysis, several studies have favoured using Mutation and Polymorphism Surveyor (MAPS) bioinformatics pipeline [25, 26]. This pipeline treats all mutants as control samples for each other, such that any recurrent polymorphisms between the mutants are recognized as non-consequential and only the unique mutations in each line are considered further. This is especially useful for polyploids, where polymorphisms between homoeologous alleles could otherwise be identified as mutations [25]. Following identification, the non-synonymous polymorphisms can further be filtered using SIFT (sorting intolerant from tolerant) and/or position-specific scoring matrix (PSSM-PareSNP) analysis. These analyses compare the mutated protein sequence against its homologs from all available proteomes to assign a SIFT score/PSSM value that denotes likelihood of the identified polymorphism to be detrimental to the protein function [28–30]. Polymorphisms with a SIFT score $\leq 0.05$ or PSSM value $>10$ are more likely to be detrimental and are preferentially selected to substantially reduce the number of candidate genes [25, 26, 31]. This can be followed by Sanger sequencing of the candidate regions to reconfirm the causal mutations and rule out the possibility of sequencing errors. Further, the effect of mutation on the candidate genes can be assessed on publicly available TILLING lines/mutant resources, or by generating transgenics with altering gene expression using CRISPR-Cas or RNA-interference methods. TILLING lines and mutant resources for wheat (http://www.wheat-tilling.com/, https://dubcovskylab.ucdavis.edu/wheat-tilling), barley (http://www.distagenomics.unibo.it/TILLMore/) [31] and rice [32] are publicly available.

## 5 Notes

1. Seed sterilization is essential if the screening of mutant seeds is to be conducted in a sterile environment (Sect. 3.3.1a). In this case, sterile solution should be maintained for all the subsequent steps till mutant screening. If the mutant seeds are screened on soil, replace seed sterilization with thorough washing of seeds with 0.01% Tween 20 in water.

2. The concentration of EMS or gamma irradiation dose is optimized by quantifying the $LD_{50}$, which is the concentration at which 50% of plants show a quantifiable trait, such as lethality. To calculate $LD_{50}$, incubate 50 seeds each with different concentrations/doses of mutagen from 0.1% to 0.9% (for 8 h) for EMS and 100 to 400 Gy for gamma rays, followed by phenotyping. Plot a line graph for phenotype versus mutagen doses. Using straight-line equation $y = mx + c$ to solve for $x$ (mutagen concentration) when 50% plants display lethality (phenotype on $y$-axis). When lower mutation frequency is desired, as for mutagenesis of an elite background where minimal background mutations are favoured, $LD_{20}$ or $LD_{30}$ is considered [33].
3. All steps involving EMS must be done in a fume hood. EMS is volatile and a toxic mutagen and therefore must be handled and discarded with utmost precautions. EMS on labware can be inactivated by soaking them in 20% sodium thiosulfate and 1% sodium hydroxide solution at 37 °C for 24 h.
4. Ideally, each mutagenized seed ($M_0$) should be propagated to subsequent generations independently. However, this can be restrained by available infrastructure. Therefore, the $M_1$ generation and the resulting $M_2$ seeds are grown and collected in batches to make the screen manageable. For example, $M_2$ seeds from 15,000 $M_1$ plants (with 50% lethality of 30,000 $M_0$ seeds) can be divided into 150 batches of 100 plants each or 300 batches of 50 plants each.
5. Before starting the mutagenesis screen, the trait to be phenotyped must be chosen with great diligence. The screen should be conducted on a phenotype that can be assessed as a clear positive or negative, for example, survival on mannitol- or herbicide-supplemented medium for drought or herbicide tolerance. If a similar selection is not possible, consider fluorescent tagging of a gene whose expression is regulated by the desired mutation. The aim of the preliminary screen in the $M_2$ generation is to allow the identification of a few strong candidates that can be reconfirmed using the same or different screening methods in subsequent generations.
6. While designing the experiment, a positive and a negative control must be decided. Seeds treated with 0% EMS solution serve as negative control, while an existing mutant or drug treatment that mimics phenotype of interest can be used as the positive control. In addition, it is advisable to gather seeds from segregating mutant plants that do not show phenotype of interest populations in $M_3$ and $M_4$ generations. These seeds are collected for each mutant line separately and can be very helpful in identifying background mutations not associated with the phenotype.

7. Mutant phenotype caused by dominant mutations would appear in the $F_1$ generation following the backcross. However, a recessive mutation-associated phenotype would only appear in the $F_2$ generation. An informed decision regarding the screening procedure could be made after considering the total number of candidate mutant lines identified/needed and the available time and resources.
8. The same or different selections can be used in subsequent generations for more stringent selection. For example, we used 10μM ABA in the $M_2$ generation, and 300μM mannitol in the $M_3$ generation for stringent selection in our screen. Similarly, if large numbers of mutants are obtained in the preliminary screen, the concentration of selection can be increased in the subsequent generations. For example, 300μM mannitol in the $M_2$ generation, followed by 350 and 400μM in subsequent generations.
9. Using mutant lines as egg donor and parental control lines as pollen donor allows for easy labelling of the plants. However, certain plant loci show paternal or maternal inheritance and phenotype patterns. Therefore, it is suggested to conduct reciprocal crosses.

## References


1. Avni R, Nave M, Barad O et al (2017) Wild emmer genome architecture and diversity elucidate wheat evolution and domestication. Science 357:93–97
2. Sidhu JS, Ramakrishnan SM, Ali S et al (2019) Assessing the genetic diversity and characterizing genomic regions conferring Tan Spot resistance in cultivated rye. PLoS One 14: e0214519
3. Velu G, Crespo Herrera L, Guzman C et al (2018) Assessing genetic diversity to breed competitive biofortified wheat with enhanced grain Zn and Fe concentrations. Front Plant Sci 9:1971
4. Martin AR, Cadotte MW, Isaac ME et al (2019) Regional and global shifts in crop diversity through the Anthropocene. PLoS One 14: e0209788
5. Liu W, Chen L, Zhang S et al (2019) Decrease of gene expression diversity during domestication of animals and plants. BMC Evol Biol 19:19
6. Vellend M, Baeten L, Becker-Scarpitta A et al (2017) Plant biodiversity change across scales during the Anthropocene. Annu Rev Plant Biol 68:563–586
7. FAO. Harvesting Nature's Diversity, http://www.fao.org/3/v1430e/V1430E04.htm
8. Sikora P, Chawade A, Larsson M, et al (2011) Mutagenesis as a tool in plant genetics, functional genomics, and breeding, https://www.hindawi.com/journals/ijpg/2011/314829/
9. Zhang P, Dundas IS, Xu SS et al (2017) Chromosome engineering techniques for targeted Introgression of rust resistance from wild wheat relatives. Methods Mol Biol 1659:163–172
10. Stadler LJ (1928) Genetic effects of X-rays in maize. Proc Natl Acad Sci U S A 14:69
11. Acquaah G (2012) Principles of plant genetics and breeding, 2nd edn | Wiley, https://www.wiley.com/en-us/Principles+of+Plant+Genetics+and+Breeding%2C+2nd+Edition-p-9780470664766
12. Chakrabarti SN (1999) PNR 519, a promising fine grained rice variety. Journal of Nuclear Agriculture and Biology 28: 116–122
13. Chakrabarti SN (1995) Mutation breeding in India with particular reference to PNR rice varieties, https://eurekamag.com/research/002/901/002901468.php

14. de Oliveira Camargo CE, Tulmann Neto A, Ferreira Filho AWP et al (2000) Genetic control of aluminum tolerance in mutant lines of the wheat cultivar Anahuac. Euphytica 114:47–53
15. Riede CR, Anderson JA (1996) Linkage of RFLP markers to an aluminum tolerance gene in wheat. Crop Sci 36:905–909
16. Gupta PK (1998) Mutation breeding in cereals and legumes. In: Jain SM, Brar DS, Ahloowalia BS (eds) Somaclonal variation and induced mutations in crop improvement. Springer Netherlands, Dordrecht, pp 311–332
17. Henikoff S, Comai L (2003) Single-nucleotide mutations for plant functional genomics. Annu Rev Plant Biol 54:375–401
18. Yang C, Mulligan BJ, Wilson ZA (2004) Molecular genetic analysis of pollen irradiation mutagenesis in Arabidopsis. 164:279–288
19. Settles AM (2020) EMS mutagenesis of maize pollen. Methods Mol Biol 2122:25–33
20. Serrat X, Esteban R, Guibourt N et al (2014) EMS mutagenesis in mature seed-derived rice calli as a new method for rapidly obtaining TILLING mutant populations. Plant Methods 10:5
21. Oladosu Y, Rafii MY, Abdullah N et al (2016) Principle and application of plant mutagenesis in crop improvement: a review. Biotechnol Biotechnol Equip 30:1–16
22. Amano E (2006) Use of induced mutants in rice breeding in Japan (INIS-XA--965). International Atomic Energy Agency (IAEA)
23. Ghosh S, Watson A, Gonzalez-Navarro OE et al (2018) Speed breeding in growth chambers and glasshouses for crop breeding and model plant research. Nat Protoc 13:2944–2963
24. Mamanova L, Coffey AJ, Scott CE et al (2010) Target-enrichment strategies for next-generation sequencing. Nat Methods 7:111–118
25. Henry IM, Nagalakshmi U, Lieberman MC et al (2014) Efficient genome-wide detection and cataloging of EMS-induced mutations using exome capture and next-generation sequencing. Plant Cell 26:1382–1397
26. Krasileva KV, Vasquez-Gross HA, Howell T et al (2017) Uncovering hidden variation in polyploid wheat. Proc Natl Acad Sci U S A 114:E913–E921
27. Mo Y, Howell T, Vasquez-Gross H et al (2018) Mapping causal mutations by exome sequencing in a wheat TILLING population: a tall mutant case study. Mol Gen Genomics 293:463–477
28. Ng PC, Henikoff S (2003) SIFT: predicting amino acid changes that affect protein function. Nucleic Acids Res 31:3812–3814
29. Sim N-L, Kumar P, Hu J et al (2012) SIFT web server: predicting effects of amino acid substitutions on proteins. Nucleic Acids Res 40: W452–W457
30. Taylor NE, Greene EA (2003) PARSESNP: a tool for the analysis of nucleotide polymorphisms. Nucleic Acids Res 31:3808–3811
31. Talamè V, Bovina R, Sanguineti MC et al (2008) TILLMore, a resource for the discovery of chemically induced mutants in barley. Plant Biotechnol 6:477–485
32. Wang N, Long T, Yao W et al (2013) Mutant resources for the functional analysis of the rice genome. Mol Plant 6:596–604
33. Maluszynski M, Szarejko I, Bhatia CR, et al (2009) Methodologies for generating variability. Part 4: Mutation techniques, FAO

# Chapter 8

# Using TILLING Lines for Functional Genomics in Wheat

## Liang Chen and Yin-Gang Hu

## Abstract

TILLING (targeting induced local lesions in genomes) combines chemical mutagenesis with high-throughput screening for point mutations in genes of interest. The TILLING lines screened can be used for both functional genomics and crop improvement. The traditional protocols for creating the mutagenized population, screening the desired mutants, and utilizing the mutants for the studies in functional genomics and genetic improvement in wheat are described. Screening mutants by sequencing is also discussed.

**Key words** TILLING, Functional genomics, Wheat improvement

## 1 Introduction

The TILLING method (targeting induced local lesions in genomes), which combines chemical mutagenesis with high-throughput genome-wide screening for point mutations in genes of interest, is useful for both functional genomics and crop improvement. TILLING can provide the practical verification needed for this type of sequence-driven hypothesis. The importance of this locus-to-phenotype method is demonstrated by the success of the *Arabidopsis* TILLING Project, a high-throughput platform for gene functional analysis, as well as in other plants. TILLING also finds application in crop improvement, as the mutants identified by TILLING can be readily utilized in conventional breeding programs since it is non-transgenic and the novel variations can be inherited stably. Particularly, wheat is especially well suited for TILLING because it tolerates a much higher mutation density than diploid plants due to its polyploid nature.

The general protocol for creation of a TILLING platform in crops includes the following steps (Fig. 1) [1]:

1. Creation of mutated populations.
2. Detection of mutations in a targeted sequence.

Andriy Bilichak and John D. Laurie (eds.), *Accelerated Breeding of Cereal Crops*, Springer Protocols Handbooks,
https://doi.org/10.1007/978-1-0716-1526-3_8, © Springer Science+Business Media, LLC, part of Springer Nature 2022

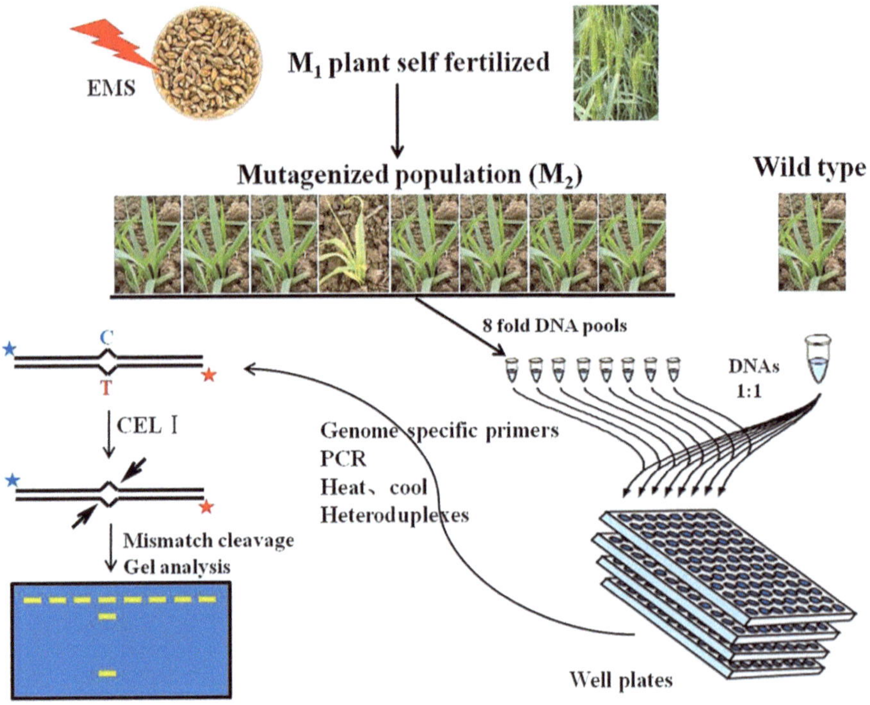


**Fig. 1** Outline of high-throughput TILLING strategy

3. Analysis of the mutant phenotype.
4. Functional genomics and crop improvement.

## 2 Materials

### 2.1 Solutions

0.2 M phosphate buffer (pH 7.0, 100 mL): 0.02 mol $Na_2HPO_4$; 1.2% EMS solution (ethyl methyl sulfone, phosphate buffer); chloroform/isoamyl alcohol; ethanol; CTAB Extraction Buffer (2% cetyltrimethylammonium bromide, 1% polyvinyl pyrrolidone, 100 mM Tris-HCl, 1.4 M NaCl, 20 mM EDTA); TE buffer (10 mM Tris-HCl, 1 mM EDTA); 10× loading buffer; 2.5% agarose gel (100 mL) (2.5 g agarose, 100 mL 1X TAE buffer); 1 μg/mL ethidium bromide buffer; HotShot Master Mix; 10 × LCGreen Plus; $ddH_2O$.

### 2.2 Equipment

Mortar, pestle, water bath, high-speed refrigerated centrifuge, low temperature refrigerator, thermal cycler, electrophoresis apparatus, and LightScanner.

## 3 Methods

### 3.1 Chemical Mutagenesis and Development of $M_1$ and $M_2$ Generations

There are many factors that may affect the frequency of induced mutations, such as mutagen concentration, exposure time, environmental effects, and the genotype of the target species [2]. A concentration of 1.2% ethyl methanesulfonate (EMS) was used in a pilot wheat TILLING study in attempting a high mutation frequency, but only ~8% of the $M_1$ seeds germinated under the treatment [3]. The EMS doses, GC content of the target regions, and also the genotypes employed should be considered before TILLING.

Approximately 3000 ~ 5000 seeds were soaked in 0.8% (v/v) EMS solution at the ratio ~ 350 grains/100 mL with gentle agitation overnight (~18 h) at room temperature (*see* **Note 1**) [3]. After EMS treatment, seeds were thoroughly washed with tap water for 3 h and then placed at 4 °C for 5 days before being transferred to room temperature. The EMS-treated seeds ($M_1$) were sown in pots with soil in a greenhouse until harvest. $M_2$ seeds were harvested from $M_1$ individuals, threshed, and packed separately. Fifteen $M_2$ seeds per line were sown in 1-meter rows in the field, three individuals per line were harvested from the $M_2$ lines, and the $M_3$ seeds from each $M_2$ individual were collected and labeled accordingly. Alternatively, it is recommended that only one seed is taken from each $M_1$ plant for laborsaving. However, at this stage, the population is still segregating, and not all $M_3$ plants will carry the mutations identified in the $M_2$ lines. The mutagenized population may be taken through further generations by single seed descent to generate near-homozygous material (~3% heterozygous mutations at $M_6$), although up to half of all mutations present in the $M_1$ are lost in the process.

Forward genetic mutant screening is performed on $M_2$ families. All visible mutant phenotypes are carefully inspected and recorded in reference to the wild type during the plant life cycle. The frequency of the distinguishable phenotypes is recorded. The novel phenotypes can also be photographed with a digital camera.

### 3.2 DNA Preparation and Pooling of Individuals

Genomic DNA from young leaves was isolated from each $M_2$ plants using the CTAB method [4]. DNA concentrations were measured using a spectrophotometer and normalized to 50 ng/μL. Then, DNA samples from eight $M_2$ plants with equivalent amounts were pooled for initial screening.

The detailed steps of genomic DNA extraction were as follows:

Plant samples can be prepared by cryogenically grinding tissue in a mortar and pestle after chilling in liquid nitrogen. In either case, a fine powder is best for extracting DNA (*see* **Note 2**).

1. For each 100 mg homogenized tissue, use 800 μL of CTAB Extraction Buffer (2% cetyltrimethylammonium bromide, 1%

polyvinyl pyrrolidone, 100 mM Tris-HCl, 1.4 M NaCl, 20 mM EDTA), mix, and thoroughly vortex. Transfer the homogenate to a 65 °C water bath for 45 min.

2. Following the incubation, add an equal volume of chloroform/isoamyl alcohol (24:1). Vortex for 5 s and then centrifuge the sample for 10 min at 12,000 g to separate the phases.
3. Transfer the aqueous upper phase to a new tube, and add an equal volume of chloroform. Centrifuge the homogenate for 10 min at 12,000 g.
4. Transfer the upper aqueous phase to a new tube. Precipitate the DNA by adding the equal volume of cold isopropanol, and incubate at −20 °C for 20 min.
5. Centrifuge the sample at 12,000 g for 5–10 min. Decant the supernatant without disturbing the pellet, and subsequently wash with 800 μL ice-cold 75% ethanol. Decant the ethanol. Remove residual ethanol by drying in a SpeedVac.
6. Dry the pellet long enough to remove alcohol but without completely drying the DNA. Dissolve DNA in 50 μL TE buffer.
7. Store DNA at 4 °C or − 20 °C.

### 3.3 Detection of Mutations in a Targeted Sequence

#### 3.3.1 Design of Genome-Specific PCR Primers

Primers should be designed to amplify a single target region. It was found that increasing the melting temperature of primers reduces low molecular weight background noise in TILLING assays, presumably arising from nonspecific primer hybridization. For designing genome-specific primers of target genes for TILLING, it is necessary to first predict the mutation hot spots and functional regions of important gene loci. Then, genome-specific primers can be designed based on variations between the three homologous genes to achieve the target genome-specific amplification [6]. Generally, CODDLE software (Codons Optimized to Discover Deleterious Lesions; http://www.proweb.org/coddle) plays an important role in predicting hot spots of genomic variation. CODDLE combined with the Primer3 software was used to define the best amplicon for TILLING screening, aiming for an expected primer Tm of 60–70 °C and of a length in the range of 20–30 nucleotides. A 100 mM solution of each primer is prepared in TE buffer, and aliquots are stored at −80 °C to avoid repeated freeze. CODDLE identifies the region(s) of a user-selected gene, which has the highest probability of affecting gene function when mutated by EMS.

Additionally, with the development of next-generation sequencing technology (NGS), the reference genome sequences have been obtained for all major crops, including wheat, maize, rice, and so on. A large number of whole-genome resequencing (WGR) and transcriptome data have been released, and many mutation sites and regions have been identified in multiple crops' genomes. Based on these studies, hot spots of genomic mutations are easy to screen. Websites for functional genomics of several crops can provide the search and screening of mutation regions, for example, wheat (*Triticeae* Multi-omics Center, http://202.194.139.32; Ensembl Plants, https://plants.ensembl.org/) and maize (Maize Genetics and Genomes Database, https://www.maizegdb.org). Moreover, multiple sequence alignment of important homologous genes among different species using BLAST (ClusterW) will also help to mine important functional regions. Thus, the combination of mutation hot spots screening within species and identification of important functional regions among species can assist the design of TILLING primers for target genes.

#### 3.3.2 PCR Amplification of Target Regions and Heteroduplex Formation

After DNA extraction and concentration normalization, the optimal amount of genomic DNA to use for PCR should be determined empirically. For wheat, according to different ploidy, 50 to 100 ng is used per PCR reaction. A concentration is chosen that provides the best signal-to-noise ratio when using a fixed amount of enzyme for heteroduplex digestion. A higher concentration may be required when using lower quality genomic DNA samples. The protocol for the TILLING targets produced through the non-fluorescence methods is similar to that used for the LI-COR detection system. The PCR reactions are performed as showed in Table 1. The PCR reaction system is adjusted according to different target gene sequence features and primer sequences. One of the key steps of TILLING technology is denaturation and renaturation after the completion of routine PCR reaction to form heteroduplex DNA fragments [7].

For increasing the specificity of the primers, touchdown PCR is favorable and conducted using a thermal cycler (Bio-Rad, DNA Engine, PTC200) as follows: 95 °C, 2 min; 5 cycles (94 °C, 20 s; 72 °C, 30 s, reduce temperature at 1 °C per cycle; 72 °C, 90 s); 37 cycles (94 °C, 20 s; 68 °C, 30 s; 72 °C, 90 s); and finally elongation at 72 °C, 5 min; a denaturing and reannealing step is included at the end of the PCR reaction (99 °C, 10 min; followed by 70 cycles of 70 °C for 20 s decreasing by 0.3 °C per cycle) to allow the formation of heteroduplexes if a mutation is present in the pool. However, the annealing temperature should be adjusted according to different primers.

#### 3.3.3 Heteroduplex Digestion

After PCR amplification, samples were digested with celery juice extract (prepared by following the extraction process of CEL I from celery as detailed in Till et al. (2006) (*see* **Note 3**) [5]). For the mixed pool containing mutant individual, the PCR products contain heteroduplex DNA fragments, which could be recognized and cut by CEL I nuclease [8]. The activity of CEL I nuclease could be determined by the cleavage efficiency of CEL I on the mismatch sites of heteroduplex DNA fragments, and factors such as working

**Table 1**
**The PCR reactions were performed in a 20 μL volume as outlined below**

| Component | Amount (μL) |
|---|---|
| Deionized water | 13.4 |
| 10× Ex-Taq buffer | 2 |
| 2.5 mM each dNTP (the stock solution contains all four dNTPs) | 1.4 |
| Forward primer (10 μM) | 0.5 |
| Reverse primer (10 μM) | 0.5 |
| Pooled DNA (8 samples) | 2 |
| Ex-Taq DNA polymerase (5 U/μL) | 0.2 |
| Total | 20 |

Keep Taq polymerase on ice, add it last, and then vortex thoroughly

**Table 2**
**Digestion of the PCR products by CEL I**

| Component | Amount (μL) |
|---|---|
| Deionized water | 6.8 |
| 10 × digestion buffer | 2 |
| Celery juice extract | 1.2 |
| PCR product | 10 |
| Total | 20 |

10 × digestion buffer: 10 mM HEPES, pH 7.5, 10 mM $MgSO_4$, 10 mM KCL, 0.002% Triton X-100, and 0.2 mg/mL BSA. Due to the variable activity of different celery juice extracts, the optimal amount of CJE for heteroduplex digestion can be determined empirically using targets with known mutations. The amount listed here is for illustration purposes only

concentration, digestion time, working temperature, and buffer solution should all be optimized. Digestion reactions are showed in Table 2, and the steps of CEL I digestion are listed below:

1. Prepare the celery juice extract reaction mix (CJE and digestion buffer), and store on ice.
2. Add 10 mL of celery juice extract reaction mix to each PCR tube. Mixing is recommended.
3. Incubate the plate for 45 min at 45 °C.
4. Stop the reaction by adding 2 μL 10× loading buffer, and the samples can be stored at −20 °C for weeks.

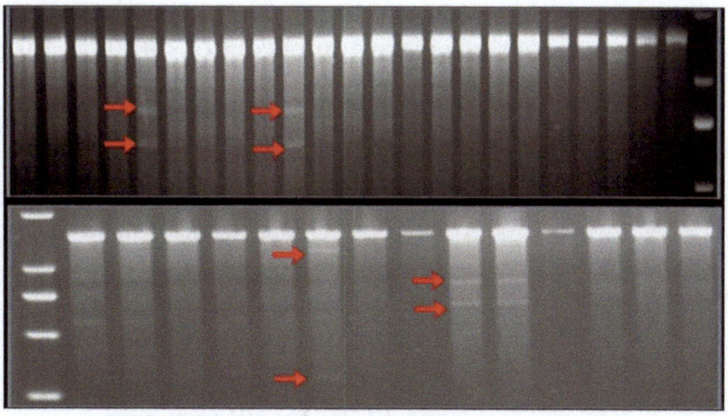


**Fig. 2** Digested bands detected with agarose gel electrophoresis. Putative mutations in the pools are identified by the presence of two bands (indicated by white arrows), with sizes adding up to the full-length polymerase chain reaction (PCR) product

*3.3.4 Detection of Digestion Products by Agarose Gel*

The traditional enzyme digestion detection system requires the use of fluorescent-labeled primers and fluorescence electrophoresis detection instrument (LI-COR genetic analyzer or ABI sequencer), which is time-consuming and expensive. Agarose gel systems eliminate the need for expensive genotyping instruments and labeled primers and are therefore especially compatible with small, low-budget laboratories [3]. High concentration of agarose gel can significantly improve the resolution of small DNA fragment (Fig. 2). The detailed steps are as follows:

1. Prepare 2.5% agarose gel (2.5 g agarose in 100 mL 1X TAE buffer).
2. Sample loading (5 ~ 8 μL).
3. 115 V running for 40 min in 1X TAE buffer.
4. Stained in an ethidium bromide buffer (1 μg/mL) for 15 min.
5. Visualized by Gel Doc XR (Bio-Rad Laboratories, Inc.).

### 3.4 Identification of Mutation in Mixed Pool by High-Resolution Melting (HRM) System

High-resolution melting (HRM) analysis is one of the techniques for mutation detection that does not rely on mismatch cleavage by endonuclease. HRM depends on the loss of fluorescence from intercalating dyes bound to dsDNA during thermal denaturation [9]. These dyes, such as LCGreen Plus, have low toxicity to PCR and can therefore be used at high concentration to saturate the dsDNA PCR product. The combination of these characteristics provides greater melt sensitivity, stability, and higher resolution melt profiles and makes it feasible to detect SNPs in PCR fragments. Accurate control of temperature and continuous monitoring of fluorescence in instruments such as the LightScanner (Idaho Technology, Salt Lake City, UT, USA) allow detection of single-base mismatches in target fragments from the mutant pools. This method is especially suited for target genes with multiple small exons separated by large introns as the detection sensitivity of HRM is optimized with amplicons of <450 bp. Moreover, HRM is a rapid assay and completed in a closed tube with no digestion and gel separation steps required, making it a good choice for screening mutations in TILLING or EcoTILLING.

For hexaploid wheat, mutation screening by HRM requires a two-step nested PCR amplification: a larger fragment containing several coding regions is first amplified using genome-specific primers, and then HRM analysis is conducted using primers specific for each exon or part thereof; an example is shown in Fig. 3. For

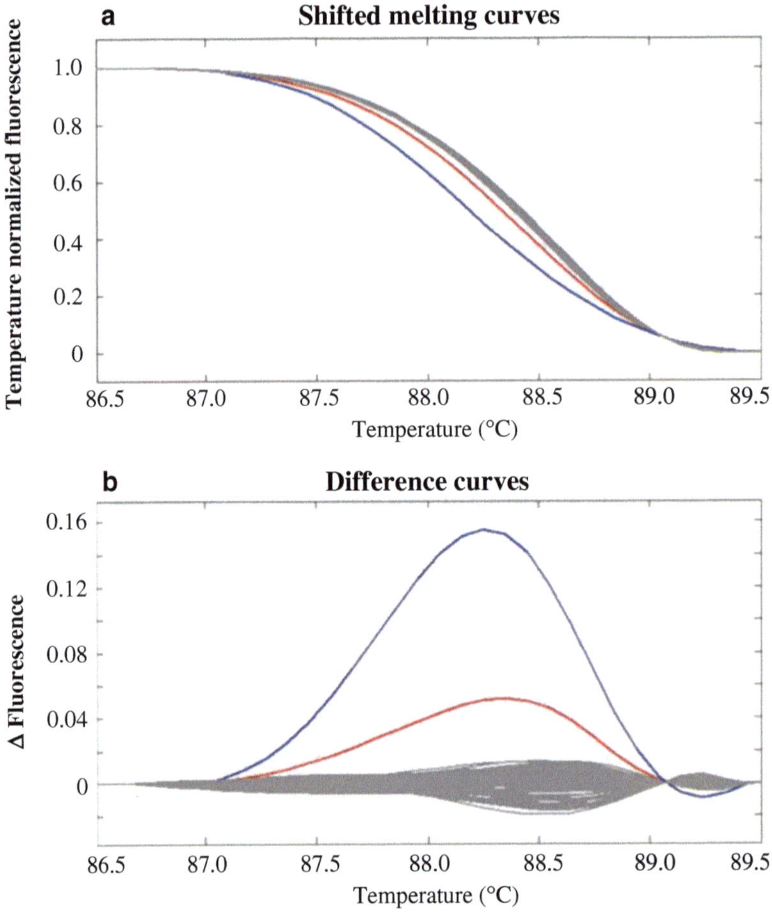


**Fig. 3** Mutation detection for dwarfing gene Rht-D1 in bread wheat using high-resolution melting (HRM) analysis. The HRM data are presented as two different parts: (A) displays normalized melting curves for the tested amplicons, whereas (B) shows different curves displaying the difference in fluorescence (DF) against temperature from melt analysis of the amplicons in relation to the wild type in a 96-well plate from an ethyl methanesulfonate (EMS)-mutagenized population of bread wheat. Wild-type samples are in gray; the putative mutant pools identified through their different melt profiles are colored. These amplicons were sent for DNA sequencing to confirm the mutant site in the gene. In this project, the first-round homoeolog-specific polymerase chain reaction (PCR) targeted the bread wheat Rht-D1 gene; the second-round amplicon was 200 bp of its first exon. LCGreen Plus dye was included in the second-round PCR for melt detection, and heteroduplex melt analysis was carried out on the LightScanner (Idaho Technology)

diploid plants, the procedure could be simplified as a one-step PCR as there is no need to distinguish the different homoeologous copies.

The detailed steps of HRM are as follows: PCR conduct using genome-specific primers as described above. Then, the PCR products are used as templates for a second-round PCR amplification to add HRM system-specific fluorescent dyes LCGreen. The PCR reactions are performed in a 10 μL volume containing HotShot Master Mix (5 μL), 10 × LCGreen Plus (1 μL), forward primer (0.3 μL, 5 M), reverse primer (0.3 μL, 5 M), DNA template (1.2 μL, 50 ng/l), and $ddH_2O$ (2.2 μL).

The PCR procedure is as follows: 95 °C, 5 min; 35 cycles (95 °C, 20 s; 68 °C, 30 s; 72 °C, 90 s); 72 °C, 10 min; 95 °C, 30 s, reduce temperature by speed 0.3 °C/s to 40 °C; store at 10 °C.

After PCR amplification, the PCR products are put into the high-resolution analysis system LightScanner (Idaho Technology Inc.) for HRM analysis. The working parameters of LightScanner are as follows: set the melting curve from 79 to 96 °C, and analyze the collected curve with LightScanner Call IT software. Finally, according to the type of dissolution curve, the DNA pool with mutation is determined, and the individuals in the pool are detected separately.

### 3.5 Analysis of the Mutant Phenotype

#### 3.5.1 Predicting Effects of Mutation on Gene Function

Generally, the mutation sites are obtained by sequencing, and the mutation type is determined. To predict the influence of mutation site on gene function, several software are used to predict the effect of site mutation, such as SIFT (Sorting Intolerant From Tolerant, http://blocks.fhcrc.org/sift/SIFT.html), PARSESNP (Project Aligned Related Sequences and Evaluate SNPs, http://blocks.fhcrc.org/proweb/parsesnp/), and Predict Protein (https://predictprotein.org). Additionally, the function of the gene where the mutation site is located on can be annotated by BLAST against Nr/Nt database.

#### 3.5.2 Mutation Phenotype Observation

TILLING in polyploids involves more effort on the downstream of mutation discovery, as the phenotype of a single homozygous mutant may be masked by the wild-type homoeolog(s) present in the other genome(s). In hexaploid wheat, for example, it is often necessary to identify mutations in the A, B, and D homoeologs and to combine these alleles by crossing and selfing over three generations to create a triple mutant that is more likely to display a phenotype. Random mutagenesis in TILLING can generate loss-of-function mutations with a relatively high probability, which provide the chance to combine mutations in different gene copies by crossing for functional genomics.

Additionally, a large number of recessive mutations in single homoeologs of genes are less likely to show a phenotype. Whereas there may be many exotic "visible" mutant phenotypes that were masked in $M_2$ (or $M_3$) plants, their values could not be observed or less attended. To see the possible hidden phenotypes in wheat TILLING population, $M_3$ mutants were chosen to backcross with the wild type. Then, the visible mutant phenotypes of each $BCF_2$ segregating population were recorded in reference to the wild type. Moreover, the mutant and nonmutant plants derived from the same line could be used to study the phenotypic change.

## 4 Functional Genomics and Crop Improvement

The newly identified alleles in TILLING could be used for elucidating gene function, while it could also be used as valuable resources in crop improvement. Previous study reported that 54 mutations were identified in wheat gene *TaSSIVb-D* using TILLING approach, and three of the missense mutations and one nonsense mutation were found to have effects on protein function [10]. TILLING was also used to identify partial loss-of-function alleles for the *TaMlo* gene, an ortholog of the barley *Mlo* gene which was known to confer durable broad-spectrum powdery mildew resistance. Sixteen missense mutations that resulted in amino acid changes were identified in three *TaMlo* homoeologs. These lines represent an important step toward the production of commercial non-transgenic, powdery-mildew-resistant bread wheat varieties [11]. Dong et al. screened the waxy genes Wx-A1 and Wx-D1 in 2348 EMS-treated M2 plants and found 121 mutants in total; then, a complete waxy wheat was successfully bred by crossing two truncation mutants (Wx-A1-truncation and Wx-D1-truncation; the lines used were already null in Wx-B1) [12, 13].

Several TILLING resources have been developed for wheat (Table 3).

## 5 Prospect of TILLING

DNA sequencing has become very affordable since the surge of next-generation sequencing (NGS) technologies, and there has been an acceleration in the integration of these methods into the TILLING approach. A number of TILLING by sequencing (TbyS) procedures have been modified to recognize point mutations from TILLING populations in an effective way that avoids the problems associated with traditional TILLING methods. At the same time, sequencing costs have rapidly decreased, contributing to a diffusion of NGS applications [30]. Only 10 years ago, the sequencing cost for a million base pairs was approximately \$1000; the cost for the same sequence is now below \$0.10 [31]. Furthermore, compared with whole-genome sequencing (WGS), if the complete genome sequence of a target species is available, then the cost of genotyping

**Table 3**
**TILLING resources developed in wheat**

| Ploidy level | Mutagen | Mutation detection technology | M2 size | Mutation frequency (1/kb) | Target traits | References |
|---|---|---|---|---|---|---|
| Wheat (6×) | EMS | HRM | 2020 | 1/26 | Powdery mildew disease resistance (TaMlo gene) | [11] |
| Wheat (6×) | EMS | TbyS | 4500 | 1/35000 | – | [14] |
| Wheat (6×) | EMS | LI-COR | 10,000 | 1/24 | Starch quality | [15, 16] |
| Wheat (6×) | EMS | PAGE, LI-COR | 1536 | 1/38 | Starch quality | [17] |
| Wheat (6×) | EMS | CEL-I, agarose gel | 2348 | 1/37, 1/23 | Starch quality and grain hardness | [12, 13] |
| Wheat (6×) | EMS | LI-COR, HRM | 4500 | 1/84 | Starch quality | [18] |
| Wheat (6×) | EMS | LI-COR | 4244 | – | Starch biosynthesis | [19] |
| Wheat (6×) | EMS | Agarose gel, PAGE | 2610 | 1/34, 1/47 | Spike development | [3] |
| Wheat (6×) | EMS | Direct sequencing | 630 | 1/3 | Grain hardness | [20] |
| Wheat (4×) | EMS | LI-COR | 8000 | 1/40 | Starch quality | [15, 16] |
| Wheat (4×) | EMS | LI-COR | 1386 | 1/51 | Starch quality | [17] |
| Wheat (4×) | EMS | HRM | 3992 | – | Starch metabolism | [21, 22] |
| Wheat (4×) | EMS | Agarose gel, dHPLC | 1140 | 1/77 | Carotenoid metabolism | [23] |
| Wheat (4×) | EMS | Exome sequencing | 733 | – | Plant height | [24] |
| Wheat (4×) | EMS | TbyS, CEL I PAGE | – | – | Thousand grain weight | [25] |
| Wheat (4×) | EMS | – | 10,000 | – | Gluten content | [26] |
| Wheat (4×) | EMS | CEL I PAGE | 1122 | – | Kernel hardness and starch | [27] |
| Wheat (2×) | EMS | – | 1400 | 1/1300 | Grain quality and lignin biosynthesis | [28] |
| Wheat (2×) | EMS | – | 1532 | 1/92 | Waxy and lignin | [29] |

*TbyS* TILLING by sequencing, *EMS* ethyl methanesulfonate, *PAGE* polyacrylamide gel electrophoresis, *HRM* high-resolution melting can be lowered 10–14-fold by using processes such as MSG (multiplexed shotgun genotyping) or GBS (genotyping by sequencing) [32]. Targeted sequencing is likely to be a better option for the exploration of large-scale markers, especially in the case of un-decoded genomes. The development of different high-throughput sequencing methods such as PacBio and Ion Torrent is helping in the improvement of TILLING platforms.

## 6 Notes

1. When determining the most suitable EMS mutagenic concentration, the concentration range can be determined by testing small batches of seed with a range of concentrations.
2. During DNA extraction, freeze-dried samples can be ground at room temperature.
3. When preparing the celery juice extract (CEL I nuclease extract), scientists should operate it with gloves, and avoid direct contact with the skin or inhalation, because PMSF is hazardous; thus, operate with gloves, and avoid direct contact with the skin or inhalation. All steps should be performed at 4 °C. Additionally, to determine the optimal amount of enzyme, perform TILLING reactions with varying amounts of enzyme, and choose the amount that provides the best signal-to-noise ratio.

### References

1. Kurowska M, Daszkowska-Golec A, Gruszka D et al (2011) TILLING—a shortcut in functional genomics. J Appl Genet 52:371–390. https://doi.org/10.1007/s13353-011-0061-1
2. Xin Z, Wang ML, Barkley NA et al (2008) Applying genotyping (TILLING) and phenotyping analyses to elucidate gene function in a chemically induced sorghum mutant population. BMC Plant Biol 8:103. https://doi.org/10.1186/1471-2229-8-103
3. Chen L, Huang L, Min D et al (2012) Development and characterization of a new TILLING population of common bread wheat (Triticum aestivum L.). PLoS One 7:e41570. https://doi.org/10.1371/journal.pone.0041570
4. Aldrich C (1993) CTAB DNA extraction from plant tissues. Plant Mol Biol Rep 11:128–141
5. Till BJ, Zerr T, Comai L et al (2006) A protocol for TILLING and Ecotilling in plants and animals. Nat Protoc 1:2465–2477. https://doi.org/10.1038/nprot.2006.329
6. Chen L, Hao L, Parry MAJ et al (2014) Progress in TILLING as a tool for functional genomics and improvement of crops. J Intergr Plant Biol 56:425–443. https://doi.org/10.1111/jipb.12192
7. Till BJ, Colbert T, Tompa R et al (2003) High-throughput TILLING for functional genomics. Plant Functional Genomics:205–220. https://doi.org/10.1385/1-59259-413-1:205
8. Till BJ, Reynolds SH, Greene EA et al (2003) Large-scale discovery of induced point mutations with high-throughput TILLING. Genome Res 13:524–530. https://doi.org/10.1101/gr.977903
9. Taheri S, Abdullah TL, Jain SM et al (2017) TILLING, high-resolution melting (HRM), and next-generation sequencing (NGS) techniques in plant mutation breeding. Mol Breed 37:40. https://doi.org/10.1007/s11032-017-0643-7
10. Guo H, Liu Y, Li X et al (2017) Novel mutant alleles of the starch synthesis gene TaSSIVb-D result in the reduction of starch granule number per chloroplast in wheat. BMC Genomics 18:358. https://doi.org/10.1186/s12864-017-3724-4

11. Acevedo-Garcia J, Spencer D, Thieron H et al (2016) Mlo-based powdery mildew resistance in hexaploid bread wheat generated by a non-transgenic TILLING approach. Plant Biotechnol J 2016:1–12. https://doi.org/10.1111/pbi.12631
12. Dong C, Dalton-Morgan J, Vincent K et al (2009) A modified TILLING method for wheat breeding. Plant Genome 2:39–47. https://doi.org/10.3835/plantgenome2008.10.0012
13. Dong C, Vincent K, Sharp P (2009) Simultaneous mutation detection of three homoeologous genes in wheat by high resolution melting analysis and mutation surveyor. BMC Plant Biol 9:1. https://doi.org/10.1186/1471-2229-9-143
14. King R, Bird N, Ramirez-Gonzalez R et al (2015) Mutation scanning in wheat by exon capture and next-generation sequencing. PLoS One 10:e0137549. https://doi.org/10.1371/journal.pone.0137549
15. Slade AJ, Fuerstenberg SI, Loeffler D et al (2005) A reverse genetic, nontransgenic approach to wheat crop improvement by TILLING. Nat Biotechnol 23:75–81. https://doi.org/10.1038/nbt1043
16. Slade AJ, McGuire C, Loeffler D et al (2012) Development of high amylose wheat through TILLING. BMC Plant Biol 12:1. https://doi.org/10.1186/1471-2229-12-69
17. Uauy C, Paraiso F, Colasuonno P et al (2009) A modified TILLING approach to detect induced mutations in tetraploid and hexaploid wheat. BMC Plant Biol 9:115. https://doi.org/10.1186/1471-2229-9-115
18. Botticella E, Sestili F, Hernandez-Lopez A et al (2011) High resolution melting analysis for the detection of EMS induced mutations in wheat SbeIIa genes. BMC Plant Biol 11:156. https://doi.org/10.1186/1471-2229-11-156
19. Sestili F, Botticella E, Bedo Z et al (2010) Production of novel allelic variation for genes involved in starch biosynthesis through mutagenesis. Mol Breed 25:145–154. https://doi.org/10.1007/s11032-009-9314-7
20. Feiz L, Martin J, Giroux M (2009) Creation and functional analysis of new Puroindoline alleles in Triticum aestivum. Theor Appl Genet 118:247–257. https://doi.org/10.1007/s00122-008-0893-1
21. Bovina R, Brunazzi A, Gasparini G et al (2014) Development of a TILLING resource in durum wheat for reverse-and forward-genetic analyses. Crop Pasture Sci 65:112–124. https://doi.org/10.1071/CP13226
22. Sestili F, Palombieri S, Botticella E et al (2015) TILLING mutants of durum wheat result in a high amylose phenotype and provide information on alternative splicing mechanisms. Plant Sci 233:127–133. https://doi.org/10.1016/j.plantsci.2015.01.009
23. Colasuonno P, Incerti O, Lozito ML et al (2016) DHPLC technology for high-throughput detection of mutations in a durum wheat TILLING population. BMC Genet 17:1. https://doi.org/10.1186/s12863-016-0350-0
24. Mo Y, Howell T, Vasquez-Gross H et al (2018) Mapping causal mutations by exome sequencing in a wheat TILLING population: a tall mutant case study. Mol Gen Genomics 293:463–477. https://doi.org/10.1007/s00438-017-1401-6
25. Wang W, Simmonds J, Pan Q et al (2018) Gene editing and mutagenesis reveal inter-cultivar differences and additivity in the contribution of TaGW2 homoeologues to grain size and weight in wheat. Theor Appl Genet 131:2463–2475. https://doi.org/10.1007/s00122-018-3166-7
26. Moehs CP, Austill WJ, Holm A et al (2018) Development of reduced gluten wheat enabled by determination of the genetic basis of the lys3a low hordein barley mutant. Plant Physiol 179:1692–1703. https://doi.org/10.1101/354548
27. Li W, Guo H, Wang Y et al (2017) Identification of novel alleles induced by EMS-mutagenesis in key genes of kernel hardness and starch biosynthesis in wheat by TILLING. Genes Genom 39:387–395. https://doi.org/10.1007/s13258-016-0504-5
28. Rothe N (2010) Validation of tilling populations in diploid and hexaploid wheat. Dissertation, Kansas State University, Kansas
29. Rawat N, Sehgal SK, Joshi A et al (2012) A diploid wheat TILLING resource for wheat functional genomics. BMC Plant Biol 12:1. https://doi.org/10.1186/1471-2229-12-205
30. Irshad A, Guo H, Zhang S et al (2020) TILLING in cereal crops for allele expansion and mutation detection by using modern sequencing technologies. Agronomy 10:405. https://doi.org/10.3390/agronomy10030405
31. Van Nimwegen KJ, van Soest RA, Veltman JA et al (2016) Is the $1000 genome as near as we think? A cost analysis of next-generation sequencing. Clin Chem 62:1458–1464. https://doi.org/10.1373/clinchem.2016.258632
32. Rothberg JM, Hinz W, Rearick TM et al (2011) An integrated semiconductor device enabling non-optical genome sequencing. Nature 475:348. https://doi.org/10.1038/nature10242

# Chapter 9

# A Transposon-Based Activation Tag System for Functional Genomics in Cereals: Detection of *Mping*-Based Activation Tag Mobilization in Wheat


Angela N. Plemmons, Amanda R. Askins, Sarah M. Mendoza, and C. Nathan Hancock



## Abstract

Mutagenesis using transposable element tags is an important resource for gene discovery in monocots. The *mPing* element has a high transposition rate in its original host rice and has shown potential as a transposon tagging system for other species. An activation tag derived from *mPing* called *mmPing20F* contains an enhancer sequence from the figwort mosaic virus promoter, allowing it to induce overexpression of nearby genes. Transgenic wheat containing both the *mmPing20F* construct and transposase expression constructs were developed. We provide an in-depth protocol for detecting *mPing20F* mobilization in wheat by PCR and gel electrophoresis.



**Key words** Transposable element, *mPing*, Activation tagging, Mutagenesis


## 1 Introduction

### *1.1 Insertional Mutagenesis*

Mutation analysis is the foundation upon which modern genetics and molecular biology are founded [1–4]. These changes to the genome can occur naturally due to DNA replication errors, sequence rearrangement, exposure to mutagens, or the insertion of sequences by bacteria (i.e., *Agrobacterium tumefaciens*), viruses, and transposable elements in the genome. This last category is referred to as insertional mutagenesis and has been widely used to intentionally produce tractable mutations because the inserted sequence acts as a tag to indicate the location of the alteration. The most notable insertional mutagenesis techniques in plants are T-DNA tagging and transposon tagging. In T-DNA tagging, large population of plants are transformed with *Agrobacterium tumefaciens,* which inserts the transfer DNA (T-DNA) portion of its tumor-inducing plasmid into the host plant's genome [5]. While this method has been highly successful in *Arabidopsis* [6–8], it has


Andriy Bilichak and John D. Laurie (eds.), *Accelerated Breeding of Cereal Crops*, Springer Protocols Handbooks, https://doi.org/10.1007/978-1-0716-1526-3_9, © Springer Science+Business Media, LLC, part of Springer Nature 2022

been harder to achieve sufficient populations of mutagenized plants in species that are more difficult to transform. In these situations, scientists have often relied on native or heterologous transposable elements as transposon tags [8–10]. To improve the effectiveness or identify specific genetic components, modifications have been made to these insertional sequence tags. The most common, an activation tag, includes an enhancer sequence to induce overexpression of nearby genes [7, 11–13]. The ability of activation tags to induce gain-of-function phenotypes is especially important in species with duplicated genomes, where simple gene knockouts rarely induce phenotypic changes.

### 1.2 Transposon Tagging in Monocots

Transposable elements are segments of DNA with the potential to mobilize or "jump" around the genome [14, 15]. Transposable elements are divided into two classes: (1) retrotransposons are mobilized by an RNA intermediate utilizing reverse transcriptase and integrase, and (2) DNA transposons are mobilized by a DNA intermediate and rely on the transposase enzyme to excise the DNA and reinsert it elsewhere. Since the discovery of transposable elements by Barbara McClintock [14], transposable elements have been an important aspect in monocot gene discovery. This is due to the fact that, before plant transformation was available, these natural insertional mutagens created tractable mutations. Since then a number of additional mutagenesis systems have been developed based on a number of elements. Below we review some of the major transposon tagging systems used in monocots.

#### 1.2.1 Ac/Ds

*Activator/Dissociator* (*Ac/Ds*) is a related set of DNA transposons that was first discovered in maize by Barbara McClintock [16, 17]. *Ac* is an autonomous element that encodes the transposase protein required for mobilization, while the *Ds* element is nonautonomous and is mobilized *in trans* by *Ac* transposase [18]. One of the distinguishing characteristics of this system is that these elements tend to jump to nearby, linked, regions of the chromosome [19–21]. Thus, these elements have often been used in a regional mutagenesis strategy [22]. In maize, this involves establishing a collection of lines with mapped *Ac* or *Ds* elements that can induce insertional mutations in the 2–3 centimorgan region near the element launchpads [22–25]. The use of an immobilized version of *Ac* allows for a two-component system, in which researchers can control the activity of *Ds* elements and prevent off-target insertions by *Ac* [26].

Since its initial characterization, the *Ac/Ds* elements have been used for gene identification in a broad range of monocot species. This requires transformation of constructs that contain both a mobile element (usually a *Ds* derivative) and the *Ac* transposase gene. For example, the rice genome has been mutagenized with a *Ds* element that carried a GUS screenable marker and a BAR selectable marker gene [27]. These markers along with a GFP reporter next to the initial *Ds* location allow for selection of plants with heritable transposition of the *Ds* element. In a related rice mutagenesis system, *Ds* elements were induced to mobilize by tissue culture treatment of F1 embryos [28, 29]. To improve gene discovery efficiency in rice, researchers have developed activation tagging [12, 30] and enhancer trapping *Ds* elements [31, 32]. In addition, *Ds* elements distributed across the rice genome have been mapped to use as launchpads for regional mutagenesis [33]. Similarly, *Ac/Ds*-based mutagenesis systems have been developed for barley [34–40], wheat [41, 42], and *Brachypodium distachyon* [43].

#### *1.2.2* Tos17

The *Tos17* element is a native retrotransposon discovered in the rice genome that was found to be activated by prolonged tissue culture treatment [44]. This element transposes by creating new copies; thus, about ten new insertions can be found in each plant produced [45]. One advantage of this system is its high frequency of genetic insertions, increasing the probability of creating phenotypes [45]. Using this element, large populations of mutagenized rice have been created and screened [45–47]. However, there are also disadvantages to this system. First, the large number of insertions and potential for other mutations to occur during tissue culture can make phenotyping challenging. Second, this system has not been conducive to producing activation or enhancer trapping tags. Finally, this system has not been successfully transferred to other plant species.

#### *1.2.3* mPing/Ping/Pong

*mPing* is a DNA transposon from rice that was initially noticed as the causative mutation for an unstable seed phenotype [48]. Sequencing of the rice genome uncovered the true nature of this element, with this nonautonomous miniature element being mobilized by the transposase protein from the closely related autonomous *Ping* or *Pong* elements [49–52]. Though widely distributed in wild and domesticated rice, *mPing* is only highly active in a limited group of japonica cultivars, where it can reach thousands of copies [53, 54]. In rice, *mPing* preferentially inserts near genes while avoiding the CG-rich exons [53, 55]. Recombinant inbred lines derived from a cultivar with thousands of copies of *mPing* have been used for genetic mapping [56], and analysis of the structural variations present in this and related populations has been used to determine the role of these miniature elements on genome evolution [55, 57].

The *mPing* element is mobilized in other species (yeast, *Arabidopsis*, and soybean), when the ORF1 and TPase proteins from either the *Pong* or *Ping* elements are expressed [50, 58, 59]. Prior efforts to use this element for transposon tagging in monocots are limited to *Setaria viridis* [60]. The generation of this population and the observations in *Arabidopsis* and soybean showed that *mPing* transposes to unlinked sites with a high percentage of insertions in promoter regions [50, 59, 60]. This insertion preference is ideal for activation tagging because almost half of the insertions should result in overexpression of nearby genes. *mPing* was developed into an activation tag by inserting figwort mosaic virus enhancer sequence [61] into a hyperactive version of *mPing* called *mmPing20* [62]. This *mPing*-based activation tag is called *mmPing20F* and is being applied to create an activation tagging library in wheat (this study).

### 1.3 Methods for Detecting Transposition

Development of a transposon tagging population requires both efficient mobilization of the element and identification of plants that have experienced transposition. Many different strategies have been utilized to detect transposition as reviewed below.

#### 1.3.1 Reporter Genes

For some DNA elements, excision of the element can be measured using a marker gene that is activated by repair of the excision site. For example, placement of elements between a promoter and GUS or GFP genes can produce a situation where expression of the reporter genes only occurs if the element is excised from the transgene [50, 63]. However, this method is element dependent, and leaky expression of the reporter genes or imprecise repair of the excision site can cause problems. A benefit of this technique is that it can indicate the specific tissues affected by transposition. If transposition occurs in somatic tissues, growth and cell division results in spotted or striped phenotypes [58, 64].

Alternatively, some tagging systems rely on the transposable element moving a selectable or screenable reporter gene to a new location [11, 12, 27]. This allows for identification of plants with mobilized elements but also may inadvertently select for unlinked insertions. This strategy is not acceptable for some elements as increasing element size has been shown to decrease the transposition frequency [62, 65]. It often relies on segregation of the T-DNA transgene, so it is primarily used in F2 populations, increasing the amount of labor needed to create and screen the populations.

#### 1.3.2 PCR Analysis

PCR analysis of plants is the most sensitive method for detecting mobilization of DNA elements as it can detect changes in a single cell. Using primers that flank the DNA element location (usually 100–300 bp away) produces different sized bands, depending on the position of the transposable element [50, 58, 59]. A larger band indicates that the transposable element is still in its original position, while a smaller band indicates that the element has excised and the excision site was repaired. This technique is especially useful for testing if transposition is occurring from a novel construct or when screening lines to determine which candidates exhibit high transposition. The sensitivity of this technique means that somatic excision that affects a portion of the cells results in two bands. Similarly, a plant that is heterozygous for the presence of the element in a location will produce two bands. A plant in which all copies of the element have excised from the original location will result in a lower band only. Thus, in a plant that is homozygous for the original transgene, detection of the lower band only indicates that both copies of the element have excised. However, not all elements that excise are reinserted into the genome [66], so this method is not a direct measurement of insertion. This method can also be labor-intensive and is not practical for extremely large populations.

Similarly, PCR can be used to identify transposable element insertion sites. This technique relies on using primers anchored to the transposable element sequence. The most common technique called TAIL-PCR uses degenerate primers to allow for amplification of the desired sequences [67]. Similarly, transposon display involves ligating adapter sequenced to fragmented DNA to allow for amplification of the desired sequences [68]. In both cases, nested PCR results in fragments with a small portion of the transposable element and some adjacent genomic DNA. Another technique, inverse PCR, relies on fragmenting the genomic DNA and ligation to produce circular products that can be amplified using primers pointing out from the element [69]. These techniques have been successful but, generally, are not able to capture all insertion sites.

#### *1.3.3 Sequencing*

As the costs of next-generation sequencing have gone down, a number of sequencing-based techniques for identifying transposon insertion sites have been established. Because scientists are only interested in the sequence that is adjacent to the transposable elements, efforts are made to enrich the libraries for the regions of interest. This can be done thorough hybridization capture [70] or PCR enrichment using element anchored primers [71, 72]. Because of the large numbers of sequences captured, these techniques tend to identify the majority of the insertions present in the genome. This method is especially critical in later stages of population development, where high-throughput characterization of many individuals is needed.

## 2 Materials

### *2.1 DNA Extraction*

CTAB buffer (2% CTAB, 1.42 M NaCl, 5 mM ascorbic acid, 2% PVP-40, 4.5 mM diethyldithiocarbamic acid (DIECA), 0.02 M EDTA, 0.1 M Tris pH 8).

Chloroform.

Isopropanol.

70% EtOH.

1xTE buffer.

Bead mill.

4.5 mm steel BBs.

2 ml screw cap tubes.

1.5 ml tubes.

### 2.2 Polymerase Chain Reaction

PCR tubes.
Filter tips.
PCR grade water.
2X Polymerase Master Mix.
Forward and reverse primers:

Detection of *mPing* excision.

*mPing* Flank 35S 2.0: AGGAAGGTGGCTCCTACAAATGCC,

GUS 3.0 Rev.: GGTCTCGGTGTTGATCGGGTACAG.

Detection of the *Pong transposase* gene.
*Pong* TPase For: ATGCAGAGTTTAGCCATCTCTCTACT.

*Pong* TPase Rev.: ATTTCCAGTTCTATGTGCAGCACCA.

## 3 Methods

A wheat *mPing20F* activation tagging population will be used below to demonstrate how PCR analysis can be used for analysis of transposon tagging populations. The plants being analyzed were created by transforming two constructs (Fig. 1) separately into spring wheat [genotype CB037, University of Nebraska Plant Transformation Core Research Facility] by *Agrobacterium*-mediated transformation [73]. Plants shown to contain the desired constructs by PCR were used in crosses to create F1 plants. PCR analysis will be used to confirm the presence of the transgenes and detect if transposition is occurring.

### 3.1 DNA Purification

1. Place two BBs and a 2.5 cm piece of leaf tissue into a labeled 2 ml screw cap tube.
2. Add 400μl of CTAB buffer.
3. Place the tubes in a bead mill and grind them for 1.5 min at max speed (*see* **Note 1**).
4. Incubate the tubes in a 55 °C water bath for 30 min.

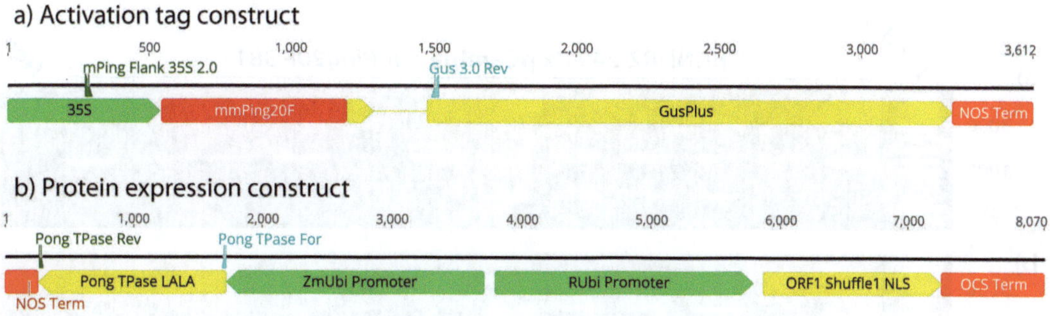


**Fig. 1** Maps of the activation tag and the protein expression constructs. The activation tag construct (**a**) has *mmPing20F* inserted between the *35S* promoter and *GusPlus* gene. The protein expression (**b**) construct has ORF1 Shuffle1 NLS [74] and *Pong* TPase L418A, L420A [58] genes expressed by the rice and maize ubiquitin promoters, respectively. The location of the primers used in this study are shown as wedges. Image was made using Geneious version 2020.0 created by Biomatters

5. Add 400μl of chloroform to each tube and mix by inverting several times.
6. Centrifuge the tubes at 9.3 × *g* for 5 min.
7. Transfer the supernatant to a new labeled tube. Be careful not to transfer any of the precipitated material or chloroform.
8. Add 300μl of isopropanol and mix by inverting several times.
9. Centrifuge for 10 min at 16.1 × *g*.
10. Pour off the supernatant and add 400μl of 70% EtOH.
11. Centrifuge for 5 min at 16.1 × *g*.
12. Carefully remove the supernatant, and air-dry the pellet in the inverted tube for 15 min.
13. Resuspend the pellet in 50μl of 1x TE.
14. Measure the genomic DNA concentration using the Qubit dsDNA broad range quantification kit or similar method.
15. Dilute the genomic DNA to a final concentration of 10 ng/μl in a total of 50 μl PCR grade water.

### 3.2 PCR Analysis

1. Label the necessary PCR tubes, including a negative control.
2. For each sample combine the following using filter tips to make a PCR mix:
   - 4 μl PCR grade purified water
   - 0.2μl 10μM forward primer
   - 0.2μl 10μM reverse primer
   - 5μl 2x Polymerase Master Mix
3. Using filter tips, transfer 9.4μl of the PCR mix into each labeled tube.

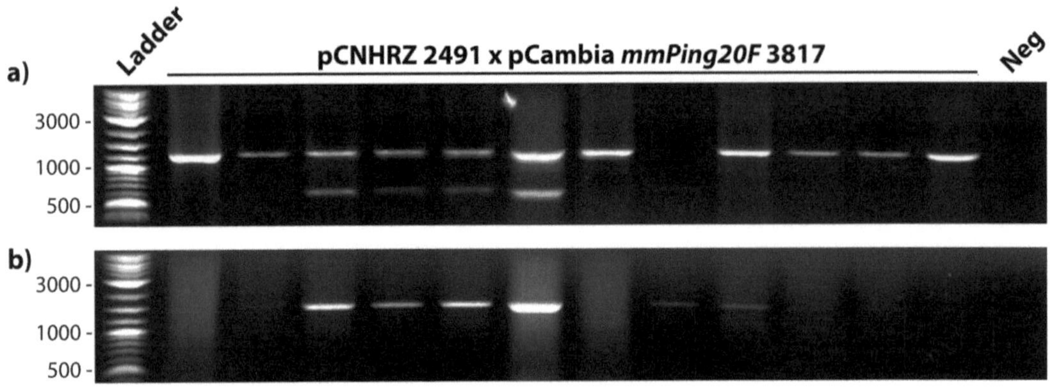


**Fig. 2** PCR analysis of F1 progeny. Using primers that flank the excision site (**a**), we can detect excision of the *mmPing20F* element from the transgene. The upper band (1244 bp) indicates the presence of *mmPing20F* in the construct, while the lower band (589 bp) indicates excision has occurred. Primers that detect the *Pong* TPase gene (**b**) produces a 1446 bp band and indicates that transposition is only occurring in plants that have the protein expression construct (see **Note 3**). Ladder, 1Kb Plus (NEB); Neg, negative control (water)

4. Using filter tips, pipette 0.6 μl of the diluted genomic DNA into the corresponding tube, and use 0.6 μl of water for the negative control.
5. Briefly vortex the tubes to mix and then centrifuge for 5 s.
6. Use a thermocycler to perform 30 cycles of amplification according to the expected product size and primer Tm (*see* **Note 2**).
7. Perform gel electrophoresis on the PCR products using a 1% agarose gel, and capture image (see Fig. 2).

## 4 Notes

1. We utilize a BeadBug™ microtube homogenizer at 4000 rpm. Leaf tissue should be ground into very small particles. Young tissue is easier to get complete grinding.
2. The amplification conditions used for the example experiment were as follows: 95 °C for 4 min; 30 cycles of 95 °C for 30 s, 58 °C for 30 s, and 72 °C for 1.5 min; 72 °C for 7 min; and 12 °C hold.
3. The PCR analysis of these F1 plants provide a lot of helpful information. First, identification of the presence of the two transgenes allows us to identify which progeny inherited both transgenes. We can also deduce the inheritance pattern of the constructs. For example, the PCR results for these plants suggest that both parents are not homozygous for the transgenes, as not all the progeny inherited it. The observation that the

protein expression construct was only inherited in 50% of the F1 plants suggests that some self-pollination occurred. Importantly, the presence of the lower band when using primers that flank the *mPing20F* insertion site indicated that mobilization of *mmPing20F* was occurring in four of the plants. As expected, transposition only occurred in the presence of the protein expression construct. However, this success rate was not observed for all combinations of transgenic events tested in this study (data not shown), suggesting that some transgenic events had poor protein expression. Thus, this is an effective method for identifying which lines are capable of transposition when establishing a transposon tagging population.

Despite the success of this technique, there are limitations on what can be determined by excision site analysis. While it does indicate that excision occurred in this tissue, it cannot confirm if the element inserted into a new location in the genome. In general, you can expect that excision of a DNA element will result in insertion, but the insertion frequency is not 100%, depending on the element being used [66]. Secondly, detection of excision in somatic tissue does not ensure that heritable mutations will be present in the next generation. To achieve heritable changes, transposition must occur in cells that become egg or pollen cells. Thus, it is critical to observe multiple generations to determine if plants that only contain the lower band (element excised) can be identified. Analysis of the 54 F2 plants derived from the four lines with transposition activity identified above showed that (1) both constructs were single-copy insertions inherited in a 3:1 ratio and (2) transposition was detected by PCR in 89% of plants that inherited both constructs. However, none of these F2 plants showed evidence of heritable transposition of the *mmPing20F* element in the F1 generation. Efforts to create heritable events in this population will include screening of additional generations and passage through tissue culture.

## Acknowledgments

Special thanks to the Thomas Clemente Lab (University of Nebraska-Lincoln) for the production of the transgenic wheat plants and Priscilla Redd for careful editing of the manuscript. This work was supported by NSF Grant #1444581, a Magellan Scholar Award (University of South Carolina) to ARA, and a USCA Connections Award to AP.

## References


1. Strauss BS (2016) Biochemical genetics and molecular biology: the contributions of George Beadle and Edward Tatum. Genetics 203(1):13
2. Beadle GW, Tatum EL (1941) Genetic control of biochemical reactions in *Neurospora*. Proc Natl Acad Sci USA 27(11):499
3. Morgan TH (1911) The origin of nine wing mutations in *Drosophila*. Science 33 (848):496–499
4. Shine I, Wrobel S (2014) Thomas Hunt Morgan: pioneer of genetics. University Press of Kentucky
5. Gelvin SB (2003) *Agrobacterium*-mediated plant transformation: the biology behind the "gene-jockeying" tool. Microbiol Mol Biol Rev 67(1):16–37
6. Alonso JM, Stepanova AN, Leisse TJ, Kim CJ, Chen HM, Shinn P, Stevenson DK, Zimmerman J, Barajas P, Cheuk R, Gadrinab C, Heller C, Jeske A, Koesema E, Meyers CC, Parker H, Prednis L, Ansari Y, Choy N, Deen H, Geralt M, Hazari N, Hom E, Karnes M, Mulholland C, Ndubaku R, Schmidt I, Guzman P, Aguilar-Henonin L, Schmid M, Weigel D, Carter DE, Marchand T, Risseeuw E, Brogden D, Zeko A, Crosby WL, Berry CC, Ecker JR (2003) Genome-wide insertional mutagenesis of *Arabidopsis thaliana*. Science 301(5633):653–657
7. Weigel D, Ahn JH, Blazquez MA, Borevitz JO, Christensen SK, Fankhauser C, Ferrandiz C, Kardailsky I, Malancharuvil EJ, Neff MM, Nguyen JT, Sato S, Wang ZY, Xia YJ, Dixon RA, Harrison MJ, Lamb CJ, Yanofsky MF, Chory J (2000) Activation tagging in *Arabidopsis*. Plant Physiol 122(4):1003–1013
8. Walbot V (1992) Strategies for mutagenesis and gene cloning using transposon tagging and T-DNA insertional mutagenesis. Annu Rev Plant Physiol Plant Mol Biol 43:49–82
9. Haring MA, Rommens CM, Nijkamp HJJ, Hille J (1991) The use of transgenic plants to understand transposition mechanisms and to develop transposon tagging strategies. Plant Mol Biol 16(3):449–461
10. Conrad LJ, Kikuchi K, Brutnell TP (2008) Transposon tagging in cereal crops. The handbook of plant functional genomics: concepts and protocols:267–290
11. Mathieu M, Winters E, Kong FM, Wan JR, Wang SX, Eckert H, Luth D, Paz M, Donovan C, Zhang ZY, Somers D, Wang K, Nguyen H, Shoemaker R, Stacey G, Clemente T (2009) Establishment of a soybean (*Glycine max* Merr. L) transposon-based mutagenesis repository. Planta 229(2):279–289. https://doi.org/10.1007/s00425-008-0827-9
12. Qu S, Desai A, Wing R, Sundaresan V (2008) A versatile transposon-based activation tag vector system for functional genomics in cereals and other monocot plants. Plant Physiol 146 (1):189–199. https://doi.org/10.1104/pp.107.111427
13. Suzuki Y, Uemura S, Saito Y, Murofushi O, Schmitz G, Theres K, Yamaguchi I (2001) A novel transposon tagging element for obtaining gain-of-function mutants based on a self-stabilizing *Ac* derivative. Plant Mol Biol 45 (2):123–131
14. McClintock B (1950) The origin and behavior of mutable loci in maize. Proc Natl Acad Sci USA 36(6):344–355
15. Craig NL (2020) Mobile DNA III. John Wiley & Sons
16. McClintock B (1945) Cytogenetic studies of maize and *Neurospora*. Carnegie Institution of Washington, p 44
17. McClintock B (1947) Mutable loci in maize. Carnegie Institution of Washington, p 47
18. Fedoroff N, Wessler S, Shure M (1983) Isolation of the transposable maize controlling elements *Ac* and *Ds*. Cell 35(1):235–242
19. Orton ER, Brink RA (1966) Reconstitution of the variegated pericarp allele in maize by transposition of *Modulator* back to the P locus. Genetics 53(1):7
20. Dooner HK, Belachew A (1989) Transposition pattern of the maize element *Ac* from the *bz-m2* (*Ac*) allele. Genetics 122(2):447–457
21. Greenblatt IM (1984) A chromosome replication pattern deduced from pericarp phenotypes resulting from movements of the transposable element, *Modulator*, in maize. Genetics 108 (2):471–485
22. Ahern KR, Deewatthanawong P, Schares J, Muszynski M, Weeks R, Vollbrecht E, Duvick J, Brendel VP, Brutnell TP (2009) Regional mutagenesis using *Dissociation* in maize. Methods 49(3):248–254
23. Vollbrecht E, Duvick J, Schares JP, Ahern KR, Deewatthanawong P, Xu L, Conrad LJ, Kikuchi K, Kubinec TA, Hall BD (2010) Genome-wide distribution of transposed *Dissociation* elements in maize. Plant Cell 22 (6):1667–1685
24. Dooner HK, Belachew A, Burgess D, Harding S, Ralston M, Ralston E (1994) Distribution of unlinked receptor sites for

transposed *Ac* elements from the *bz-m2* (*Ac*) allele in maize. Genetics 136(1):261–279

25. Kolkman JM, Conrad LJ, Farmer PR, Hardeman K, Ahern KR, Lewis PE, Sawers RJ, Lebejko S, Chomet P, Brutnell TP (2005) Distribution of *Activator* (*Ac*) throughout the maize genome for use in regional mutagenesis. Genetics 169(2):981–995
26. Conrad LJ, Brutnell TP (2005) *Ac*-immobilized, a stable source of *Activator* transposase that mediates sporophytic and gametophytic excision of *Dissociation* elements in maize. Genetics 171(4):1999–2012
27. Kolesnik T, Szeverenyi I, Bachmann D, Kumar CS, Jiang S, Ramamoorthy R, Cai M, Ma ZG, Sundaresan V, Ramachandran S (2004) Establishing an efficient *Ac/Ds* tagging system in rice: large-scale analysis of *Ds* flanking sequences. Plant J 37(2):301–314
28. Kim CM, Piao HL, Park SJ, Chon NS, Je BI, Sun B, Park SH, Park JY, Lee EJ, Kim MJ (2004) Rapid, large-scale generation of *Ds* transposant lines and analysis of the *Ds* insertion sites in rice. Plant J 39(2):252–263
29. Park SH, Jun NS, Kim CM, Oh TY, Huang J, Y-h X, Park SJ, Je BI, Piao HL, Park SH (2007) Analysis of gene-trap *Ds* rice populations in Korea. Plant Mol Biol 65(4):373–384
30. Kota VR, Gundra SR, Vudem DR, Pulugurtha BK, Khareedu VR (2020) Development of a large population of activation-tagged mutants in an elite indica rice variety. Plant Breed 139 (2):328–343
31. Upadhyaya NM, Zhou X-R, Zhu Q-H, Ramm K, Wu L, Eamens A, Sivakumar R, Kato T, Yun D-W, Santhoshkumar C (2002) An *iAc/Ds* gene and enhancer trapping system for insertional mutagenesis in rice. Funct Plant Biol 29(5):547–559
32. Ito Y, Eiguchi M, Kurata N (2004) Establishment of an enhancer trap system with *Ds* and *GUS* for functional genomics in rice. Mol Gen Genomics 271(6):639–650
33. Upadhyaya NM, Zhu Q-H, Zhou X-R, Eamens AL, Hoque MS, Ramm K, Shivakkumar R, Smith KF, Pan S-T, Li S (2006) *Dissociation* (*Ds*) constructs, mapped *Ds* launch pads and a transiently-expressed transposase system suitable for localized insertional mutagenesis in rice. Theor Appl Genet 112(7):1326–1341
34. Ayliffe MA, Pallotta M, Langridge P, Pryor AJ (2007) A barley activation tagging system. Plant Mol Biol 64(3):329–347
35. Ayliffe MA, Pryor AJ (2011) Activation tagging and insertional mutagenesis in barley. In: Plant reverse genetics. Springer, pp 107–128
36. Randhawa HS, Singh J, Lemaux PG, Gill KS (2009) Mapping barley *Ds* insertions using wheat deletion lines reveals high insertion frequencies in gene-rich regions with high to moderate recombination rates. Genome 52 (6):566–575
37. Zhao T, Palotta M, Langridge P, Prasad M, Graner A, Schulze-Lefert P, Koprek T (2006) Mapped *Ds*/T-DNA launch pads for functional genomics in barley. Plant J 47(5):811–826
38. Lazarow K, Lütticke S (2009) An *Ac/Ds*-mediated gene trap system for functional genomics in barley. BMC Genomics 10(1):55
39. Singh S, Singh J (2017) Sequence-based genetic mapping of *Ds*-tagged insertions to characterize malting-related traits in barley. Crop J 5(1):11–20
40. Bregitzer P, Brown RH, Dahleen LS, Singh J, Cooper LD, Hayes PM, Lemaux PG (2019) Registration of the barley transposon-tagged population II: sixty-one lines each characterized for *Ds* insertion sites. J Plant Regist 13 (1):109–112
41. Takumi S, Murai K, Mori N, Nakamura C (1999) Trans-activation of a maize *Ds* transposable element in transgenic wheat plants expressing the *Ac* transposase gene. Theor Appl Genet 98(6–7):947–953
42. Pastori GM, Huttly A, West J, Sparks C, Pieters A, Luna CM, Jones HD, Foyer CH (2007) The maize *Activator/Dissociation* system is functional in hexaploid wheat through successive generations. Funct Plant Biol 34 (9):835–843
43. Wu H, Xue X, Qin C, Xu Y, Guo Y, Li X, Lv W, Li Q, Mao C, Li L (2019) An efficient system for *Ds* transposon tagging in *Brachypodium distachyon*. Plant Physiol 180(1):56–65
44. Hirochika H, Sugimoto K, Otsuki Y, Tsugawa H, Kanda M (1996) Retrotransposons of rice involved in mutations induced by tissue culture. Proc Natl Acad Sci 93 (15):7783–7788
45. Miyao A, Tanaka K, Murata K, Sawaki H, Takeda S, Abe K, Shinozuka Y, Onosato K, Hirochika H (2003) Target site specificity of the *Tos17* retrotransposon shows a preference for insertion within genes and against insertion in retrotransposon-rich regions of the genome. Plant Cell 15(8):1771–1780
46. Miyao A, Iwasaki Y, Kitano H, Itoh J-I, Maekawa M, Murata K, Yatou O, Nagato Y, Hirochika H (2007) A large-scale collection of phenotypic data describing an insertional mutant population to facilitate functional analysis of rice genes. Plant Mol Biol 63 (5):625–635

47. Piffanelli P, Droc G, Mieulet D, Lanau N, Bès M, Bourgeois E, Rouvière C, Gavory F, Cruaud C, Ghesquière A (2007) Large-scale characterization of *Tos17* insertion sites in a rice T-DNA mutant library. Plant Mol Biol 65 (5):587–601
48. Teraishi M, Okumoto Y, Hirochika H, Horibata A, Yamagata H, Tanisaka T (1999) Identification of a mutable slender glume gene in rice (*Oryza sativa* L.). Mol Gen Genet MGG 261(3):487–494
49. Jiang N, Bao ZR, Zhang XY, Hirochika H, Eddy SR, McCouch SR, Wessler SR (2003) An active DNA transposon family in rice. Nature 421(6919):163–167
50. Yang GJ, Zhang F, Hancock CN, Wessler SR (2007) Transposition of the rice miniature inverted repeat transposable element *mPing* in *Arabidopsis thaliana*. Proc Natl Acad Sci USA 104(26):10962–10967
51. Kikuchi K, Terauchi K, Wada M, Hirano HY (2003) The plant MITE *mPing* is mobilized in anther culture. Nature 421(6919):167–170
52. Nakazaki T, Okumoto Y, Horibata A, Yamahira S, Teraishi M, Nishida H, Inoue H, Tanisaka T (2003) Mobilization of a transposon in the rice genome. Nature 421 (6919):170–172
53. Naito K, Cho E, Yang GJ, Campbell MA, Yano K, Okumoto Y, Tanisaka T, Wessler SR (2006) Dramatic amplification of a rice transposable element during recent domestication. Proc Natl Acad Sci USA 103 (47):17620–17625
54. Chen J, Lu L, Benjamin J, Diaz S, Hancock CN, Stajich JE, Wessler SR (2019) Tracking the origin of two genetic components associated with transposable element bursts in domesticated rice. Nat Commun 10(1):1–10
55. Naito K, Zhang F, Tsukiyama T, Saito H, Hancock CN, Richardson AO, Okumoto Y, Tanisaka T, Wessler SR (2009) Unexpected consequences of a sudden and massive transposon amplification on rice gene expression. Nature 461(7267):1130–1134. https://doi.org/10.1038/Nature08479
56. Monden Y, Naito K, Okumoto Y, Saito H, Oki N, Tsukiyama T, Ideta O, Nakazaki T, Wessler SR, Tanisaka T (2009) High potential of a transposon *mPing* as a marker system in japonica× japonica cross in rice. DNA Res 16 (2):131–140
57. Chen J, Lu L, Robb SM, Collin M, Okumoto Y, Stajich JE, Wessler SR (2020) Genomic diversity generated by a transposable element burst in a rice recombinant inbred population. bioRxiv 117:26288
58. Hancock CN, Zhang F, Wessler S (2010) Transposition of the *Tourist*-MITE *mPing* in yeast: an assay that retains key features of catalysis by the Class 2 *PIF/Harbinger* superfamily. Mob DNA 1(5):5
59. Hancock CN, Zhang F, Floyd K, Richardson AO, LaFayette P, Tucker D, Wessler SR, Parrott WA (2011) The rice miniature inverted repeat transposable element *mPing* is an effective insertional mutagen in soybean. Plant Physiol 157(2):552–562. https://doi.org/10.1104/pp.111.181206
60. Kikuchi K, Shyu C, Brutnell TP (2017) Transposon tagging in *Setaria viridis*. In: Genetics and genomics of *Setaria*. Springer, pp 323–342
61. Maiti IB, Gowda S, Kiernan J, Ghosh SK, Shepherd RJ (1997) Promoter/leader deletion analysis and plant expression vectors with the figwort mosaic virus (FMV) full length transcript (FLt) promoter containing single or double enhancer domains. Transgenic Res 6 (2):143–156
62. Johnson A, Mcassey E, Diaz S, Reagin J, Redd P, Parrilla DR, Nguyen H, Stec A, McDaniel LA, Clemente TE, Stupar RM, Parrott WA, Hancock CN (2021) Development of *mPing*-based activation tags for crop insertional mutagenesis. Plant direct 5(1):e00300
63. McElroy D, Louwerse JD, McElroy SM, Lemaux PG (1997) Development of a simple transient assay for *Ac/Ds* activity in cells of intact barley tissue. Plant J 11(1):157–165
64. Scheres B, Wolkenfelt H, Willemsen V, Terlouw M, Lawson E, Dean C, Weisbeek P (1994) Embryonic origin of the *Arabidopsis* primary root and root meristem initials. Development 120(9):2475–2487
65. Way J, Kleckner N (1985) Transposition of plasmid-borne *Tn10* elements does not exhibit simple length-dependence. Genetics 111 (4):705–713
66. Lazarow K, Du M-L, Weimer R, Kunze R (2012) A hyperactive transposase of the maize transposable element *Activator* (*Ac*). Genetics 191(3):747–756
67. Singer T, Burke E (2003) High-throughput TAIL-PCR as a tool to identify DNA flanking insertions. In: Plant functional genomics. Springer, pp 241–271
68. Van den Broeck D, Maes T, Sauer M, Zethof J, De Keukeleire P, D'Hauw M, Van Montagu M, Gerats T (1998) Transposon display identifies individual transposable elements in high copy number lines. Plant J 13(1):121–129
69. Ochman H, Ajioka JW, Garza D, Hartl DL (1989) Inverse polymerase chain reaction. In: PCR technology. Springer, pp 105–111

70. Williams-Carrier R, Stiffler N, Belcher S, Kroeger T, Stern DB, Monde R-A, Coalter R, Barkan A (2010) Use of Illumina sequencing to identify transposon insertions underlying mutant phenotypes in high-copy *Mutator* lines of maize. Plant J 63(1):167–177
71. Kanizay LB, Jacobs TB, Gillespie K, Newsome JA, Spaid BN, Parrott WA (2015) HtStuf: high-throughput sequencing to locate unknown DNA junction fragments. Plant Genome 8(1):1–10
72. McCarty DR, Mark Settles A, Suzuki M, Tan BC, Latshaw S, Porch T, Robin K, Baier J, Avigne W, Lai J (2005) Steady-state transposon mutagenesis in inbred maize. Plant J 44 (1):52–61
73. Clemente T, Mitra A (2004) Genetic engineering of wheat: protocols and use to enhance stress tolerance. Genetically modified crops-their development, uses and risks:131–163
74. Payero L, Outten G, Burckhalter C, Hancock CN (2016) Alteration of the *Ping* and *Pong* ORF1 proteins allows for hyperactive transposition of *mPing*. J S C Acad Sci 14(2):3

# Chapter 10

# A Modular Cloning Toolkit for Genome Editing in Cereals

**Florian Hahn, Laura Sanjurjo Loures, and Vladimir Nekrasov**


## Abstract

The CRISPR/Cas technology has recently become the tool of choice for gene editing applications in plants. Here we describe the method of assembling CRISPR/Cas DNA constructs, using the modular cloning (MoClo), for deployment in cereal species. For this purpose, we use a cloning toolkit that consists of a library of Golden Gate (GG) modules, which encode genetic elements, such as promoters, coding sequences of CRISPR/Cas nucleases (e.g. Cas9), terminators, guide RNA backbones, etc. A part of the toolkit are modules that allow assembly of tRNA-sgRNA polycistronic units without a PCR step involved. The tRNA-sgRNA system has proved to be efficient in a number of plant species, including cereals (rice, wheat). The cloning toolkit we present can be used for a variety of gene editing applications, including gene deletions and single nucleotide substitutions. It can also be easily expanded by adding new standardised parts.


**Key words** CRISPR, Cas9, Plant, Genome editing, Golden Gate, MoClo, Cereal, Wheat

## 1 Introduction

Over the last 7 years, the CRISPR/Cas technology has been widely adopted by the plant research community as a versatile tool for genome editing applications. As the genome editing field is highly dynamic, new molecular tools (e.g. sequence-specific nucleases) become regularly available. In order to be able to easily apply such new tools, one needs a cloning system where genetic elements, such as promoters, coding sequences (CDSs) or terminators, can be readily interchanged. The modular cloning (MoClo) system is based on Golden Gate (GG) cloning and makes possible hierarchical assembly of multigene constructs using standard cloning parts [1]. The MoClo system has been successfully used for CRISPR/Cas-based genome editing applications in plants [2–4].

Here we describe a detailed procedure for assembling CRISPR/Cas constructs for gene editing in cereals based on the modular cloning toolkit by Hahn et al. [5]. In particular, we present strategies for expressing guide RNAs both under individual

Andriy Bilichak and John D. Laurie (eds.), *Accelerated Breeding of Cereal Crops*, Springer Protocols Handbooks, https://doi.org/10.1007/978-1-0716-1526-3_10, © Springer Science+Business Media, LLC, part of Springer Nature 2022

RNA polymerase III (Pol III) promoters and as part of a polycistronic tRNA-sgRNA transcript.

## 2 Materials

### 2.1 Cut-Ligations

GG cloning modules and recipient vectors are available from Addgene (www.addgene.org) as separate plasmids (figshare: https://doi.org/10.6084/m9.figshare.12998861) or whole toolkits (catalogue numbers #1000000159 and #1000000044). DNA oligos can be ordered from Merck. Required enzymes include T4 DNA ligase (New England Biolabs), BpiI (Thermo Fisher Scientific) and BsaI-HFv2 (New England Biolabs). Cut-ligation reactions are supplemented with BSA (New England Biolabs).

### 2.2 E. coli Transformation

Cut-ligations are transformed into One Shot™ TOP10 chemically competent *E. coli* (Thermo Fisher Scientific) according to the supplier's instructions. *E. coli* cells are plated onto LB (10 g/L tryptone, 5 g/L yeast extract, 5 g/L NaCl and 15 g/L agar) agar supplemented with one of the following antibiotics: 100μg/mL spectinomycin (Sigma-Aldrich), 100μg/mL carbenicillin (Sigma-Aldrich) or 50μg/mL kanamycin (Sigma-Aldrich). For the blue/white screening, agar is supplemented with 40μg/mL X-Gal (Promega) and 100μM IPTG (Promega).

### 2.3 Plasmid Extraction and Verification

Plasmid DNA is extracted from *E. coli* overnight cultures (grown in the LB medium supplemented with respective antibiotics) using the QIAprep Spin Miniprep Kit (Qiagen). DNA constructs are verified by restriction digest and Sanger sequencing (Eurofins Genomics).

## 3 Methods

### 3.1 General Golden Gate Cut-Ligation Reaction Protocol

1. Set up GG cut-ligations using the following recipe:

| | |
|---|---|
| 1μL | Vector backbone (adjusted to 100 ng/μL) |
| 1μL | Each assembly piece (vectors, adjusted to 100 ng/μL; annealed primers, 10μM each) |
| 1.5μL | 10x T4 DNA Ligase buffer (NEB catalogue # B0202S) |
| 1.5μL | BSA (50μg/mL; diluted 1:20 in ddH$_2$O from the 1 mg/mL stock, NEB catalogue # B9000S) |
| 1μL | BpiI (for assembly of level 0 and level 2 constructs; 10 U/μL, Thermo Fisher Scientific catalogue # ER1011) **or** BsaI-HFv2 (for assembly of level 1 constructs; 20 U/μL, NEB catalogue # R3733S) |

(continued)

| | |
|---|---|
| 1μL | T4 DNA Ligase (400,000 U/μL, NEB catalogue # M0202S) |
| to 15μL | $ddH_2O$ |

2. Perform GG cut-ligation reaction in a thermocycler using the following program :

| | | |
|---|---|---|
| 37 °C | | 20 s |
| 37 °C<br>16 °C | × 50 | 3 min<br>4 min |
| 50 °C | | 5 min |
| 80 °C | | 5 min |
| 16 °C | | Store |

3. Transform 2μL of the ligation reaction into 10μL of competent *E. coli* cells, and plate on the selection media (supplemented with X-Gal and IPTG in the case of the blue/white screen).
4. The following day, pick white colonies and set up overnight cultures.
5. Extract plasmid DNA using the miniprep kit, and verify constructs by restriction digest and Sanger sequencing.

### 3.2 Assembly of CRISPR/Cas Constructs with sgRNAs/crRNAs Expressed under Individual Pol III Promoters

A timeline for assembly of a generic level 2 CRISPR/Cas construct can be found in Fig. 1. Below we describe a detailed procedure for assembling an L2 construct containing a selectable marker cassette, a transcriptional unit for a CRISPR/Cas nuclease and transcriptional unit(s) for the corresponding sgRNA/crRNA (Figs. 2 and 3). For a general overview of how GG cloning works, please refer to Marillonnet and Werner (2020) [1].

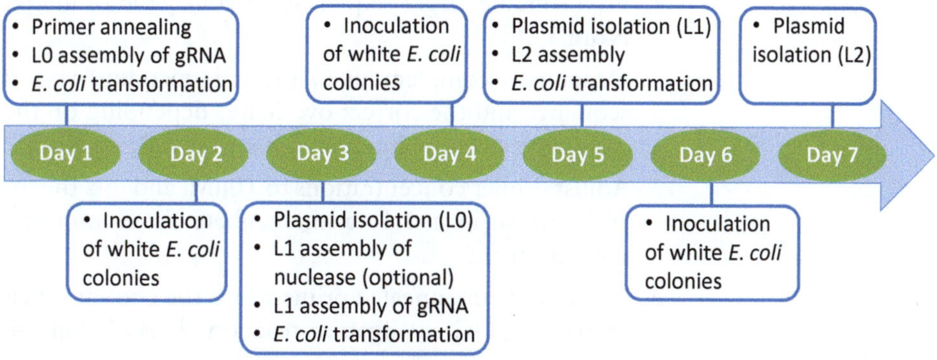


**Fig. 1** A timeline for assembly of a level 2 CRISPR/Cas construct. (Adapted from Hahn et al. (2020) [5])

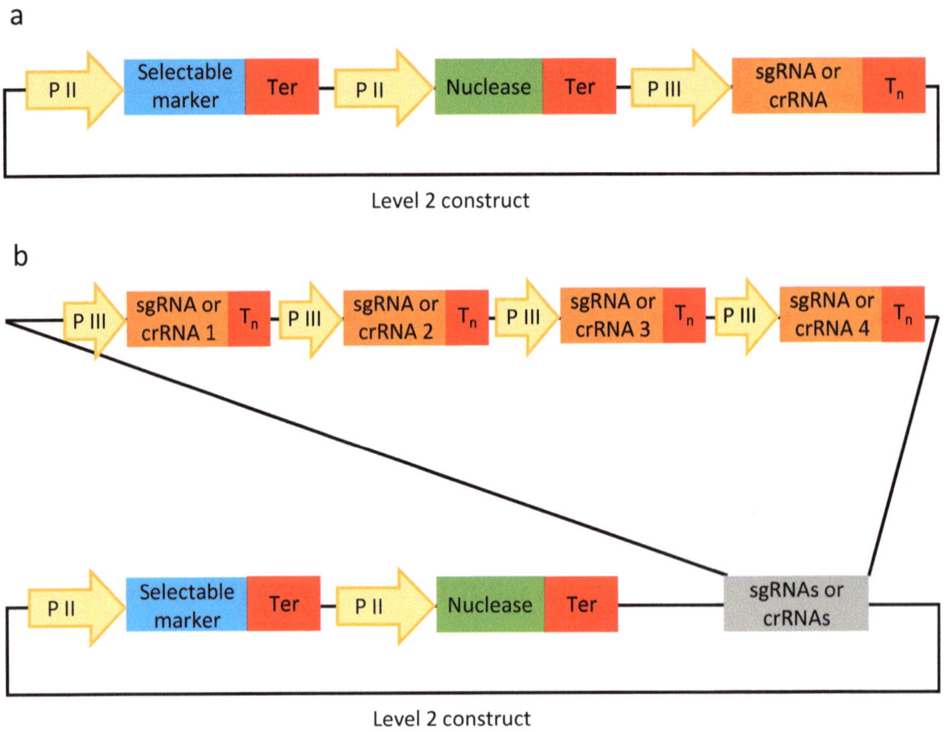


**Fig. 2** Cartoons of level 2 CRISPR/Cas constructs for targeted mutagenesis using guide RNAs expressed under individual Pol III promoters (adapted from Hahn et al. (2020) [5]). (**a**) represents a construct with one and (**b**) with four guide RNAs. P II and P III are Pol II and Pol III promoters, respectively; Ter and $T_n$ are Pol II and Pol III terminators, respectively

#### *3.2.1 Subcloning of the Guide Sequence into the Guide RNA Backbone Acceptor Vector*

1. Choose the correct guide RNA backbone vector from Table 1 for insertion of a guide sequence, which matches the CRISPR/Cas target in the cereal genome of interest. The choice would depend on the Pol III promoter (*TaU3*p, *OsU6-2p* or *OsU3p*) and the CRISPR/Cas nuclease one intends to use.
   1. Find suitable target sequences in your gene of interest (*see* **Note 1**).
   2. Order two complementary primers encoding your target sequence and the correct overhangs depending on the subcloning vector one intends to use (Table 2; *see* **Note 2**).
   3. Adjust primer concentrations to 10μM, and mix the forward and reverse in a 1:1 ratio; incubate for 5 min at room temperature.
   4. Integrate your annealed primers into the chosen subcloning vector in a GG cut-ligation reaction via BpiI (Figs. 4a and 5a).

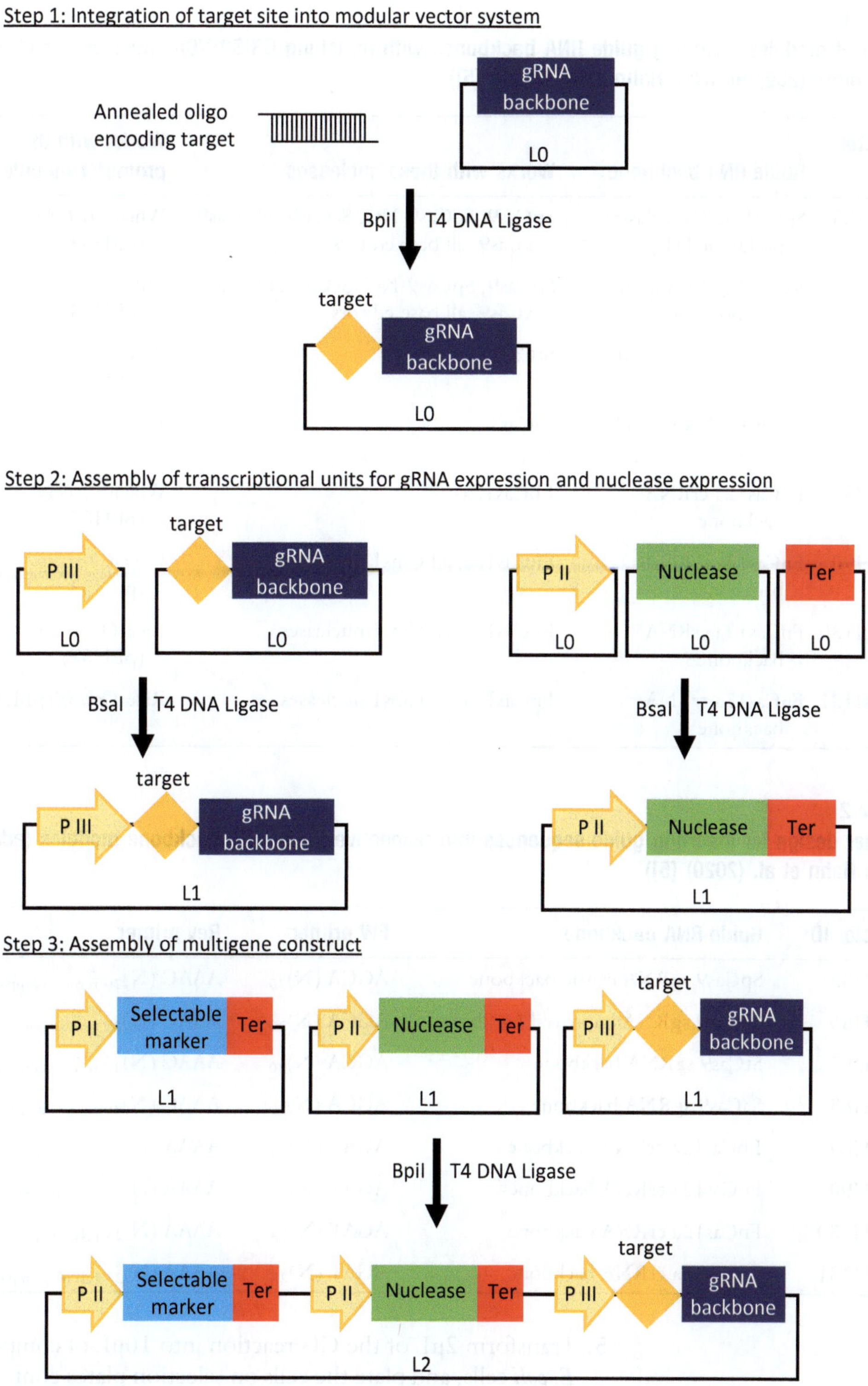


**Fig. 3** A flow chart diagram illustrating assembly of level 2 constructs with gRNAs expressed under individual Pol III promoters (adapted from Hahn et al. (2020) [5]). P II and P III are Pol II and Pol III promoters, respectively; Ter is a Pol II terminator

**Table 1**
**A list of modules encoding guide RNA backbones with matching CRISPR/Cas nucleases and Pol III promoters (adapted from Hahn et al. (2020) [5])**

| Vector ID | Guide RNA backbone | Works with these nucleases | Works with this promoter module |
|---|---|---|---|
| pFH85 | SpCas9 sgRNA classic backbone [10] | SpCas9, SpCas9-NG, ScCas9, nScCas9, xCas9, all base editors | Wheat *TaU3p* (pFH33) |
| pFH86 | SpCas9 sgRNA improved backbone [9] | SpCas9, SpCas9-NG, ScCas9, nScCas9, xCas9, all base editors | Wheat *TaU3p* (pFH33) |
| pFH87 | StCas9 sgRNA backbone | StCas9 | Wheat *TaU3p* (pFH33) |
| pFH88 | SaCas9 sgRNA backbone | SaCas9 | Wheat *TaU3p* (pFH33) |
| pFH89 | LbCas12a crRNA backbone | LbCas12a | Wheat *TaU3p* (pFH33) |
| pFH90 | FnCas12a crRNA backbone | FnCas12a, all Cms1 nucleases | Wheat *TaU3p* (pFH33) |
| pFH130 | FnCas12a crRNA backbone | FnCas12a, all Cms1 nucleases | Rice *OsU6-2p* (pFH37) |
| pFH131 | FnCas12a crRNA backbone | FnCas12a, all Cms1 nucleases | Rice *OsU3p* (pFH39) |

**Table 2**
**Primer design for inserting guide sequences into respective guide RNA backbone modules (adapted from Hahn et al. (2020) [5])**

| Vector ID | Guide RNA backbone | FW primer | Rev primer |
|---|---|---|---|
| pFH85 | SpCas9 sgRNA classic backbone | AGCA $(N)_{20}$ | AAAC $(N)_{20\ \text{reverse complement}}$ |
| pFH86 | SpCas9 sgRNA improved backbone | AGCA $(N)_{20}$ | AAAC $(N)_{20\ \text{reverse complement}}$ |
| pFH87 | StCas9 sgRNA backbone | AGCA $(N)_{20}$ | AAAC $(N)_{20\ \text{reverse complement}}$ |
| pFH88 | SaCas9 sgRNA backbone | AGCA $(N)_{20}$ | AAAC $(N)_{20\ \text{reverse complement}}$ |
| pFH89 | LbCas12a crRNA backbone | AGAT $(N)_{23}$ | AAAA $(N)_{23\ \text{reverse complement}}$ |
| pFH90 | FnCas12a crRNA backbone | AGAT $(N)_{23}$ | AAAA $(N)_{23\ \text{reverse complement}}$ |
| pFH130 | FnCas12a crRNA backbone | AGAT $(N)_{23}$ | AAAA $(N)_{23\ \text{reverse complement}}$ |
| pFH131 | FnCas12a crRNA backbone | AGAT $(N)_{23}$ | AAAA $(N)_{23\ \text{reverse complement}}$ |

5. Transform 2μL of the GG reaction into 10μL of competent *E. coli* cells, and plate the cells on selection plates containing kanamycin, IPTG and X-Gal.

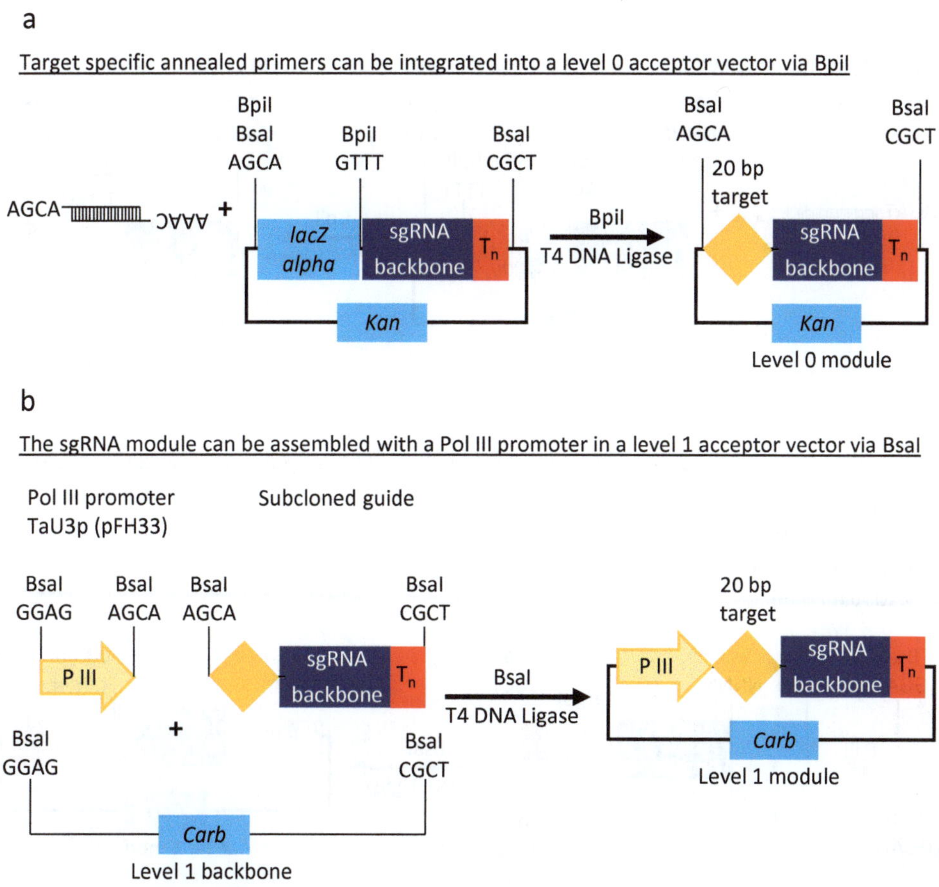


**Fig. 4** Assembly of level 1 sgRNA expression modules to be used with respective Cas9 nucleases (adapted from Hahn et al. (2020) [5]). During the first step (**a**), annealed complementary oligos encoding the guide are inserted into a level 0 acceptor with the sgRNA backbone using BpiI. During the second step (**b**), sgRNA is fused with the respective Pol III promoter using BsaI. P III is a Pol III promoter; $T_n$ is a Pol III terminator

6. Inoculate a white colony, incubate overnight, extract plasmid DNA and verify correct integration of annealed primers by sequencing with the primer FH32 (Table 3).

*3.2.2 Generation of a Transcription Unit for Guide RNA Expression*

1. Combine the assembled guide module with the correct Pol III promoter module (see Table 1) into a suitable L1 backbone (figshare: https://doi.org/10.6084/m9.figshare.12998861) using a GG cut-ligation reaction via BsaI-HFv2 (Figs. 4b and 5b; see **Note 3**).
2. Transform 2μL of the GG reaction into 10μL of competent *E. coli* cells, and plate the cells on selection media containing carbenicillin, IPTG and X-Gal.
3. Inoculate a white colony, incubate overnight, extract plasmid DNA and verify constructs by sequencing with primers Lvl1_F (0229) and Lvl1_R(0230) (Table 3).

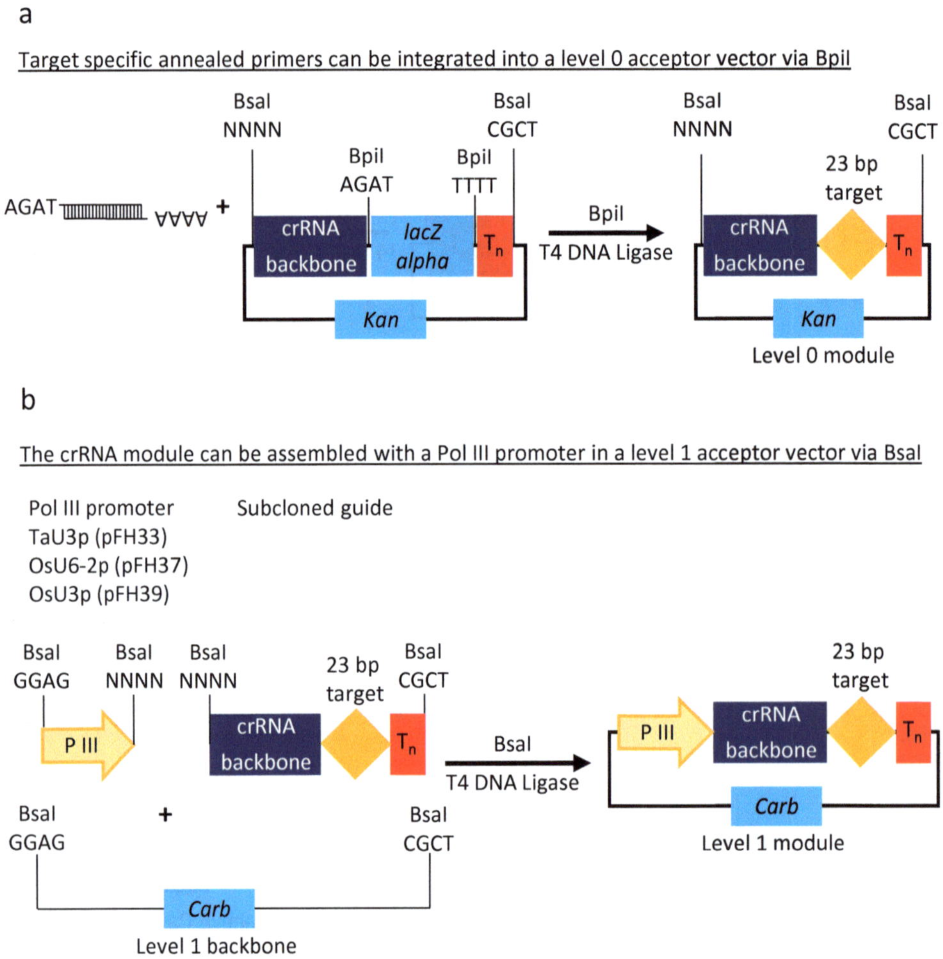


**Fig. 5** Assembly of level 1 crRNA expression modules to be used with respective Cas12a (Cpf1) and Cms1 nucleases (adapted from Hahn et al. (2020) [5]). During the first step (**a**), annealed complementary oligos encoding the guide are inserted into a level 0 acceptor carrying the crRNA backbone using BpiI. During the second step (**b**), crRNA is fused with the respective Pol III promoter using BsaI. P III is a Pol III promoter; $T_n$ is a Pol III terminator

**Table 3**
**Primers used for Sanger sequencing (adapted from Hahn et al. (2020) [5])**

| Primer name | Sequence (5′-3′) |
|---|---|
| FH32 | GCAATGTAACATCAGAGATTTTGA |
| Lvl1_F(0229) | GAACCCTGTGGTTGGCATGCACATAC |
| Lvl1_R(0230) | CTGGTGGCAGGATATATTGTGGTG |

*3.2.3 Generation of a Nuclease Transcription Unit (Optional)*

1. Many nuclease transcription units are already available as assembled L1 modules (figshare: https://doi.org/10.6084/m9.figshare.12998861). This step is only necessary if one needs to use alternative promoter or terminator sequences for the nuclease expression (otherwise, proceed to Chap. 3.2.4).
2. Cut-ligate an L0 promoter +5'UTR module with an L0 nuclease CDS module and an L0 3'UTR + Term module, using BsaI-HFv2, into a suitable L1 backbone (in most cases this would be pICH47742, which places the nuclease transcription unit at the position 2 in the final L2 construct; figshare: https://doi.org/10.6084/m9.figshare.12998861) (Fig. 3).
3. Transform 2μL of the GG reaction into 10μL of competent *E. coli* cells, and plate the cells on selection media containing carbenicillin, IPTG and X-Gal.
4. Inoculate a white colony, incubate overnight, extract plasmid DNA and verify constructs by Sanger sequencing with primers Lvl1_F(0229) and Lvl1_R(0230) (Table 3).

*3.2.4 Generation of a Multi-Expression Unit L2 Construct*

1. Combine your L1 sgRNA/crRNA construct(s) with an L1 resistance marker transcription unit, a suitable L1 nuclease module and the correct endlinker into the pICSL4723 L2 backbone (figshare: https://doi.org/10.6084/m9.figshare.12998861) using a GG cut-ligation reaction via BpiI (Fig. 3; *see* **Note 4**).
2. Transform 2μL of the GG reaction into 10μL of competent *E. coli* cells, and plate the cells on selection media (*see* **Note 5**).
3. Inoculate a white colony, incubate overnight, extract plasmid DNA and verify constructs by suitable restriction digest and Sanger sequencing (*see* **Note 6**).

### 3.3 CRISPR/Cas Construct Assembly for Multiple Targets Using the Polycistronic tRNA-sgRNA System

This section describes a detailed procedure for assembling a L2 construct containing a resistance cassette, a transcriptional unit for a CRISPR/Cas nuclease and a set of up to six sgRNAs using the polycistronic tRNA-sgRNA multiplex system [6] (Figs. 6 and 7).

*3.3.1 Subcloning of CRISPR/Cas Guide Sequences into Acceptor Vectors*

1. Choose the correct subcloning vector for the sgRNA guide sequence, which one intends to clone in at the 5′-end of the polycistronic tRNA-sgRNA (position 1), depending on the Pol III promoter that you intend to use in your experiment and the sgRNA backbone you prefer according to Table 4 and Fig. 8b.
2. Find suitable target sequences in your gene of interest (*see* **Note 1**).
3. Order two complementary primers for each of your target sequences as follows:

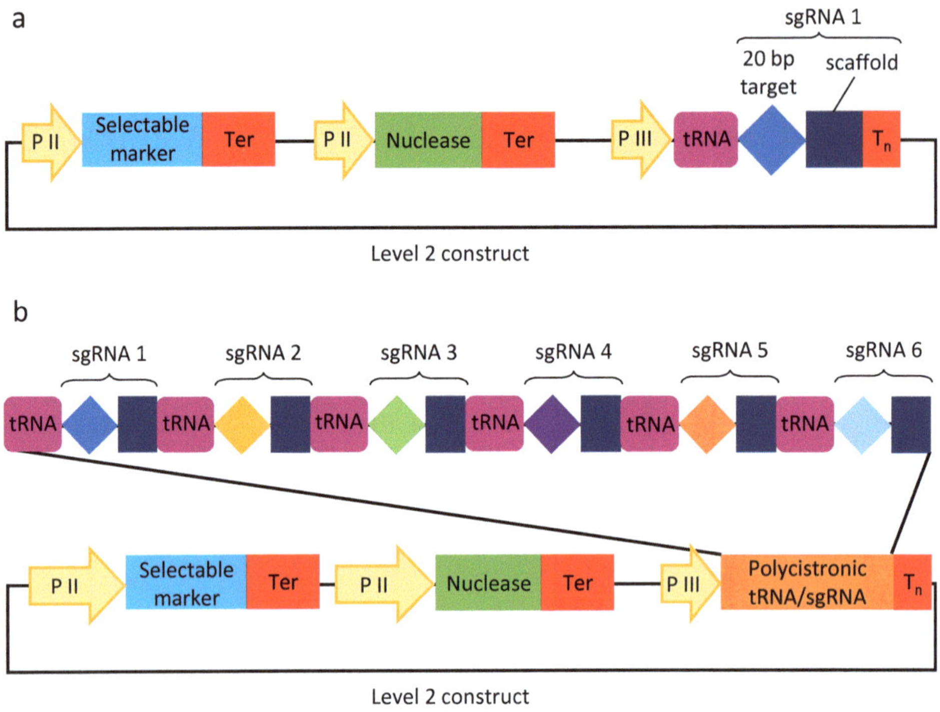


**Fig. 6** Cartoons of level 2 CRISPR/Cas constructs for targeted mutagenesis using the polycistronic SpCas9 tRNA-sgRNA system. (**a**) represents a construct with one and (**b**) with six sgRNAs (adapted from Hahn et al. (2020) [5]). P II and P III are Pol II and Pol III promoters, respectively; Ter and $T_n$ are Pol II and Pol III terminators, respectively

5′ TGCA $(N)_{20}$ 3′

5′ AAAC $(N)_{20\ \text{reverse complement}}$ 3′.

4. Adjust primer concentrations to 10μM, and mix the forward and reverse in a 1:1 ratio; incubate for 5 min at room temperature.
5. Integrate the annealed primers into respective L0 acceptor vectors, according to the position, at which one wants their guide sequence and the choice of the sgRNA backbone using a GG cut-ligation reaction via BpiI (Fig. 8).
6. Transform 2μL of the GG reaction into 10μL of competent *E. coli* cells, and plate the cells on selection media containing kanamycin, IPTG and X-Gal.
7. Inoculate a white colony, incubate overnight, extract plasmid DNA and verify correct integration of the annealed primers by sequencing with the primer FH32 (Table 3).

#### *3.3.2 Generation of a Polycistronic tRNA-sgRNA Transcription Unit*

1. Combine modules carrying sgRNAs with the correct promoter module (see Table 4 and Fig. 8b) and a suitable endlinker (only necessary, if less than six gRNAs are cloned; Fig. 9) into an L1

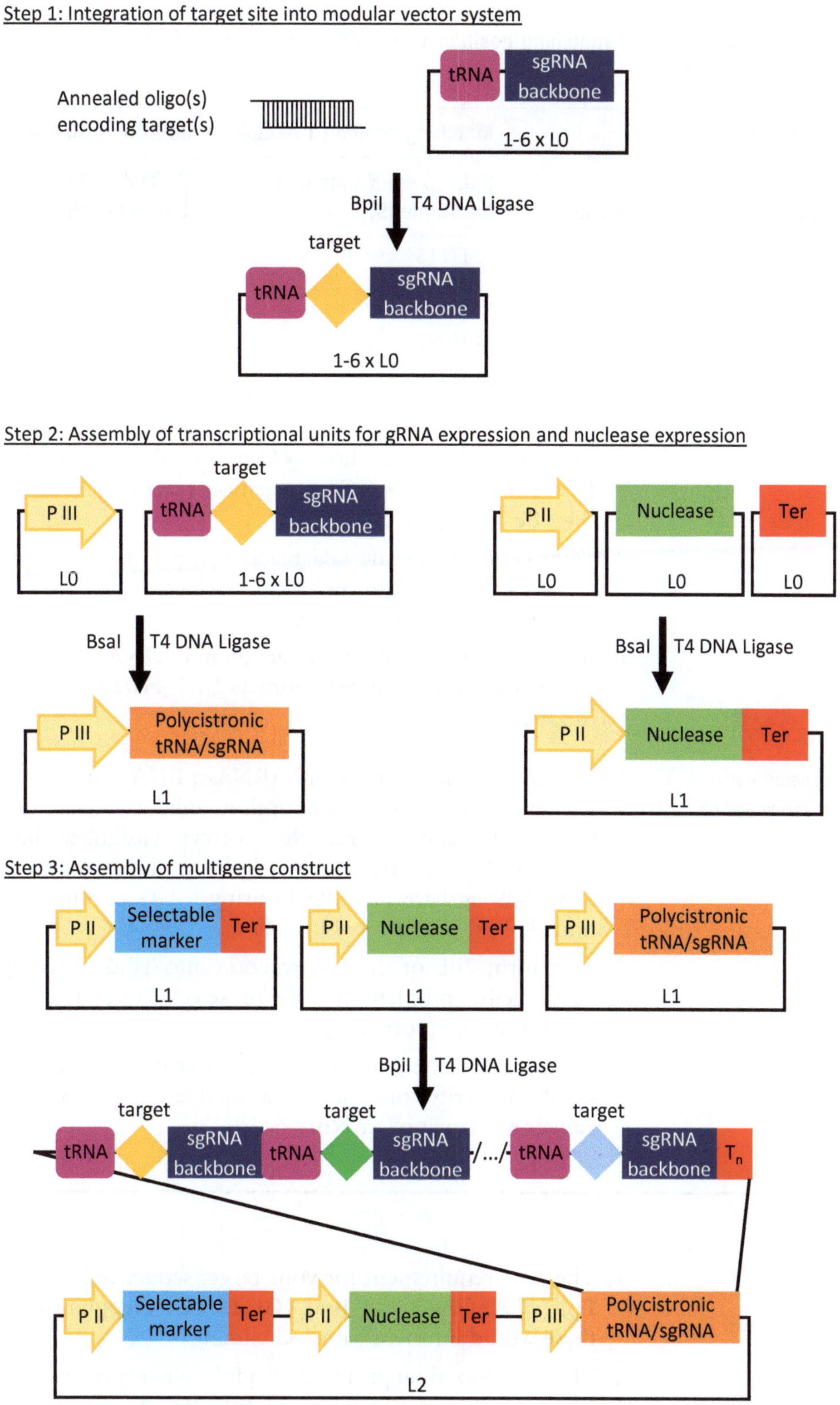


**Fig. 7** A flow chart diagram illustrating assembly of level 2 constructs with gRNAs expressed as tRNA-sgRNA polycistronic units (adapted from Hahn et al. (2020) [5]). P II and P III are Pol II and Pol III promoters, respectively; Ter and $T_n$ are Pol II and Pol III terminators, respectively

**Table 4**
**Pol III promoter modules and matching position 1 modules in the tRNA-sgRNA array (adapted from Hahn et al. (2020) [5])**

| Pol III promoter module | | Matching position 1 module in the tRNA-sgRNA array | |
|---|---|---|---|
| **Vector ID** | **Description** | **SpCas9 sgRNA improved backbone [9]** | **SpCas9 sgRNA classic backbone [10]** |
| pFH31 | Wheat *TaU3p* | pFH113 | pAK007 |
| pICSL90003 [11] | Wheat *TaU6p* | pFH71 | pFH75 |
| pFH36 | Rice *OsU6-2p* | pFH50 | pFH73 |
| pFH38 | Rice *OsU3p* | pFH51 | pFH74 |

backbone (figshare: https://doi.org/10.6084/m9.figshare.12998861) using a GG cut-ligation reaction via BsaI-HFv2 (Fig. 8b and 9; see **Note 7**).

2. Transform 2μL of the GG reaction into 10μL of competent *E. coli* cells, and plate the cells on selection plates containing carbenicillin, IPTG and X-Gal.
3. Inoculate a white colony, extract plasmid DNA and verify constructs by sequencing with primers Lvl1_F(0229) and Lvl1_R (0230) (Table 3).

*3.3.3 Generation of a Multi-Expression Unit L2 Construct*

1. Combine the L1 polycistronic tRNA-sgRNA construct with an L1 resistance marker transcription unit, a suitable SpCas9-based L1 module and the correct endlinker into the pICSL4723 L2 backbone (figshare: https://doi.org/10.6084/m9.figshare.12998861) using a GG cut-ligation reaction via BpiI (Fig. 7; *see* **Note 4**).
2. Transform 2μL of the GG reaction into 10μL of competent *E. coli* cells, and plate the cells on selection media containing kanamycin (*see* **Note 5**).
3. Inoculate a white colony, incubate overnight, extract plasmid DNA and verify constructs by a suitable restriction digest and Sanger sequencing (*see* **Note 6**).

## 4 Notes

1. The only requirement for your target sequence is that it lacks BpiI (GAAGAC) and BsaI (GGTCTC) recognition sites as these would interfere with GG cloning.
2. The corresponding promoter modules already contain the correct transcriptional start nucleotide for expression of guide

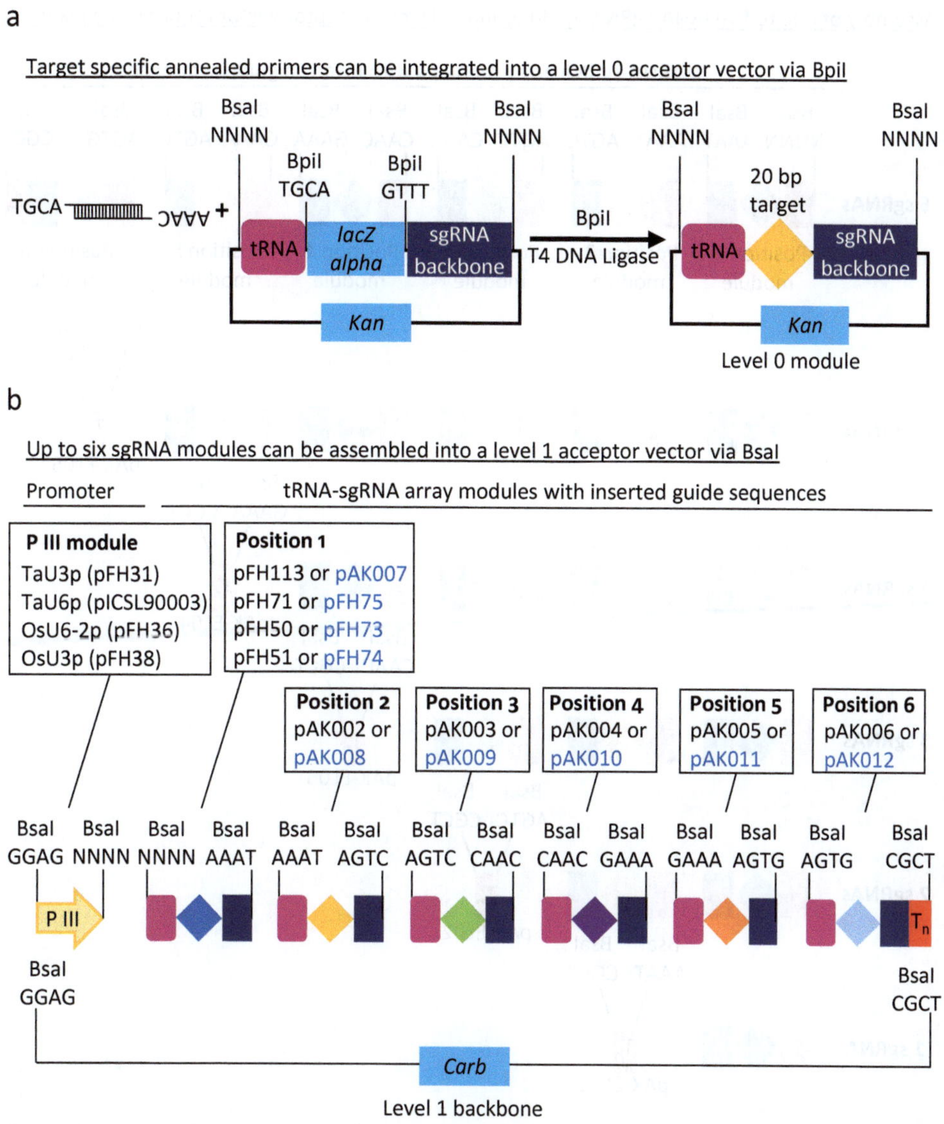


**Fig. 8** Assembly of level 1 polycistronic tRNA-sgRNA expression modules to be used with SpCas9 (adapted from Hahn et al. (2020) [5]). During the first step (**a**), annealed complementary oligos encoding the guide are inserted into a level 0 acceptor with the tRNA-sgRNA backbone using BpiI. During the second step (**b**), tRNA-sgRNA modules are fused with the respective Pol III promoter using BsaI. pFH50, pFH51, pFH71, pFH113 and pAK002 to pAK006 constructs (shown in black font) carry the improved sgRNA backbone [9], while pFH73 to pFH75, and pAK007 to pAK012 (shown in blue font)—the classic sgRNA backbone [10]. P III is a Pol III promoter; $T_n$ is a Pol III terminator

RNAs ('A' in the case of U3 promoters and 'G' in the case of U6 promoters) at their 3′-end to allow transcription of all guide RNAs independent of their first nucleotide. If the CRISPR/Cas target sequence already starts with the correct transcriptional start nucleotide, one should consider not to add

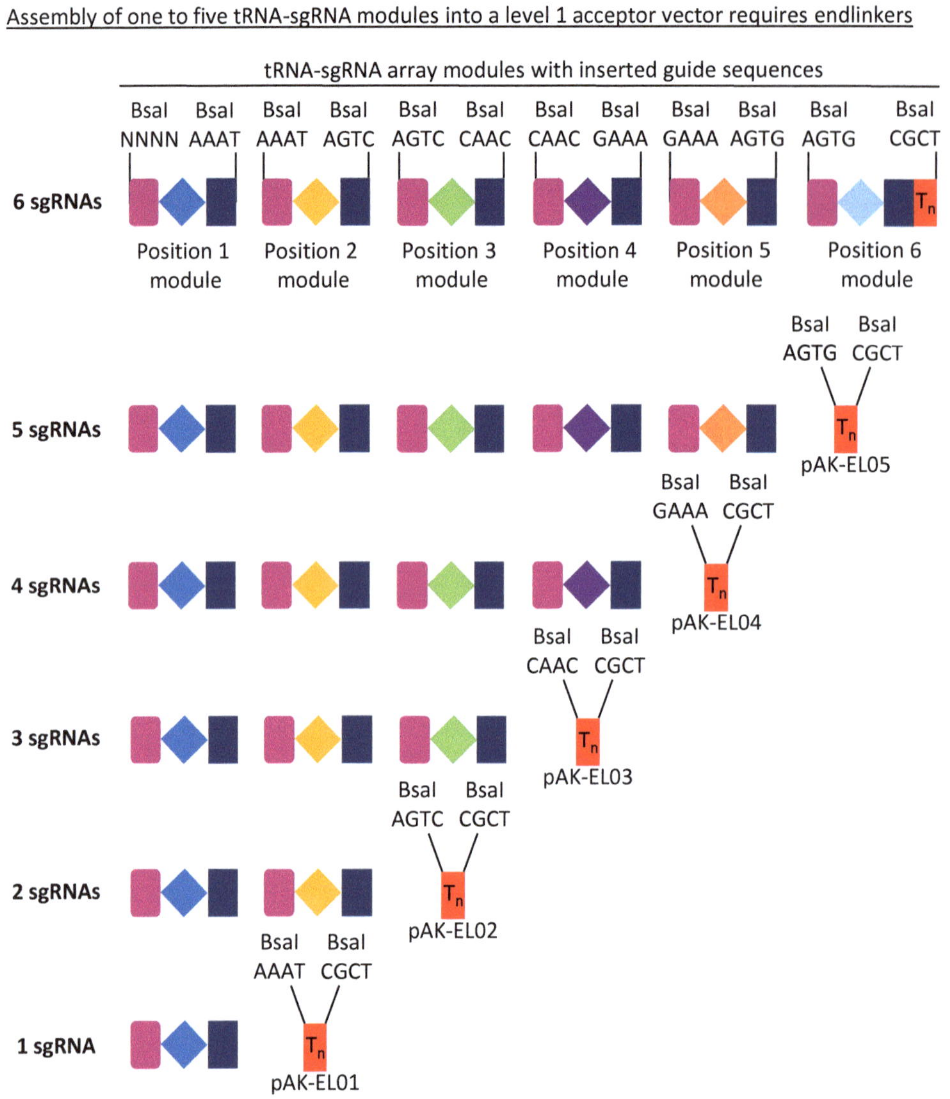


**Fig. 9** Assembly of less than six tRNA-sgRNA modules into a level 1 vector requires respective endlinkers (adapted from Hahn et al. (2020) [5]). $T_n$ is a Pol III terminator

it to the primer sequence (e.g. for pFH85, one would just order two primers: AGCA $(N)_{2\text{-}20}$ and AAAC $(N)_{20\text{-}2\ \text{reverse complement}}$). This does not apply to the Cas12a/Cms1 crRNAs, as in this case the target sequence is anyway located at the 3′-end of the crRNA backbone.

3. In the majority of cases, the most suitable L1 backbone is pICH47751 [7] (Addgene ID: 48002), which will place the promoter-gRNA unit in the position 3 in the final L2 construct. If one wants to assemble several separate promoter-gRNA units into the final L2 construct, they would need to clone them separately into different L1 backbones at this step

(e.g. position 3, 4 and 5; figshare: https://doi.org/10.6084/m9.figshare.12998861).

4. For one gRNA expression unit, use endlinker pICH41766; for two, endlinker pICH41780; for three, endlinker pICH41800; for four, endlinker pICH41822; and for five, endlinker pICH50866 (figshare: https://doi.org/10.6084/m9.figshare.12998861). If using six sgRNA expression units, no endlinker is necessary.
5. If using the pICSL4723 L2 vector [8], the red/white screening is done. No additional reagents need to be added to agar for the red colour to appear in this case.
6. One does not have to Sanger sequence the whole insert in the assembled L2 construct as it tends to be very large. One should, however, sequence the overlaps between the assembled parts to verify correct assembly (in addition to a restriction digest). As the cloning procedure does not involve PCR steps, mutations within assembled parts are unlikely at this step.
7. In the majority of cases, the most suitable L1 backbone is pICH47751 [7] (Addgene ID: 48002), which will place the promoter::polycistronic tRNA-sgRNA transcription unit in the position 3 in the final L2 construct. If one wants to add several separate polycistronic tRNA-sgRNA units to the final L2 construct (e.g. for expression of more than six guides), one would need to clone them separately into different L1 backbones at this step (e.g. position 3, 4 and 5; figshare: https://doi.org/10.6084/m9.figshare.12998861).

## Acknowledgement

Vladimir Nekrasov (Rothamsted Research) receives grant-aided support from the Biotechnology and Biological Sciences Research Council (BBSRC) Designing Future Wheat (DFW) programme.

### References

1. Marillonnet S, Werner S (2020) Assembly of multigene constructs using the modular cloning system MoClo. In: Chandran S, George KW (eds) DNA cloning and assembly: methods and protocols. Springer US, New York, NY, pp 125–141. https://doi.org/10.1007/978-1-0716-0908-8_8
2. Nekrasov V, Staskawicz B, Weigel D, Jones JD, Kamoun S (2013) Targeted mutagenesis in the model plant Nicotiana benthamiana using Cas9 RNA-guided endonuclease. Nat Biotechnol 31:691–693
3. Nekrasov V, Wang C, Win J, Lanz C, Weigel D, Kamoun S (2017) Rapid generation of a transgene-free powdery mildew resistant tomato by genome deletion. Sci Rep 7:482
4. Brooks C, Nekrasov V, Lippman ZB, Van Eck J (2014) Efficient gene editing in tomato in the first generation using the clustered regularly interspaced short palindromic repeats/CRISPR-Associated9 system. Plant Physiol 166:1292–1297
5. Hahn F, Korolev A, Sanjurjo Loures L, Nekrasov V (2020) A modular cloning toolkit for

genome editing in plants. BMC Plant Biol 20:179

6. Xie K, Minkenberg B, Yang Y (2015) Boosting CRISPR/Cas9 multiplex editing capability with the endogenous tRNA-processing system. Proc Natl Acad Sci U S A 112:3570–3575
7. Weber E, Engler C, Gruetzner R, Werner S, Marillonnet S (2011) A modular cloning system for standardized assembly of multigene constructs. PLoS One 6:e16765
8. Castel B, Tomlinson L, Locci F, Yang Y, Jones JDG (2019) Optimization of T-DNA architecture for Cas9-mediated mutagenesis in Arabidopsis. PLoS One 14:e0204778
9. Dang Y, Jia G, Choi J, Ma H, Anaya E, Ye C et al (2015) Optimizing sgRNA structure to improve CRISPR-Cas9 knockout efficiency. Genome Biol 16:280
10. Mali P, Yang L, Esvelt KM, Aach J, Guell M, DiCarlo JE et al (2013) RNA-guided human genome engineering via Cas9. Science 339:823–826
11. Lawrenson T, Shorinola O, Stacey N, Li C, Østergaard L, Patron N et al (2015) Induction of targeted, heritable mutations in barley and Brassica oleracea using RNA-guided Cas9 nuclease. Genome Biol 16:258

# Chapter 11

# Genome Editing and Identification of Targeted Heritable Mutations in Wheat

**Matthew J. Milner and Emma J. Wallington**

## Abstract

The CRISPR revolution has enabled greater manipulation of many genomes to help elucidate gene function and drive biological gains. This chapter will describe in detail the steps required to target and edit either a single gene or multiple loci in the wheat genome to generate knockouts in the target gene(s) of interest. It will walk you through how to design the construct, select guide sequences, identify mutations, induce additional edits if this is required and finally to segregate out the T-DNA bearing the Cas9 components to generate clean homozygous edited lines. This is a fast-evolving field, and there are several variations of Cas nuclease which can be used and numerous ways to design and carry out genome editing experiments. The steps described are those which we regularly used for wheat gene editing.

**Key words** CRISPR/Cas9, sgRNA, PAM, Non-homologous end-joining repair

## 1 Introduction

CRISPR technology is based on RNA-programmed DNA cleavage systems, which were first discovered in bacteria and archaea [1, 2]. Since the first reports of CRISPR editing, this technology has been demonstrated to be readily transferable to a multitude of organisms across different kingdoms, leading to advancements in our understanding of biological processes as well as commercial applications [3–5]. The CRISPR/Cas9 system has allowed researchers to edit genomes in a precise targeted manner to broaden our understanding of gene function as well as achieve more precise molecular plant breeding strategies. The CRISPR/Cas9 system has two components: a DNA endonuclease (e.g. Cas9) and an RNA molecule that confers targeting specificity, known as single-guide RNA (sgRNA). DNA double-strand breaks (DSBs) are created by the Cas9 nuclease at the target site and are repaired by the endogenous repair pathways. In higher plants the error-prone non-homologous end-joining pathway is prevalent, leading to the creation of insertions or deletions (indels), which may result

Andriy Bilichak and John D. Laurie (eds.), *Accelerated Breeding of Cereal Crops*, Springer Protocols Handbooks, https://doi.org/10.1007/978-1-0716-1526-3_11, © Springer Science+Business Media, LLC, part of Springer Nature 2022

in a frameshift mutation in the target and abolish gene function. If a donor template with homology to the adjacent sequence to the DSB is also present, the homologous recombination pathway may repair the break by incorporation or "knock-in" of the donor sequence.

The only prerequisite for applying CRISPR to a given target is the presence of a protospacer adjacent motif (PAM) sequence near the site of interest. In the case of SpCas9, originally identified from *S. pyogenes*, this is NGG, where N is any nucleotide and G is guanosine. Directing CRISPR to gene targets thus only requires different spacer sequences to specifically target a genomic sequence; hence, it is simple, rapid, efficient, inexpensive and versatile. To take advantage of this technology, a researcher only needs two things: (1) a robust transformation system to deliver the Cas9 enzyme and sgRNA to the plant cell and (2) sequence information for the target gene homoeologues in the transformable germ plasm to ensure that the sgRNA will bind to the correct sequence(s) and to facilitate identification of the mutations.

As CRISPR-mediated genome engineering is now a very active area of research across the breadth of biological systems, new nucleases are being identified or created and rapidly deployed across many different species. There are a choice of nucleases, from the original SpCas9 to other CRISPR-type enzymes, such as CpfI, also known as Cas12a, which uses an alternative PAM sequence (5'-TTN PAM) and facilitates targeting of AT-rich regions in the genome [6]. Other versions of Cas9 include a nickase version, containing the D10A mutation within the nuclease domain, which creates single-strand breaks and shows increased specificity but has lower overall editing rates [7]. Editing a single nucleotide in a predicted manner, termed base editing, utilises a catalytically inactive or "dead" Cas9 (dCas9) containing D10A and H840A mutations fused to a cytidine deaminase for conversions of cytosine to thymidine [8, 9]. Adenosine to guanosine transitions can also be generated using a dCas9 fused to adenine deaminase [10]. Base editing can therefore be used to introduce specific point mutations at a target locus. Other nucleases with alternative PAM recognition sites include xCas9, which targets NG, GAA and GAT PAM sequences, and SpCas9-NG, a variant that recognises the NG PAM [11, 12]. These variants also differ in their catalytic activity or in off-target effects [13].

Several strategies for the guide scaffolds to generate the sgRNA have also been described. At its simplest, a single guide scaffold is transcribed from a Pol III promoter, such as OsU3 or OsU6 [14]. Several guides may be stacked in parallel on a single T-DNA, with each expressed from separate Pol III promoters to increase the likelihood of editing the target gene or to create a specific whole gene or region deletion [15–19]. Alternatively, guides may be multiplexed within a polycistronic-tRNA-gRNA

(PTG) array and the endogenous tRNA processing system used to efficiently express multiple sgRNAs from a single Pol III promoter [20, 21]. A number of algorithms are available online to identify potential guide target sites in the gene of interest. However, in practice, sgRNAs with a higher predicted efficiency probability in silico fail to target the Cas9 to the targeted sequence and generate mutations [22, 23].

With the multitude of nuclease and guide scaffold options available, it can be tempting to use components from heterologous systems. Wherever possible, it is prudent to test these components first to edit, for example, a "simple" target gene, such as phytoene desaturase, either in an in vitro transient system or stable transformation experiment [24]. Changes to the components such as the promoter used for nuclease expression, codon optimisation of the nuclease or changes to the Pol III promoter or guide scaffold can then be made prior to embarking on a more complex target [25].

The efficiency of editing in hexaploid wheat has until recently lagged behind diploid cereal species such as rice, maize and barley. Certainly, the polyploid nature of the genome increases the number of target sites, which may need to be edited to achieve a knockout. For a single rice gene, such as phytoene desaturase, three homoeologous genes are present in wheat. All six wheat alleles must therefore be edited to create a full null knockout, compared with two in a diploid genome. Strategies to increase the efficiency of editing or introduce a second round of edits to create the full knockout can therefore be used [22, 26].

This chapter covers the specific stages required for CRISPR/Cas9 experiments in wheat: construct components, verification of transgenic lines, mutation identification, introduction of additional edits and progression to T-DNA-free "clean" homozygous edited lines (Fig. 1). The stages described here are also relevant to the many different nuclease and guide systems which are available. Wheat transformation per se is outside of the scope of this chapter, and several methods for *Agrobacterium*-mediated transformation or biolistic bombardment and tissue culture of wheat are available. Whilst many elite varieties of wheat can be transformed with differing levels of efficiency, a method appropriate to facilities available and specific wheat germplasm should be used. Transformation of wheat is also available on a service provision basis in many countries worldwide for both academic researchers and commercial entities. Gene editing using CRISPR/Cas in a polyploid crop, such as wheat, is now a reality and, as it becomes available in many laboratories, will undoubtedly accelerate the rate of gene characterisation and drive advances in molecular breeding.

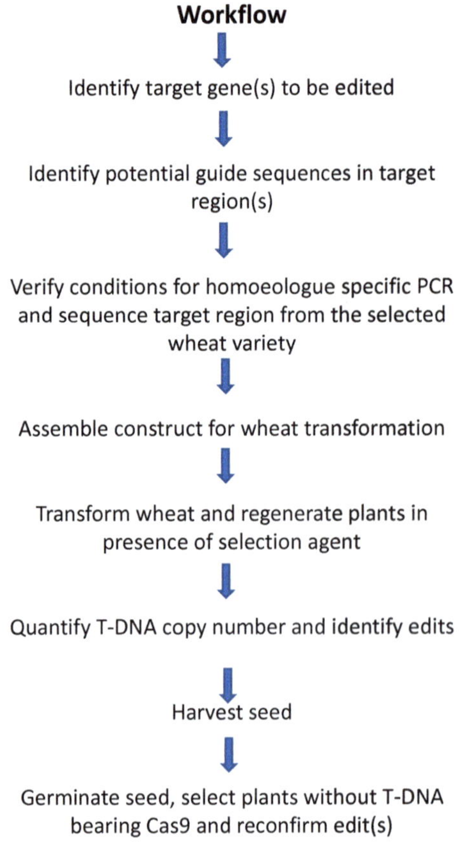


**Fig. 1** Workflow for designing and performing CRISPR-mediated knockout of targeted sequences in wheat

## 2 Materials

**Crude DNA extraction buffer:** 200 mM Tris pH 7.5, 250 mM NaCl, 25 mM EDTA, 0.5% SDS (*see* **Note 1**).

**10X TE buffer:** 100 mM Tris HCl, 10 mM EDTA

**PCR reactions:** 1μM of each primer, 2 mM $MgCl_2$, 1 mM dNTPs, 0.4 U Taq DNA polymerase, 1× PCR reaction buffer provided by manufacturer.

**PCR clean-up:** 10X ExoI buffer (67 mM glycine-KOH, 6.7 mM $MgCl_2$, 10 mM β-mercaptoethanol, pH 9.5), Exonuclease I (2 U per reaction), shrimp alkaline phosphatase (SAP) (0.1 U per reaction).

**Growth media for culture of *E. coli* and *A. tumefaciens* (2xYT):** Tryptone 16 g/L, yeast extract 10 g/L, NaCl 5 g/L, pH 7.0, plus appropriate antibiotic(s) for plasmid and strain selection.

## 3 Methods

### 3.1 Overall Construct Design

The design of the construct for stable wheat transformation can be quite varied. A typical example using a wheat optimised Cas9 is suggested below for *Agrobacterium*-mediated transformation. The construct has three expression cassettes:

1. A suitable selection gene cassette for selection of transformed cells in tissue culture, for example, *nptII*, *hph* or the *bar* gene expressed from a constitutive promoter.
2. A wheat codon optimised SpCas9 expressed from a strong constitutive promoter such as *ZmUbi*. Other Cas enzymes that could also be used as a general toolkit is now publicly available [27]. We would recommend the use of ZmUbi promoter for expression of the Cas9 as substantially higher levels of editing will be achieved if the ZmUbi promoter is used, rather than other promoters, such as OsActin, to express Cas9 in wheat.
3. A guide sequence targeting a desired sequence or sequences in wheat. The genomic target can be any ~20 nucleotide DNA sequence, provided it meets the condition that the target is present immediately adjacent to a protospacer adjacent motif (PAM). In the case of SpCas9, this is NGG at the 3′ end of the sequence. This sequence is not included in the guide itself but is necessary for editing in vivo. The guide sequence is usually expressed from a Pol III promoter, such as *TaU3* and *TaU6*, or the *OsU3* and *OsU6* promoters, which have also been shown to work in wheat [25, 27–29]. Multiple guides can also be stacked in series using either domesticated sequences and Golden Gate cloning, overlap PCR or a PTG system.

### 3.2 Wheat Target Gene Sequence Verification

1. The sequence of the target gene in the wheat germ plasm for transformation should be obtained prior to proceeding with guide design. This may mean design of primers in silico using an available wheat genome sequences, such as Chinese Spring, Cadenza or others, at Ensembl (http://plants.ensembl.org/Triticum_aestivum/Info/Index) and then amplifying and resequencing the region from the germ plasm available for transformation, e.g. Fielder, Cadenza or a more current wheat germplasm.
2. Ensure that sequence for the three homoeologues is included. Placement of primers in the flanking promoter region or intron regions, etc. may assist in homoeologue specificity for the PCR. If editing is required in only one or two homoeologues for the experiment, sequence for the third homoeologue should also be obtained to verify that no off-target edits have been introduced.

3. A region of the target gene towards the 5′ of the coding sequence should be chosen, if possible, to increase the likelihood of a frameshift mutation abolishing gene function. Ideally up to 1 kb of sequence is required for design of three to four guides plus sufficient sequence to design homoeologue-specific PCRs. Ensure that homoeologue-specific primers and PCR conditions have been validated for this region, to simplify analysis of potential edits in Sect. 3.7. To optimise PCR conditions, run PCRs with wheat gDNA using a gradient PCR setup and a range of annealing conditions (e.g. $T_a$ from 53 to 63 °C) with and without 2% DMSO (final concentration) using Taq DNA polymerase and reaction components supplied by the manufacturer (*see* **Note 2**).

### 3.3 Guide Design and Selection

1. Guide selection is one of the most important parts of designing the CRISPR/Cas experiment. There are a number of websites available to identify guide sequences and which provide a relative rating or likelihood of the guide working. Examples include CCTop (https://cctop.cos.uni-heidelberg.de/) [30], CRISPR-P 2.0 (http://crispr.hzau.edu.cn/CRISPR2/) [31], WheatCrispr (https://crispr.bioinfo.nrc.ca/WheatCrispr/) [32] and CHOPCHOP v3 (http://chopchop.cbu.uib.no/) [33]. In our experience these sites do a good job at identifying the guides themselves but a poor job in predicting the likelihood of a mutation occurring in the transformed plants (*see* **Notes 3–6**). This has been recently published in both wheat and other model species, showing the comparison of prediction versus actual editing rates [22, 23].
2. The Pol III promoters have specific requirements for the first nucleotide transcribed. Transcription from U3 promoters must initiate with A, and U6 promoters must initiate with G. The gRNA selected must therefore be compatible with this precondition. This would make the guide sequence 5'-ANNNNNNNNNNNNNNNNNNNN<u>NGG</u> for guides expressed from a U3 promoter and 5'-GNNNNNNNNNNNNNNNNNNNN<u>NGG</u> for guides expressed from a U6 promoter. If a PTG array is used, only the first guide in the system needs to follow this design requirement.
3. Many of the guide design websites will also provide potential off-target sequences, which could be edited. Alternatively, a BLAST search can be performed (http://blast.ncbi.nlm.nih.gov/) with the 20 nt sgRNA-binding sequence plus the GG in the NGG PAM—to make sure that the target sequence is unique. However, as previously noted, the reference sequence used is often Chinese Spring, which may not be relevant to the germplasm for transformation, but sequences of more wheat varieties are increasingly becoming available.

### *3.4 Construct Design and Construction*

1. Decide on the gene target region(s) for genome modification by CRISPR and whether this is a single locus or there is a requirement to target the presumed three wheat homoeologues. Use one of the many sites listed in the Guide Design and Selection section to identify sequences which fit the guide selection parameters for the number of targets desired. The same guide can be used to edit multiple sites in the wheat genome. The homoeologue sequences should be downloaded from a public repository, such as Ensembl (http://plants.ensembl.org/Triticum_aestivum/Info/Index), and aligned to compare conserved and unique portions of the targeted sequence(s).
2. Design primers specific to the genomic region for targeting, leaving at least 50 bp on either side of the potential target site for sequencing the PCR amplicons later. Use the alignment, from **step 2** of the methods, of the three (or more) putative homoeologues to design primers either unique to a particular homoeologue or targeting all potential sites, and optimise the PCR conditions, as described in the Wheat Target Gene Sequence Verification section, to ensure amplicons are specific, with only a single band of the correct size amplified from the wheat variety for transformation. Sequencing this amplicon will confirm that the guide target is correct in the wheat transformation germplasm.
3. Construct your guide, or guide stack, to target your genomic location(s), depending on your method of cloning. Examples include Golden Gate, restriction enzyme and ligation cloning, MultiSite Gateway and Gibson assembly. Publicly available components, such as those published by Hahn et al., use annealed primers and the Golden Gate construction method, which is inexpensive [27]. Alternatively, commercial synthesis of the guide stack is recommended but is more expensive.
4. Construct the final binary construct containing the promoter-Cas9, guide(s) and selection marker for selection in tissue culture. Sequence any regions of the final binary construct, which have not previously been verified.
5. Transform the verified binary construct into *Agrobacterium tumefaciens* strain AglI, EHA101, EHA105 or LBA4404pSB1, and select three or four colonies to grow a liquid culture for 2–3 days at 28 °C. Re-isolate the plasmid from the *Agrobacterium* cultures to confirm that the plasmid is correct by restriction enzyme digestion or sequencing before starting the plant transformation experiment [34]. Make 4–5 mini-glycerol stocks, each containing 100μL per tube, and store at −80 °C until the wheat germplasm is ready for transformation.

### 3.5 Wheat Transformation

1. An *Agrobacterium*-mediated or biolistic bombardment transformation method can be employed for stable wheat transformation [24, 25, 35–39]. DNA-free systems have also recently been published, albeit with low efficiency [40].
2. The efficiency of editing may be increased in the transformed lines regenerated by using a temperature-inducible promoter to drive expression of Cas9 or a higher temperature for the callus induction and selection stages of tissue culture, where Cas9 is expressed from the ZmUbi promoter [22, 26].

### 3.6 Verification of Transformed $T_0$ Plants

1. Extract genomic DNA from small rooted plantlets using either crude or CTAB method [24, 41].
2. Use a quantitative PCR assay to determine T-DNA copy number. We routinely determine copy number based on the selectable marker gene as a proxy for the T-DNA copy number [42]. This type of assay may also be available from commercial providers. Southern hybridisation of DNA extracted with a CTAB method would provide additional useful information on the number of T-DNA integration loci present [41, 43].
3. Amplification of the Cas9 gene and/or guide stack by PCR. Typically for an *A. tumefaciens* binary construct, the selectable marker gene is placed towards the T-DNA left border sequence and is integrated last into the genome. A PCR to verify that both the Cas9 gene and guide stack are present will confirm that a full-length T-DNA has been inserted.

### 3.7 Identification of Edits in Transformed Plants

Several methods are available for identification of edits in transformed plants based on PCR amplification of the target gene region. Considerable time is saved if homoeologue-specific primers and PCR conditions have been developed in advance, as described above. The ability to detect and track the mutation is in our experience the most important aspect to consider before attempting CRISPR-mediated modification of wheat, as most gene targets have at least three but sometimes more copies within the genome.

1. An initial screen by homoeologue-specific PCR using Taq polymerase may identify plants with smaller amplicons compared with wild-type control, indicating that a region between two guides has been deleted. An aliquot of the PCR is added to loading dye and run on an agarose gel to visualise PCR product sizes. Two bands may indicate a heterozygous mutation. However, if only full-length amplicons are observed by gel electrophoresis, the amplicons should be sequenced directly to determine whether smaller indels have been created.
2. Cleavage of amplified polymorphic sequences (CAPS assay) can be used if a restriction site is adjacent to the PAM. Again, only an aliquot of the PCR product is digested and run on the

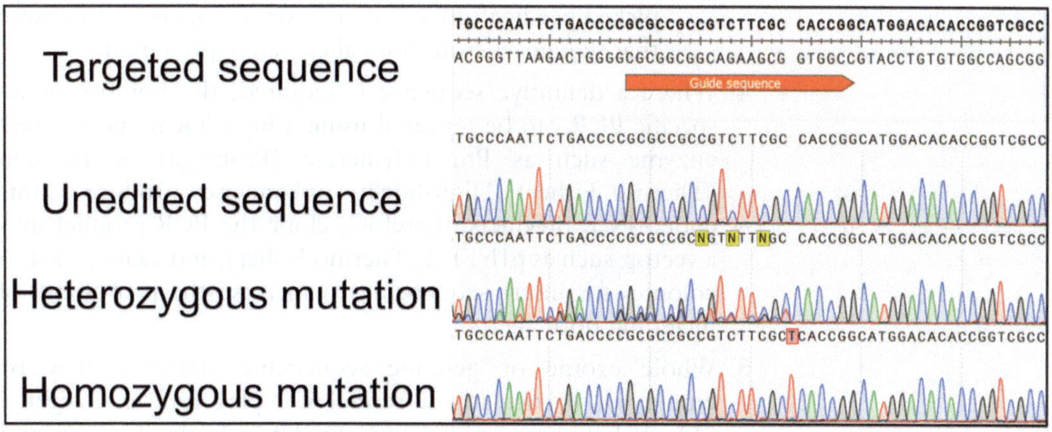


**Fig. 2** Examples of chromatographs from Sanger-sequenced amplicons of a homoeologue-specific targeted mutation in wheat. Chromatograms from unedited control plant and plants with heterozygous or homozygous mutations are shown. The sequence reaction proceeds right to left in the chromatogram and demonstrates that the inserted nucleotide in the heterozygous mutation chromatogram alters the ability to clearly call the desired sequence

agarose gel. If only wild-type amplicon digestion fragments are observed by gel electrophoresis, the remaining undigested amplicons should be sequenced directly to determine whether indels have been created outside the enzyme recognition sequence.

3. Direct Sanger sequencing of homoeologue-specific PCR products may be the first indication that editing has been successful.
   (a) Clean up the PCR reaction using ExoSAP (Thermo Fisher) or using 0.1μL ExoI, 0.1μL shrimp alkaline phosphatase (SAP) and 1x ExoI buffer in a 20μL reaction volume. Incubate tubes/plate for 30 mins at 37 °C and 10 mins at 80 °C to inactivate the enzymes.
   (b) Use 5μL of the cleaned-up PCR reaction for a Sanger sequencing reaction using an appropriate primer, which binds within the amplified PCR product. If the sequencing chromatogram is of good quality but then fails, or produces double peaks at each position, three to four bases into the guide sequence, this may indicate that a heterozygous mutation is present (Fig. 2). Sites such as TIDE (https://tide.nki.nl/) can be used to deconvolute the sequence and identify the mutations. However, this provides a probability of a few mutations and not an absolute confirmation of the sequence and mutation itself.
   (c) PCR products can be cloned into a suitable cloning vector, e.g. pGEM-T Easy (Promega) for Taq polymerase PCR products. Multiple colonies, usually 4–6, should be

picked per ligation and their inserts sequenced to ensure that sequences from both alleles are represented.

4. Where a definitive sequence is required, the homoeologue-specific PCR can be repeated using a high-fidelity polymerase enzyme such as Pfu polymerase (Promega) or Phusion (Thermo Fisher). High-fidelity polymerases produce blunt-ended PCR products; therefore, clone the PCR product into a vector, such as pJET1.2 (Thermo Fisher), and again pick 4–6 colonies for insert sequencing to increase the probability of including both alleles.
5. Whole exome or genome sequencing strategies may be employed, particularly if several complex loci are targeted [44, 45]. Exome sequencing can detect both on-target and off-target mutations, which may have been induced, but will not detect mutations in regulatory or non-coding regions. Whole genome sequencing will provide the best coverage of mutations throughout the genome but will be most costly to carry out.

### 3.8 Plant Growth and Induction of Additional Edits in Transformed Lines

Edited plants expressing the Cas9 nuclease from the ZmUbi promoter can produce additional editing under elevated temperatures. This could cause secondary edits in somatic tissues, which may confound genotyping, or could be utilised to produce additional heritable edits. Care should therefore be taken that plant growth is undertaken under well-controlled temperature conditions, unless a deliberate strategy of edit induction through increased temperature is being employed [22]. Segregation of the T-DNA bearing the Cas9 and guide should be undertaken as soon as possible. Useful additional steps include:

1. Embryo rescue for reduction in the seed-to-seed generation time. Immature embryos are isolated aseptically at 12–18 days after anthesis and germinated in vitro. After 5–7 days the germinated shoots are transferred to compost in 40-well trays and then genotyped. This can shorten the generation time under standard growth conditions by at least 8 weeks [24]. Speed breeding may also be used to reduce generation time if these specialist facilities are available [46].
2. Germ-line secondary edit induction. Seed from wheat lines transformed with a ZmUbi-Cas9 can be germinated at an elevated temperature, 28–37 °C, to induce secondary editing [22]. Seedlings are grown for 7 days in a petri dish with moistened filter paper at the elevated temperature, before transferring seedlings to compost and growth under standard temperature and growth conditions (20/15 °C day/night temperature).

### 3.9 Generation of T-DNA-Free "Clean" Homozygous Lines

It is important to segregate out the T-DNA bearing Cas9 and guides at the earliest opportunity and identify "clean" lines containing homozygous mutations in the target gene (*see* **Note 7**). This is straightforward if edits in the target gene have been identified in a line with 1–2 copies of the T-DNA.

1. Typically, 20–40 seeds are germinated and the plantlets genotyped by selection gene copy number qPCR assay and Cas9 PCR. The observed segregation ratios for the selection gene copy number and the Cas9 amplicon are compared with the expected segregation ratio for Mendelian inheritance [24].
2. Edits in the target gene(s) should be reconfirmed in these clean lines by resequencing PCR amplicons as above. Full genome resequencing could also be considered for selected lines if budget permits.

## 4 Notes

1. Commercial plant DNA extraction kits can be used, but we do not find them necessary and add expense.
2. A $MgCl_2$ gradient may also be used to optimise PCR conditions using a final concentration of 1–4 mM $MgCl_2$ to ensure that a specific homoeologue is amplified.
3. WheatCrispr enables selection of guides, which target all three homoeologous coding regions simultaneously and off-target effects to be assessed.
4. Programs such as Vector NTI, SnapGene or Geneious can be used to identify the target region. Note that if the region selected for gene editing is larger than 1 kb, the cost of identifying the mutations will increase as multiple PCRs and sequencing reactions will be needed to cover the region.
5. When using multiple guides in one region, it is prudent to leave more than 50 bp between guides, so that if multiple mutations occur, you can detect these individual mutations.
6. In our experience one or two guides out of the three or four will cause a mutation in a given $T_0$ plant—but this is sufficient for the disruption of the targeted sequence. See Figs. 3 and 4 for a realistic outcome of mutations in 20 transgenic plants using a four-guide stack with a wheat optimised Cas9 [25].
7. We would strongly recommend segregating away the Cas9 as quickly as possible, once the desired mutations in the target gene are observed.

| Plant No. | RPG1 A genome guide | | | | RPG1 B genome guide | | | | RPG1 D genome guide | | | | Phenotype |
|---|---|---|---|---|---|---|---|---|---|---|---|---|---|
| | 1 | 2 | 3 | 4 | 1 | 2 | 3 | 4 | 1 | 2 | 3 | 4 | |
| RPG 1 | Wt/Wt | Wt/Wt | Wt/-1 | Wt/Wt | Wt/Wt | Wt/Wt | Wt/Wt | Wt/Wt | Wt/Wt | Wt/Wt | Wt/Wt | Wt/Wt | Fertile |
| RPG 2 | Wt/Wt | Wt/-119 | Wt/-119 | Wt/Wt | Wt/Wt | Wt/Wt | -3/-3 | Wt/Wt | Wt/Wt | Wt/Wt | Wt/-4 | Wt/Wt | Fertile |
| RPG 3 | Wt/Wt | Wt/Wt | Wt/-3 | Wt/Wt | Wt/Wt | Wt/Wt | Wt/Wt | Wt/Wt | Wt/Wt | Wt/Wt | Wt/Wt | Wt/Wt | Fertile |
| RPG 4 | Wt/Wt | Wt/Wt | Wt/Wt | Wt/Wt | Wt/Wt | Wt/Wt | Wt/Wt | Wt/Wt | Wt/Wt | Wt/Wt | Wt/Wt | Wt/Wt | Fertile |
| RPG 5 | Wt/Wt | Wt/Wt | Wt/+1 | Wt/Wt | Wt/Wt | Wt/Wt | Wt/+12 rearangement | Wt/Wt | Wt/Wt | Wt/Wt | -1/-4 | Wt/Wt | Fertile |
| RPG 6 | Wt/Wt | Wt/Wt | Wt/+1 | Wt/Wt | Wt/Wt | Wt/Wt | Wt/+19 rearangement | Wt/Wt | Wt/Wt | Wt/Wt | Wt/-4 | Wt/Wt | Fertile |
| RPG 7 | Wt/Wt | Wt/Wt | Wt/Wt | Wt/Wt | Wt/Wt | Wt/Wt | Wt/Wt | Wt/Wt | Wt/Wt | Wt/Wt | Wt/Wt | Wt/Wt | Fertile |
| RPG 8 | Wt/Wt | Wt/Wt | +1/+1 | Wt/Wt | Wt/Wt | Wt/-4 | Wt/-5 | Wt/Wt | Wt/Wt | Wt/Wt | Wt/Wt | Wt/Wt | Fertile |
| RPG 9 | Wt/Wt | -532/-459 | -532/-459 | Wt/Wt | Wt/Wt | -532/Wt | -532/-5 | Wt/Wt | Wt/Wt | -532/Wt | -532/ -2 | Wt/Wt | Sterile |
| RPG 10 | Wt/Wt | Wt/Wt | -23/-23 | Wt/Wt | Wt/Wt | -532/-4 | -532/-5 | Wt/Wt | Wt/Wt | 532/Wt | -532/-2 | Wt/Wt | Sterile |
| RPG 11 | Wt/Wt | Wt/Wt | -5/Wt | Wt/Wt | Wt/Wt | Wt/Wt | +1/+1 | Wt/Wt | Wt/Wt | Wt/Wt | -4/-5 | Wt/Wt | Fertile |
| RPG 12 | Wt/Wt | -119/Wt | -119/-5 | Wt/Wt | Wt/Wt | +1/Wt | -3/5 | Wt/Wt | Wt/Wt | Wt/Wt | -4/-4 | Wt/Wt | Sterile |
| RPG 13 | Wt/Wt | Wt/Wt | +1/+1 | Wt/Wt | Wt/Wt | -4/-5 | -5/-6 | Wt/Wt | Wt/Wt | Wt/Wt | -2/-3 | Wt/Wt | Sterile |
| RPG 14 | Wt/Wt | -558/Wt | -558/+1 | Wt/Wt | Wt/Wt | Wt/-4 | -5/-6 | Wt/Wt | Wt/Wt | -19/Wt | -3/-2 | Wt/Wt | Sterile |
| RPG 15 | Wt/Wt | Wt/Wt | +1/+1 | Wt/Wt | Wt/Wt | -34/Wt | Wt/-469 | Wt/Wt | Wt/Wt | Wt/Wt | -1/-1 | Wt/Wt | Sterile |
| RPG 16 | Wt/Wt | Wt/Wt | +1/1 | Wt/Wt | Wt/Wt | Wt/Wt | -1/-1 | Wt/Wt | Wt/Wt | Wt/Wt | -2/-3 | Wt/Wt | Sterile |
| RPG 17 | Wt/Wt | Wt/Wt | -5/-4 | Wt/Wt | Wt/Wt | Wt/Wt | -1/-1 | Wt/Wt | Wt/Wt | Wt/Wt | -4/-4 | Wt/Wt | Sterile |
| RPG 18 | Wt/Wt | Wt/Wt | -3/-4 | Wt/Wt | Wt/Wt | Wt/Wt | -10/-5 | Wt/Wt | Wt/Wt | Wt/Wt | +1/-3 | Wt/Wt | Fertile |
| RPG 19 | Wt/Wt | Wt/Wt | -4/-3 | Wt/Wt | Wt/Wt | Wt/Wt | -5/-9 | Wt/Wt | Wt/Wt | Wt/Wt | +1/ -3 | Wt/Wt | Fertile |
| RPG 20 | Wt/Wt | Wt/Wt | -5/-5 | Wt/Wt | Wt/Wt | Wt/Wt | -1/-4 | Wt/Wt | Wt/Wt | Wt/Wt | -4/-5 | Wt/Wt | Sterile |

**Fig. 3** Examples of edits in 20 transgenic wheat plants transformed with a construct expressing a wheat codon optimised Cas9 and 4 guides targeted to the RPG1 gene. Wt indicates wild type sequence, +/- indicates number of inserted or deleted nucleotides. In this example full knockout of the 6 alleles results in a male sterile phenotype. Reproduced from [25]

| Plant No. | CalS5 A genome guide | | | | CalS5 B genome guide | | | | CalS5 D genome guide | | | | Phenotype |
|---|---|---|---|---|---|---|---|---|---|---|---|---|---|
| | 1 | 2 | 3 | 4 | 1 | 2 | 3 | 4 | 1 | 2 | 3 | 4 | |
| CalS 1 | Wt/Wt | +1/+4 | Wt/Wt | Wt/Wt | Wt/Wt | -936/Wt | -936/Wt | -936/Wt | Wt/Wt | Wt/Wt | -30/Wt | -30/Wt | Fertile |
| CalS 2 | Wt/Wt | +1/Wt | Wt/Wt | Wt/Wt | Wt/Wt | -61/-936 | Wt/-936 | Wt/-936 | Wt/Wt | Wt/Wt | Wt/Wt | Wt/Wt | Fertile |
| CalS 3 | Wt/Wt | -1/-1 | -1/Wt | Wt/Wt | Wt/Wt | -3/-936 | Wt/936 | Wt/-936 | Wt/Wt | -940/Wt | -940/-4 | -940/Wt | Fertile |
| CalS 4 | Wt/Wt | Wt/Wt | Wt/Wt | Wt/Wt | Wt/Wt | +1/Wt | Wt/Wt | Wt/Wt | Wt/Wt | Wt/Wt | Wt/Wt | Wt/Wt | Fertile |
| CalS 5 | Wt/Wt | +1/+1 | Wt/Wt | Wt/Wt | Wt/Wt | Wt/Wt | Wt/Wt | -1/-1 | Wt/Wt | -17/Wt | -62/+1 | -62/Wt | Sterile |
| CalS 6 | Wt/Wt | Wt/Wt | Wt/Wt | Wt/Wt | Wt/Wt | -154/Wt | Wt/Wt | Wt/Wt | Wt/Wt | Wt/Wt | +1/Wt | Wt/Wt | Fertile |
| CalS 7 | Wt/Wt | -1/-1029 | Wt/-1029 | Wt/-1029 | Wt/Wt | -938/Wt | -938/Wt | -938/Wt | Wt/Wt | Wt/Wt | Wt/Wt | Wt/Wt | Fertile |
| CalS 8 | Wt/Wt | -4/+1 | Wt/Wt | Wt/Wt | Wt/Wt | Wt/Wt | Wt/Wt | Wt/Wt | Wt/Wt | Wt/Wt | Wt/Wt | Wt/Wt | Fertile |
| CalS 9 | Wt/Wt | -4/-5 | Wt/Wt | Wt/Wt | Wt/Wt | -2/-542 | Wt/Wt | Wt/Wt | Wt/Wt | Wt/Wt | +1/+1 | Wt/Wt | Sterile |
| CalS 10 | Wt/Wt | -1/-33 | Wt/Wt | Wt/Wt | Wt/Wt | -17/-17 | Wt/Wt | Wt/Wt | Wt/Wt | -5/-5 | Wt/Wt | Wt/Wt | Sterile |
| CalS 11 | Wt/Wt | -1/Wt | Wt/Wt | Wt/Wt | Wt/Wt | Wt/Wt | Wt/Wt | Wt/Wt | Wt/Wt | Wt/Wt | Wt/Wt | -1/Wt | Fertile |
| CalS 12 | Wt/Wt | -4/-7 | +1/-61 | Wt/-61 | Wt/Wt | 4/-17 | Wt/Wt | Wt/Wt | Wt/Wt | -1/-1 | Wt/Wt | Wt/Wt | Sterile |
| CalS 13 | Wt/Wt | -1/Wt | Wt/Wt | Wt/Wt | Wt/Wt | +1/Wt | Wt/Wt | Wt/Wt | Wt/Wt | Wt/Wt | Wt/Wt | Wt/Wt | Fertile |
| CalS 14 | Wt/Wt | Wt/Wt | Wt/Wt | Wt/Wt | Wt/Wt | Wt/Wt | Wt/Wt | Wt/Wt | Wt/Wt | Wt/Wt | Wt/Wt | Wt/Wt | Fertile |
| CalS 15 | Wt/Wt | +118/Wt | Wt/Wt | Wt/Wt | Wt/Wt | missense/-946 | Wt/-946 | Wt/-946 | Wt/Wt | Wt/Wt | Wt/Wt | Wt/Wt | Fertile |
| CalS 16 | Wt/Wt | -1029/Wt | -1029/+1 | -1029/Wt | Wt/Wt | -5/-5 | +1/Wt | Wt/Wt | Wt/Wt | -5/-5 | Wt/Wt | Wt/Wt | Sterile |
| CalS 17 | Wt/Wt | -2/-5 | Wt/Wt | Wt/Wt | Wt/Wt | -5/Wt | Wt/Wt | Wt/ -1 | Wt/Wt | -933/-939 | -933/-939 | -933/-939 | Sterile |
| CalS 18 | Wt/Wt | -963/-963 | -963/-963 | Wt/Wt | Wt/Wt | Wt/Wt | Wt/Wt | Wt/Wt | Wt/Wt | Wt/Wt | Wt/Wt | Wt/Wt | Fertile |
| CalS 19 | Wt/Wt | -963/-965 | -963/-965 | Wt/Wt | Wt/Wt | Wt/Wt | Wt/Wt | Wt/Wt | Wt/Wt | -963/Wt | -963/Wt | Wt/Wt | Fertile |
| CalS 20 | Wt/Wt | -24/+1 | Wt/Wt | Wt/Wt | Wt/Wt | Wt/Wt | -1/+1 | Wt/Wt | Wt/Wt | -1/-1 | Wt/Wt | Wt/Wt | Sterile |

**Fig. 4** Examples of edits in 20 transgenic wheat plants transformed with a construct expressing a wheat codon optimised Cas9 and 4 guides targeted to the CalS gene. Wt indicates wild type sequence, +/- indicates number of inserted or deleted nucleotides. In this example full knockout of the 6 alleles results in a male sterile phenotype. Reproduced from [25]

## Acknowledgements

This work was funded by BBSRC grant BB/R014876/1.

### References

1. Mojica FJM, Juez G, Rodriguez-Valera F (1993) Transcription at different salinities of Haloferax mediterranei sequences adjacent to partially modified PstI sites. Mol Microbiol 9:613–621
2. Ishino Y, Shinagawa H, Makino K et al (1987) Nucleotide sequence of the iap gene, responsible for alkaline phosphatase isoenzyme conversion in Escherichia coli, and identification of the gene product. J Bacteriol 169:5429–5433

3. Cong L, Ran FA, Cox D et al (2013) Multiplex genome engineering using CRISPR/Cas systems. Science (80- ) 339:819–823
4. Shan Q, Wang Y, Li J et al (2013) Targeted genome modification of crop plants using a CRISPR-Cas system. Nature biotechnology, 31(8):686–688
5. Miao J, Guo D, Zhang J, et al (2013) Targeted mutagenesis in rice using CRISPR-Cas system. Cell research, 23(10):1233–1236
6. Kleinstiver BP, Tsai SQ, Prew MS et al (2016) Genome-wide specificities of CRISPR-Cas Cpf1 nucleases in human cells. Nat Biotechnol 34:869–874
7. Mikami M, Toki S, Endo M (2016) Precision targeted mutagenesis via Cas9 paired Nickases in Rice. Plant Cell Physiol 57:1058–1068
8. Zong Y, Wang Y, Li C et al (2017) Precise base editing in rice, wheat and maize with a Cas9-cytidine deaminase fusion. Nat Biotechnol 35:438–440
9. Shimatani Z, Kashojiya S, Takayama M et al (2017) Targeted base editing in rice and tomato using a CRISPR-Cas9 cytidine deaminase fusion. Nat Biotechnol 35:441–443
10. Li C, Zong Y, Wang Y et al (2018) Expanded base editing in rice and wheat using a Cas9-adenosine deaminase fusion. Genome Biol 19:59
11. Hu JH, Miller SM, Geurts MH et al (2018) Evolved Cas9 variants with broad PAM compatibility and high DNA specificity. Nature 556:57–63
12. Wang M, Wang Z, Mao Y et al (2019) Optimizing base editors for improved efficiency and expanded editing scope in rice. Plant Biotechnol J 17:1697–1699
13. Kim HK, Lee S, Kim Y et al (2020) High-throughput analysis of the activities of xCas9, SpCas9-NG and SpCas9 at matched and mismatched target sequences in human cells. Nat Biomed Eng 4:111–124
14. Mikami M, Toki S, Endo M (2015) Comparison of CRISPR/Cas9 expression constructs for efficient targeted mutagenesis in rice. Plant Mol Biol 88:561–572
15. Essletzbichler P, Konopka T, Santoro F et al (2014) Megabase-scale deletion using CRISPR/Cas9 to generate a fully haploid human cell line. Genome Res 24:2059–2065
16. Zhou H, Liu B, Weeks DP et al (2014) Large chromosomal deletions and heritable small genetic changes induced by CRISPR/Cas9 in rice. Nucleic Acids Res 42:10903–10914
17. Brooks C, Nekrasov V, Lipppman ZB et al (2014) Efficient gene editing in tomato in the first generation using the clustered regularly interspaced short palindromic repeats/CRISPR-associated9 system. Plant Physiol 166:1292–1297
18. Sánchez-León S, Gil-Humanes J, Ozuna CV et al (2018) Low-gluten, nontransgenic wheat engineered with CRISPR/Cas9. Plant Biotechnol J 16:902–910
19. Jouanin A, Schaart JG, Boyd LA et al (2019) Outlook for coeliac disease patients: towards bread wheat with hypoimmunogenic gluten by gene editing of α- and γ-gliadin gene families. BMC Plant Biol 19:333
20. Qi W, Zhu T, Tian Z et al (2016) High-efficiency CRISPR/Cas9 multiplex gene editing using the glycine tRNA-processing system-based strategy in maize. BMC Biotechnol 16:58
21. Kapusi E, Corcuera-Gómez M, Melnik S et al (2017) Heritable genomic fragment deletions and small Indels in the putative ENGase gene induced by CRISPR/Cas9 in barley. Front Plant Sci 8:540
22. Milner MJ, Craze M, Hope M et al (2020) Turning up the temperature on CRISPR: increased temperature can improve the editing efficiency of wheat using CRISPR/Cas9. Front Plant Sci 11:1780
23. Naim F, Shand K, Hayashi S et al (2020) Are the current gRNA ranking prediction algorithms useful for genome editing in plants? PLoS One 15:e0227994
24. Howells RM, Craze M, Bowden S et al (2018) Efficient generation of stable, heritable gene edits in wheat using CRISPR/Cas9. BMC Plant Biol 18:215
25. Milner M J, Craze M, Bowden L, et al (2020) Identification of genes involved in male sterility in wheat (Triticum aestivum L.) which could be used in a genic hybrid breeding system. Plant direct, 4(3), e00201
26. Nandy S, Pathak B, Zhao S, et al (2019) Heat-shock-inducible CRISPR/Cas9 system generates heritable mutations in rice. Plant direct, 3 (5), e00145
27. Hahn F, Korolev A, Sanjurjo Loures L et al (2020) A modular cloning toolkit for genome editing in plants. BMC Plant Biol 20:179
28. Wang Y, Cheng X, Shan Q et al (2014) Simultaneous editing of three homoeoalleles in hexaploid bread wheat confers heritable resistance to powdery mildew. Nat Biotechnol 32:947–951
29. Shan Q, Wang Y, Li J et al (2014) Genome editing in rice and wheat using the CRISPR/Cas system. Nat Protoc 9:2395–2410
30. Stemmer M, Thumberger T, del Sol Keyer M et al (2015) CCTop: an intuitive, flexible and

reliable CRISPR/Cas9 target prediction tool. PLoS One 10:e0124633

31. Liu H, Ding Y, Zhou Y et al (2017) CRISPR-P 2.0: an improved CRISPR-Cas9 tool for genome editing in plants. Molecular plant, 10 (3): 530–532

32. Cram D, Kulkarni M, Buchwaldt M et al (2019) WheatCRISPR: a web-based guide RNA design tool for CRISPR/Cas9-mediated genome editing in wheat. BMC Plant Biol:19

33. Labun K, Montague TG, Krause M et al (2019) CHOPCHOP v3: expanding the CRISPR web toolbox beyond genome editing. Nucleic Acids Res 47:W171–W174

34. Bates R, Craze M, Wallington EJ (2017) Agrobacterium -mediated transformation of oilseed rape ( Brassica napus ). In: Current protocols in plant biology, Hoboken, NJ, pp 287–298

35. Ishida Y, Tsunashima M, Hiei Y, et al (2015) Wheat (Triticum aestivum L.) transformation using immature embryos, in: methods in molecular biology (Clifton, N.J.), pp. 189–198

36. Richardson T, Thistleton J, Higgins TJ et al (2014) Efficient agrobacterium transformation of elite wheat germplasm without selection. Plant Cell Tissue Organ Cult 119:647–659

37. Risacher T, Craze M, Bowden S, et al (2009) Highly efficient agrobacterium-mediated transformation of wheat via in planta inoculation, in: methods in molecular biology (Clifton, N.J.), pp. 115–124

38. Jones HD, Doherty A, and Wu H (2005), Review of methodologies and a protocol for the agrobacterium-mediated transformation of wheat

39. Hamada H, Linghu Q, Nagira Y et al (2017) An in planta biolistic method for stable wheat transformation. Sci Rep 7:1–8

40. Liang Z, Chen K, Li T et al (2017) Efficient DNA-free genome editing of bread wheat using CRISPR/Cas9 ribonucleoprotein complexes. Nat Commun 8:1–5

41. Stacey J, Isaac PG (1994) Isolation of DNA from plants. Methods Mol Biol 28:9–15

42. Milner MJ, Howells RM, Craze M et al (2018) A PSTOL-like gene, TaPSTOL, controls a number of agronomically important traits in wheat. BMC Plant Biol 18:115

43. Glenn G, Andreou LV (2013) Analysis of DNA by southern blotting. In: Methods in enzymology. Academic Press Inc, pp 47–63

44. Hussain M, Iqbal MA, Till BJ et al (2018) Identification of induced mutations in hexaploid wheat genome using exome capture assay. PLoS One 13:e0201918

45. Jouanin A, Borm T, Boyd LA et al (2019) Development of the GlutEnSeq capture system for sequencing gluten gene families in hexaploid bread wheat with deletions or mutations induced by γ-irradiation or CRISPR/Cas9. J Cereal Sci 88:157–166

46. Ghosh S, Watson A, Gonzalez-Navarro OE et al (2018) Speed breeding in growth chambers and glasshouses for crop breeding and model plant research. Nat Protoc 13:2944–2963

# Chapter 12

# A Genotyping Protocol to Identify CRISPR/Cas9-Edited Events in Hexaploid Wheat

Johann Schernthaner, Margaret Balcerzak, Jhadeswar Murmu, and Rajagopal Subramaniam

## Abstract

Targeted genome editing using the clustered regularly interspaced short palindromic repeats (CRISPR)/Cas9 system has become an established method to introduce changes into a plant genome. While traditional genetic modification allows us to follow transgenic events with selection markers as they are often co-integrated, this is not always feasible when Cas9/guide RNA-based methods are applied. In cases where genome editing does not result in a selectable phenotype, the only option to analyze editing events is by sequencing the target region through either conventional or high-throughput sequencing, both of which can be tedious and costly. In this chapter, we present a rapid, economic genotyping method that was designed to identify editing events in the three wheat sub-genomes. The method is not only economical but is in general very useful when dealing with polyploid organisms where identification and characterization of edited events can be challenging.

**Keywords** CRISPR, Cas9, Guide RNA, Genotyping, Universal primer, Fluorescence-labeled primer, FAM, NED, VIC

## 1 Introduction

Since its first appearance in 1989 [1], genotyping has been an indispensable genetic research and diagnostic tool where analysis of small differences at a nucleotide level between sequences is required. At present, capillary electrophoresis fluorescence detection systems in combination with sophisticated software are able to precisely determine the length of a PCR product, provided that the PCR product size is within the manageable range of the electrophoresis system. Genome editing by the CRISPR/Cas9 system and the subsequent repair by the nonhomologous end-joining repair mechanism typically results in small deletions and, on rarer occasions, in small insertions [2]. Editing events can be analyzed in a number of different ways. Typically, methods such as high-throughput sequencing (HTS) are highly successful with haploid

Andriy Bilichak and John D. Laurie (eds.), *Accelerated Breeding of Cereal Crops*, Springer Protocols Handbooks, https://doi.org/10.1007/978-1-0716-1526-3_12, © Springer Science+Business Media, LLC, part of Springer Nature 2022

or diploid organisms; however, this approach can make sequence analysis very cumbersome with higher ploidy genomes such as in hexaploid wheat. Amplicon sequencing has been used in characterizing editing in polyploidy genomes, but it can be expensive where a larger number of samples need to be analyzed.

In contrast, genotyping is a fairly simple and economical method for analyzing a large number of progenies due to the reduced costs of separation by capillary electrophoresis. Based on a previously reported method [3], we demonstrate how to use a differentially fluorescence-labeled universal primer in combination with homoeolog-specific amplification primers to analyze editing events in a hexaploid wheat [4]. We utilize single nucleotide polymorphisms (SNPs) and/or indels to design amplification primers for each sub-genome. The resulting PCR products from the A, B, and D genome (wild type and edited) are then re-amplified with universal primers labeled with either the fluorescence compound FAM, NED, or VIC (Fig. 1). The PCR products are then combined and separated on a capillary electrophoresis system along a size standard. The resulting chromatograms show the precise length of the fragments, thus revealing the presence or absence of editing events in each of the three sub-genomes. We demonstrate this method on the T1 generation of wheat cv. Fielder in which deletions within the *Golden2-like (GLK2)* gene have been generated by CRISPR/Cas9.

## 2 Materials

1. Stock buffer solutions (*see* **Note 1**). pH-adjusted Tris buffers were ordered from Sigma-Aldrich, Canada, in 1 M stock solutions and stored at room temperature.
2. Water was prepared using ultrapure water (MilliQ) at a sensitivity of 18 MΩ-cm at 25 °C.
3. Potting soil and containers to grow wheat seedlings. We used 18-well trays with PRO-MIX MPV potting mixture (Premier Tech Horticulture, Rivière-du-Loup, QC, Canada).
4. Plant growth chamber (or equivalent). We used Conviron PGC 20 CMP model 6050 cabinets (Winnipeg, Manitoba, Canada) at 16–20 °C, 40–60% humidity, under long-day conditions (16 h light/8 h dark cycle), and a fluorescent light intensity of 130–160 μmol photons $m^{-2}$ $s^{-1}$.
5. Plant DNA extraction kit (*see* **Note 2**). In the example given, DNA was extracted using the Illustra Nucleon Phytopure genomic DNA extraction kit (GE Healthcare).
6. Liquid nitrogen for freezing plant tissue.

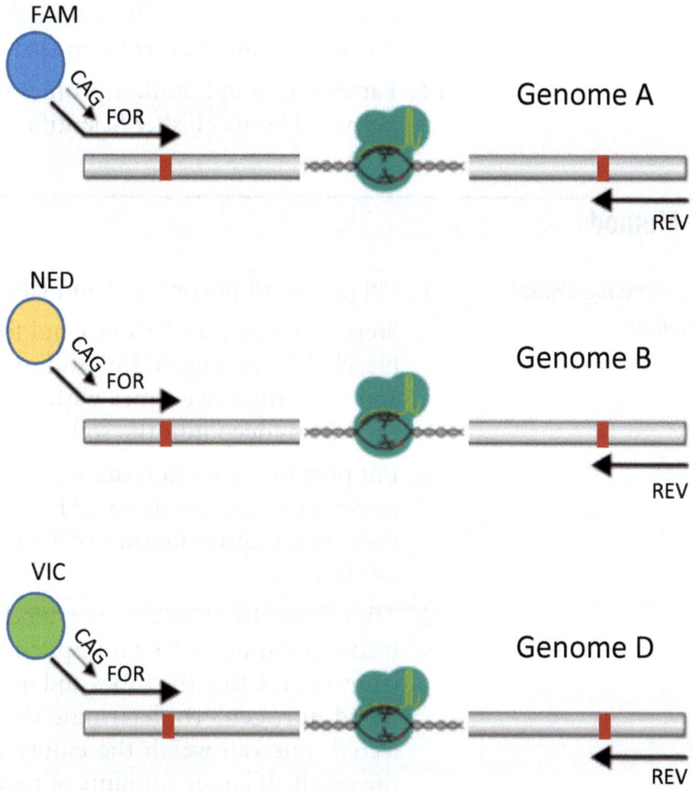


**Fig. 1** The general outline of the two-step PCR amplification is shown. In the first round, forward and reverse primers specific for target gene in either the A, B, or D genome are used to generate a product from each sub-genome. In the second round, a universal primer labeled with either the FAM, NED, or VIC fluorescence dye together with the reverse primer from the first round are used to re-amplify the first-round PCR product. The three PCR products are then pooled and injected together with a LIZ-labeled size standard into the capillary electrophoresis system

7. Blue pestles (VWR Canada) that fit the bottom of a 1.5 ml Eppendorf tube for grinding frozen plant tissue.
8. PCR enzymes, buffers, and dNTPs were from New England Biolabs Inc. (*see* **Note 3**).
9. PCR thermocycler.
10. Generic primers were ordered from IDT at 100 μM concentration and diluted to 10 μM with 10 mM Tris, pH 8.
11. Universal primers with fluorescence dyes FAM, NED, and VIC were from Thermo Fisher Scientific. The universal primer sequence CAGTCGGGCGTCATCA was reported in [5] (*see* **Note 4**).
12. Capillary electrophoresis was performed in-house on the IBI 3500 Genetic Analyzer with 8-capillary array.

13. GeneScan 600 LIZ (20–600 nucleotides, Applied Biosystems) was used as internal size standard (*see* **Note 5**).
14. Patterns of amplification profiles were analyzed by GeneMapper v5 (Thermo Fisher Scientific).

## 3 Methods

### 3.1 Growing Wheat Seedlings

1. Fill pots with potting soil and label the containers.
2. Sterilize seeds with 75% ethanol for 1 min followed by 2% v/v bleach (containing 8.25% sodium hypochlorite) for 10 min, and then rinse five times with sterile water. Place 1 seed/pot ca. 0.5 cm deep into the soil and water (*see* **Note 6**).
3. Put pots in a growth chamber at 16–20 °C, 40–60% humidity, under long-day conditions (16 h light/8 h dark cycle), and a fluorescent light intensity of 130–160 μmol photons $m^{-2}$ $s^{-1}$ (*see* **Note 7**).
4. After 2 weeks, excise young tender leaves with a clean razor blade, and transfer to 2 ml Eppendorf tubes with screw caps or equivalent. Close the tubes and immediately submerge them in liquid nitrogen. To determine the exact amount of tissue collected, one can weigh the empty tubes beforehand, and note the result. If larger amounts of tissue are collected, fill a Falcon tube, or a suitable freezer bag, with liquid nitrogen, and quick-freeze the leaves. Keep the tubes/bags on dry ice to prevent quick evaporation of $N_2$. Make sure there are enough leaves remaining on seedlings to allow for continued growth.
5. Store the frozen samples in a −80 °C freezer.

### 3.2 DNA Extraction

1. In order to obtain consistent results, it is advisable to use a dedicated plant DNA extraction kit. Although these are offered by a number of different suppliers, they all work in a similar fashion (*see* **Note 2**).
2. Process the amount of tissue recommended in the kit's manual. In our experiments, leaf tissues were homogenized under liquid nitrogen to powder in 1.5 ml Eppendorf tubes, with sterile blue pestles that fit the rounded shape of the tube bottom. This can be done either manually or with a tissue homogenizer (IKA-Eurostar power basic) (*see* **Note 8**).
3. Measure the DNA concentration using a suitable instrument, e.g., NanoDrop (Thermo Fisher Scientific). The 260/280 nm ratio should be between 1.7 and 2.0, which indicates good quality DNA (*see* **Note 9**).

4. Check the integrity of the DNA by running 50–100 ng DNA on an agarose gel. A single band above a 20 kbp marker with no visible smear indicates good integrity.
5. Dilute the DNA with 10 mM Tris, pH 8, to a final concentration of 60 ng/μl.

### 3.3 Primer Design

1. Design primers such that the targeted mutation site (the binding sequence for the gRNA(s)) is in the middle of a 400 to 500 bp region (*see* **Note 10**). One hundred bp shifts to the left or right are likely still within range. The idea is to avoid potential overlap of the primer site with the edited site, the extent of which is not known yet but is typically within a few nucleotides where a single gRNA is used. If two gRNAs are used in close proximity, that distance has to be taken into consideration. Primer design tools are freely available, often from oligo suppliers.
2. Align the three target sequences from each sub-genome using suitable software. Make sure the sequence differences such as SNPs and indels are highlighted.
3. Pick a SNP or indel on one sub-genome that is not in the other two, and anchor the 3′ end of the primer to this nucleotide. The more sequence differences in the primer site compared to the other homoeologs can be found, the better (*see* **Note 11**). Do this with all the forward and reverse primers. To the forward primers, add the universal primer sequence so that the final sequence looks like the following: 5′CAGTCGGGCGT CATCA.....forward primer sequence...0.3′ (see Fig. 1).
4. Try to achieve a primer Tm of 58–60 °C for the forward (not including the universal primer part) and reverse primer.
5. Design two to three primers for each priming site in order to have an alternative should one primer not give the desired results.
6. Order the universal primer sequence CAGTCGGGCGT CATCA with either FAM, NED, or VIC modifications on their 5′ end. We ordered our universal primers from Thermo Fisher Scientific.
7. In a laminar flow hood or designated PCR prep hood, prepare 500 μl volume working solutions for each primer by diluting the original stock with 10 mM Tris, pH 7.6, to a 10 mM final concentration. Store the primers in a −20 °C freezer.

### 3.4 First Round of PCR

1. In an area designated only for PCR work, set up the reactions for the first round of PCR.
2. Prepare a master mix consisting of PCR reaction buffer, dNTP, water (if needed), and polymerase in an Eppendorf tube on ice

(*see* **Note 12**). Concentrations are usually given in the supplier's manual.

3. Distribute the required master mix volume to each sample tube/plate. The tubes or plates should be kept on ice. For plates, cooling racks that can be kept at −20 °C are commercially available.
4. Add the primers and sample DNA to each tube and close the tubes, or if plates are used, cover them with a suitable foil or mat. Always process Cas9-edited and unedited samples together.
5. Start the PCR cycler, start the program, and wait until it has reached the temperature for the first step denaturation.
6. Insert the tubes/plates, close the lid, and let the thermocycler do its work(*see* **Note 13**).
7. In our example, we used the following setup and cycling conditions: 25 μl of the PCR reaction included 5 μl of 5 × Q5 Reaction Buffer, 1.0 μl of 10 mM dNTPs, 1.0 μl of 10 μM of each primer, 0.25 μl of Q5 High-Fidelity DNA Polymerase (*see* **Note 14**), 15.75 μl of PCR grade water, and 4.0 μl of 60 ng/μl genomic DNA (*see* **Note 15**). Cycling conditions for the touchdown PCR were as follows: initial denaturing at 98 °C for 3 min followed by 10 cycles of 98 °C for 10 seconds, 68 °C (cycle 1) for 30 seconds, and 72 °C for 30 seconds. The annealing temperature was reduced by 1 °C per cycle until it reached 58 °C. This touchdown phase was followed by 35 cycles of 98 °C for 10 s, 58 °C for 30 s, 72 °C for 30 s, and a final extension step at 72 °C for 5 min (*see* **Note 16**).
8. Evaluate the PCR with 5 μl of the product on an agarose gel to confirm the integrity of the product such as right size and presence of a single band. If multiple bands occur, alternative primers should be tested (*see* **Note 17**). Clean up successful reactions with a suitable system to get rid of excess primers that could be competing with the labeled universal primer. We routinely use ExoSAP-IT PCR Product Cleanup Reagent (Thermo Fisher Scientific) according to the manufacturer protocol. This method is simple to use and saves time when many samples have to be processed. PCR products were cleaned up by mixing 3 μl of a PCR reaction product with 1 μl of ExoSAP-IT.

### 3.5 Second Round of PCR

The concept of the two-round amplification is illustrated in Fig. 1.

1. Dilute the first-round PCR product a hundredfold in water (*see* **Note 18**).
2. Set up the PCR reactions in the same way using the same components as for the first round except replace the forward

primer used in the first round with the corresponding universal FAM, NED, or VIC-labeled primer.

3. Cycling parameters are the same as for the first round except that after the touchdown phase, only 25 cycles are added.

### *3.6 Sample Preparation for Capillary Electrophoresis*

1. Determine the right sample concentration for capillary electrophoresis by preparing dilutions of second-round PCR products (each color of one edited and one unedited sample). Suggested dilution factors are 10×, 50×, 100×, and 200× with water as diluent. The objective is to obtain a clearly readable chromatogram with sharp peaks.
2. Mix 0.2 μl of LIZ-labeled size standard and 8.8 μl high-grade formamide (both available from Applied Biosystems) with 1 μl of diluted second-round PCR product, and submit samples to the genotyping service provider.
3. When the proper dilution factor has been established, process the remaining samples accordingly (*see* **Note 19**).
4. Analyze the results and chose the seedlings to propagate further. An example of a genotyping chromato gram is shown in Fig. 2.

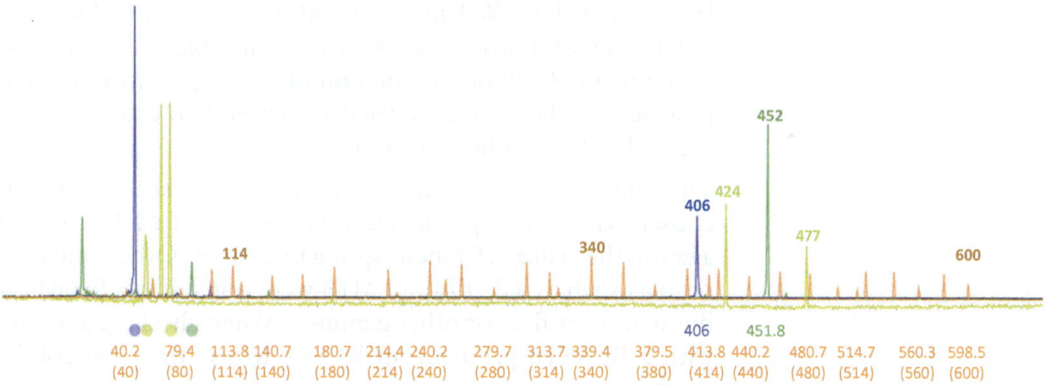


**Fig. 2** The electropherogram resulting from the capillary electrophoresis shows the Cas9-edited fragment sizes belonging to each sub-genome. Blue depicts the PCR product from genome A, yellow from genome B, and green from genome D. The LIZ-labeled size standards are in brown. Integrated software determines the exact fragment size. The size of the Cas9-generated deletions can be inferred from the electropherogram of the unedited samples (not shown). Shown here is a sample with a single bp deletion in genome A, a 12 bp deletion in genome D, and a 53 bp deletion in genome B. The fragment sizes are indicated. The respective sizes for the unedited fragments in this experiment are 407 bp for A, 477 bp for B, and 464 bp for D. PCR of the target in genome B resulted in two products: one edited and one unedited. The explanation for this is that only one of the two parental chromosomes was edited in the T0 and carried forward to the T1 shown here. The peaks at the left represent leftover primers and partial extension products

### 3.7 Finishing up

1. Once suitable homozygous lines have been chosen, determine the final target DNA sequence by dye-terminator sequencing (Applied Biosystems) using the same homoeolog-specific primers that were used for the first-round PCR. Both strands should be sequenced to verify the exact positions and sequences of the changes introduced by the Cas9/gRNA.

## 4 Notes

1. Stock buffer solutions can be prepared in the lab, but if the solutions go into the PCR process, it is recommended to obtain stock solutions from a supplier in order to avoid any contamination that can occur through in-lab preparation.
2. We use kits to extract plant DNA which work fine for DNA extractions from wheat seedlings. Depending on the tissue source, extracted DNA may be contaminated with polysaccharides or other compounds that can inhibit PCR. In this case, DNA concentration should be stepwise diluted with water until a PCR product emerges. If that is not successful, DNA extraction protocols involving CTAB [6] will most likely solve the problem.
3. We used PCR enzymes, buffers, and dNTPs from New England Biolabs Inc. We had very good outcomes with Q5 High-Fidelity DNA Polymerase as this enzyme seemed to be less prone to PCR-inhibiting compounds. Non-proofreading Taq polymerases should not be used as they add an extra dATP to the end of the synthesized strand.
4. The universal primer sequence used in these experiments was chosen based on it producing a "no hit" in a BLAST search against the wheat (Chinese spring) genome. Other universal primers such as T3, T7, and M13 were not suitable for wheat but may work fine for other genomes. When checking a primer with a BLASTn protocol, make sure to use the protocol for short sequences.
5. We used the GeneScan 600 LIZ (20–600 nucleotides range, Applied Biosystems) as an internal size standard. The choice of the fluorescence label of the standard (in this case LIZ) and of the universal primers is depending on the capillary electrophoresis equipment used and what channels are available for reporting. Different equipment may have different options, so it will be important to determine up front what combination of fluorescence colors can be used on the primers and standard.
6. Wheat seeds need to be vernalized for 2 weeks before planting.
7. If needed, wheat seeds can also be grown under sterile conditions. In this case, seeds are sterilized with 75% ethanol for

1 min followed by 50% bleach (containing 8.25% sodium hypochlorite) for 10 min and then rinsed five times with sterile water. The plants are grown in sterilized Magenta™ boxes (e.g., W × L × H: 77 mm × 77 mm × 97 mm, Sigma-Aldrich, MO) containing MS medium (4.2 g/l Murashige and Skoog Salts, 10 g/l sucrose, 3 g/l phytagel, pH 5.8), in a growth chamber in the same conditions as given in Sect. 3.1.

8. **Important:** Do not close the containers tightly before the $N_2$ is evaporated, or else they might rupture or even explode!
9. A 260/280 nm absorbance ratio outside the 1.7–2 range may indicate contamination with some buffer components in the extraction kit, such as Guanidine-HCl or similar, or tissue components, such as proteins, polysaccharide, and/or polyphenols. Often it is an indication of insufficient cleanup in the washing step of the kit protocol. Another quality indicator is the 260/230 nm absorbance ratio, which should be in the 2.0–2.2 range. In general, chemical contamination from nucleic acid extraction procedures may result in an overestimation of the nucleic acid concentration and could potentially influence downstream analysis.
10. From our experience with Cas9 editing in wheat, it is highly advisable to obtain sequences of the target regions beforehand if one is using a cultivar different from Chinese spring, for which a complete genome sequence is available. Other cultivars will invariably have SNPs or other mutations even within a comparatively small region like a single gene. Sequencing the target gene serves two purposes: (a) to ensure that the gRNA matches completely and (b) to confidently pick primer sites, which allow for homoeolog-specific primer design and PCR amplification. A possible strategy for sequencing target regions would be to set primers based on available genome sequence data and amplifying a ~ 1000 bp region, subcloning the PCR fragments into a suitable plasmid and sequencing the inserts.
11. It is advisable to choose a different PCR product size for each homoeolog in order to obtain spaced-out peaks in the chromatogram. The product size, however, should not exceed 500 bp so as to not compromise the resolution of the capillary gel. In most cases this will not be an issue since the deletions from the editing events often are greater than 5 bp; however, we have observed deletions/insertions of only one bp on more than one occasion. With PCR fragments >500 bp, a 1 bp difference may not be resolved properly by the capillary electrophoresis.
12. A PCR master mix should in general contain as many components as possible in order to reduce pipetting steps. Calculate the volume of the master mix such that the amount is sufficient

for one to two more samples than needed. This will avoid having not enough master mix in the last sample if, for instance, the pipette is not drawing the exact volume.

13. Putting the samples into the hot plate of the thermocycler helps prevent false priming/extension during the warm-up phase. This is not necessary for so-called "Hot Start" polymerases.
14. We always use high-fidelity polymerases; although generic, non-proofreading polymerases may work as well. As the protocol requires 2 PCR amplifications, switching from proofreading in the first round to non-proofreading enzymes in the second round should be avoided since non-proofreading polymerases add an additional nucleotide (dATP) to the end of the synthesized strand, which could lead to split peaks in the chromatogram.
15. This is just a suggested setup. To save on components, smaller reaction volumes can of course be used. It is best to stick to the recommendations of the PCR enzyme supplier. In our hands though, the first round of PCR on wheat samples could only be reliably repeated by using a touchdown protocol, in this case, lowering the primer annealing temperature by 1 °C each round for ten rounds and then continuing with the annealing temperature required for the primer. This approach helped to ensure correct priming if a SNP was used at the 3′ end of the primer to obtain homoeolog-specific amplification. This step may not be necessary with diploid genomes.
16. The number of cycles for the first round may have to be determined empirically. For hexaploid wheat, it is very much depending on the DNA quality. We have been successful with 30 cycles but also had samples which required 40 cycles of amplification.
17. It is not unusual to see a double band in the T1 generation analysis. This can happen in a case where only one of the two parental chromosomes is edited, resulting in a large enough gap that produces two visible bands with one band co-migrating with the unedited control (see Fig. 2). However, in most cases, the differences between edited and unedited events are too small to be resolved on an agarose gel, and samples which gave a PCR product that co-migrates with the wild-type product result should be genotyped regardless.
18. The dilution factor is likely not critical. We have also successfully used undiluted first-round PCR samples. A general rule, however, is to keep the sample concentration in a PCR as low as possible.
19. The example of sample preparation for capillary electrophoresis given here was in accordance with our service provider and

adjusted to the capillary electrophoresis system used. They are not generally applicable. The proper sample dilution and pooling methods should be established with the service provider's recommendations to obtain best results.

## References


1. Litt M, Luty JA (1989) A hypervariable microsatellite revealed by in vitro amplification of a dinucleotide repeat within the cardiac muscle actin gene. Am J Hum Genet 44:397
2. Bhaya D, Davison M, Barrangou R (2011) CRISPR-Cas systems in bacteria and archaea: versatile small RNAs for adaptive defense and regulation. Annu Rev Genet 45:273–297
3. Schuelke M (2000) An economic method for the fluorescent labeling of PCR fragments. Nat Biotechnol 18:233–234
4. Cui X, Balcerzak M, Schernthaner J, Babic V, Datla R, Brauer EK, Labbé N, Subramaniam R, Ouellet T (2019) An optimised CRISPR/Cas9 protocol to create targeted mutations in homoeologous genes and an efficient genotyping protocol to identify edited events in wheat. Plant Methods 15:1–12
5. Shirk RY, Glenn TC, Chang S, Hamrick JL (2013) Development and characterization of microsatellite primers in Geranium carolinianum (Geraniaceae) with 454 sequencing. Appl Plant Sci 1:1300006
6. Porebski S, Bailey LG, Baum BR (1997) Modification of a CTAB DNA extraction protocol for plants containing high polysaccharide and polyphenol components. Plant Mol Biol Report 15:8–15

# Chapter 13

# qPCR Methods for the Quantification of Transgene Insert Copy Number and Zygosity Using the Comparative Ct Method in Transgenic *Sorghum bicolor L.* Moench

**Joel Hague, Kimberly Nelson, Alex Yonchak, and Albert P. Kausch**


## Abstract

The development and analysis of transgenic plants is a costly and time-consuming process. A major focus of this analysis is a determination of the structure and number of T-DNA insertions in each transgenic event. Traditionally, this has been accomplished via a combination of standard end-point PCR and laborious Southern blotting techniques. Whereas end-point PCR can generally only yield presence/absence information regarding transgenes, Southern blot analysis can yield highly accurate information regarding the structure of transgene insertions as well as the number of insertions. While this process is reliable and routine, it requires relatively large amounts of high-quality high-molecular-weight plant genomic DNA (gDNA), several days to obtain a result, and sometimes multiple probes overlapping the sequence(s) of interest. In contrast, the quantitative real-time polymerase chain reaction (qPCR) requires relatively little gDNA, which does not need to be high-molecular weight; is easy to set up, requiring less labor than blotting techniques; and can deliver a result inside of 1 h. Herein, we describe a simple and rapid method using a fluorogenic 5′ nuclease "TaqMan®" assay in conjunction with the comparative ($2^{-\Delta\Delta Ct}$) quantification method to determine transgene copy number in transgenic *Sorghum bicolor*. The same assay can be used to determine the zygosity of homozygous and hemizygous transgenic segregants in a $T_1$ population. A method to extract large amounts of good-quality high-molecular-weight gDNA suitable for Southern blotting is also provided for additional validation of the qPCR assay.


**Key words** Transgene copy number, Quantitative real-time PCR, Zygosity

## 1 Introduction

Stable genetic transformation of plants results in the insertion and integration of transgenes in one or more copies into the native plant's genome. After the production of primary transformants ($T_0$ events), molecular analysis is used to determine which plants contain the transgenes and how many copies are integrated. Plants with simple transgene insertions (one or two intact transgene copies) compared with complex insertions (three or greater with possible truncations and/or rearrangements) are preferred for transgene

Andriy Bilichak and John D. Laurie (eds.), *Accelerated Breeding of Cereal Crops*, Springer Protocols Handbooks, https://doi.org/10.1007/978-1-0716-1526-3_13, © Springer Science+Business Media, LLC, part of Springer Nature 2022

expression analysis and stable inheritance without rearrangements. Thus, identifying single copy, hemizygous plants as early as possible simplifies the production and identification of transgene homozygotes in subsequent generations. Southern analysis [1] has been the primary method of determining transgene copy number [2–4]. This involves restriction enzyme digestion of genomic plant DNA, often overnight, followed by gel electrophoresis, and transfer of the DNA to a nitrocellulose or nylon membrane, which is in turn hybridized with a labelled DNA probe complementary to the transgene to produce an informative band pattern. With the appropriate choice of restriction enzyme and probe sequence, the number of bands equals the number of transgene copies per transgenic event [2]. Although Southern analysis is reliable, it is time-consuming and laborious and requires large amounts of plant material. Post-gDNA extraction, a typical Southern blot procedure takes several days to reach a result. If a restriction site is lost or if a transgenic event carries concatemers of T-DNA inserts, the number of bands will not correspond to the copy number. Furthermore, it is often not easy to distinguish hemizygotes from homozygotes by Southern analysis as they will have the same banding pattern. In contrast, a quantitative real-time polymerase chain reaction (qPCR) can take just a few hours to set up and run, and it can produce quantitative data in both digital and graphic formats depending upon the qPCR system used. In addition, qPCR does not require high-quality, high-molecular-weight gDNA for analysis, only PCR-quality DNA preparations. Therefore, an accurate, rapid, and low-labor method for screening $T_0$ transformants and $T_1$ segregants using TaqMan chemistry and comparative ($2^{-\Delta\Delta Ct}$) quantification [5] is described here, which can be used for the identification of events with one or two copies of the integrated transgene. Many transgenic plant species require significant resources including greenhouse space, inputs, personnel, and time, and therefore it is preferable to rapidly identify single-copy transgene homozygotes for subsequent phenotypic analysis and breeding. These events are more likely to exhibit stable transgene expression, Mendelian inheritance without rearrangements or gene silencing issues [6–10]. This assay also allows for copy number estimation in plants carrying three or greater transgene inserts.

## 2 Materials

### *2.1 Miscellaneous Supplies and Equipment*

1. Bead mill ("Beater") that can accommodate 15 mL tubes, e.g., a Harbil 5G-HD Heavy Duty Shaker (*see* **Note 1**).
2. Bead mill ("Beater") that can accommodate 2 mL tubes, e.g., a Qiagen TissueLyser II (*see* **Note 1**).
3. 15 mL RNase-/DNase-free polypropylene centrifuge tubes.

4. 4.8 mm stainless steel beads (Next Advance: nextadvance.com).
5. 10 mg/mL RNase A solution (Qiagen).
6. 65 °C water bath.
7. 2-Propanol (isopropanol).
8. Chloroform. This is a highly toxic reagent that needs to be handed carefully. Read the SDS/MSDS prior to handling, follow a Safe Standard Operating Procedure (SOP), wear appropriate personal protective equipment (PPE), and dispose of all associated wastes properly.
9. Equilibrated phenol, pH 8.0. This is a highly toxic reagent that needs to be handed carefully. Read the SDS/MSDS prior to handling, follow a Safe Standard Operating Procedure (SOP), wear appropriate personal protective equipment (PPE), and dispose of all associated wastes properly.
10. Benchtop centrifuge that can accommodate 15 mL centrifuge tubes, e.g., an Eppendorf 5810 R.
11. Benchtop microcentrifuge that can accommodate 1.5 mL microcentrifuge tubes, e.g., an Eppendorf 5424.
12. Benchtop microcentrifuge for routine spin downs of 2.0 mL microcentrifuge tubes and 0.2 mL PCR tubes.

### 2.2 Urea-Based gDNA Extraction Buffer (Derived and Modified from [1, 2, 11]) for High-Molecular-Weight gDNA Extraction

All solutions should be prepared using water of at least type II quality; type I is preferable and should be used if possible. Analytical grade reagents should be used to prepare stock and working solutions for all procedures, taking precautions to prevent any cross-contamination of the gDNA samples, oligonucleotide primers and TaqMan probes, and qPCR master mixes. Safe and Standard Operating Procedures (SOPs) should be developed and followed to ensure staff safety as well as proper disposal of hazardous wastes.

#### 2.2.1 Stock Solutions

1. PVP stock: Place 400 mL water in a 1000 mL glass beaker and set stirring with a stir bar on a hot plate. Gradually add 10 g PVP-10 (polyvinylpyrrolidone). Heat gently on the hot plate until dissolved. Dispense to a 1000 mL graduated cylinder and bring the volume up to 650 mL with water. Transfer to a 1 L Pyrex™ bottle and autoclave at 120 °C for 15 min.
2. 1 M Tris, pH 8: Place 8.88 g Tris-HCL and 5.30 g Tris base into 100 mL graduated cylinder. Add water to the graduated cylinder to approximately 70 mL and spin until well-mixed and then add water to 100 mL. Transfer to a 250 mL Pyrex bottle and autoclave at 120 °C for 15 min.
3. 0.5 M EDTA, pH 8: Dissolve 18.6 g NaEDTA·2 $H_20$ in 70 mL in a 100 mL beaker. Adjust pH with approximately 4.5 mL 10 N NaOH, and then add NaOH dropwise until pH 8 is

achieved. Add water to 100 mL and mix well. Dispense 50 mL each to 2 × 100 mL Pyrex bottles, and autoclave at 120 °C for 15 min.

4. 50 × TE: Place 50 mL 1 M Tris, pH 8, into a suitable graduated cylinder and add 10 mL of 0.5 M EDTA pH 8 into same graduated cylinder. Add water until 100 mL and mix well. Dispense to a 250 mL Pyrex bottle and autoclave at 120 °C for 15 min.
5. 10 N sodium hydroxide: Dissolve 4 g NaOH in 50 mL water in a 100 mL graduated cylinder. Add water until 100 mL; carefully mix well! Transfer to a 1 × 100 mL plastic bottle. Do not autoclave.
6. 1.25 M sodium sulfate: Place 17.7 g $Na_2SO_4$ into a 100 mL grad cylinder. Add 70 mL and mix well until dissolved, and then add water to 100 mL. Transfer to a 250 mL Pyrex bottle and autoclave at 120 °C for 15 min.
7. 10% N-lauroylsarcosine: Add 50 mL water to a 100 mL grad cylinder. Gradually add 1 g N-lauroylsarcosine (sodium salt) and mix well, adding water in small increments as you go to dissolve the NLS. Bring the water to 100 mL, transfer to a 250 mL Pyrex bottle, and autoclave at 120 °C for 15 min.
8. 5 M ammonium acetate: Add 70 mL of water to a 100 mL graduated cylinder. Gradually add 38.54 g $NH_4AC$ and mix well until dissolved, and then add water to 100 mL. Transfer to a 250 mL Pyrex bottle and autoclave at 120 °C for 15 min.

#### 2.2.2 Working Solutions

The following solutions are most conveniently dispensed during the procedures using "bottle top" dispensers, such as the Brinkmann® Bottle-Top Dispenser.

1. Urea extraction buffer (UEB): Place 650 mL of PVP stock into a 1000 mL beaker with a stir bar. Set stirring, and then gradually add 180 g urea powder. Spin this until dissolved, and then add 50 mL 1 M Tris, pH 8, 50 mL 0.5 M EDTA, pH 8, 50 mL 1.25 M $Na_2SO_4$, and 100 mL 10% NLS. Transfer to a 1 L Pyrex bottle and add an autoclaved; clean bottle-top dispenser. This can be stored on the bench at room temperature for up to 6 months.
2. 5 × TE + acetate + RNase: Add 500 mL of water to a clean, autoclaved 1 L Pyrex bottle. Add 50 mL of 50 × TE, 50 mL of 5 M $NH_4AC$, and 100μL of RNase (100 mg/mL) for a total volume of 600 mL. Do not autoclave. Store this solution at 4 °C for up to 3 months.
3. 1 × TE: Add 20 mL 50 × TE to a 1 L Pyrex bottle; then, add water to 1 L and mix well.

4. 1:1 phenol-chloroform. These reagents are highly toxic and should only be handled in fume hood. 500 mL equilibrated phenol, pH 8.0, should be transferred safely to a 1 L Pyrex bottle in a fume hood. 500 mL chloroform can then be added to the bottle and a bottle-top dispenser added. Carefully wrap the bottle in aluminum foil, and store refrigerated at 4 °C.

### 2.3 Rapid gDNA Extraction for qPCR

1. DNeasy Plant Mini Kit (Qiagen).
2. Tungsten Carbide Beads, 3 mm (Qiagen).
3. 1.7 mL Safe-Lock tubes.
4. Harris Uni-Core sampling tool, 4.00 mm (*see* **Note 2**).

### 2.4 Serial Dilution of gDNA for qPCR Efficiency

1. Accurately calibrated PIPETMAN with appropriate sterile filter tips.
2. Ice.
3. Fisher Vortex Genie 2 or equivalent.
4. Suitable sterile plastic ware (i.e., 0.2 mL PCR tubes) for dilution of gDNA samples and preparation of the master mix (i.e., 1.5 or 2.0 mL microcentrifuge tubes).
5. Benchtop microcentrifuge for routine spin downs of 2.0 mL microcentrifuge tubes and 0.2 mL PCR tubes.
6. A qPCR system – we use the Applied Biosystems ViiA7™—or equivalent qPCR system that can collect raw Ct values and associated data.
7. Compatible DNA/RNase/PCR inhibitor-free plastic ware for the qPCR system such as the Applied Biosystems MicroAmp® Fast Reaction Tube strips and caps.
8. A NanoDrop 2000 spectrophotometer (Thermo Fisher Scientific) or fluorometer (such as a Qubit® 3.0, Life Technologies) for determining DNA concentrations.
9. Endogenous control gene *enol-2* (*Sorghum bicolor* L.) [12, 13] 20 × Custom TaqMan® Gene Expression Assay (Applied Biosystems) or equivalent primer/probe working solutions, where the forward primer sequence (5′-3′) is GTTCATGGGCAT CATTGCAA, the reverse primer sequence (5′-3′) is TGA CAAGTAAATCATCACCCACAA, and the probe sequence (5′-3′) is CCTCAGTCGATATCCA. The primer and probe working concentrations in the qPCR reaction are 900 nM and 250 nM, respectively.
10. *bar* transgene 20 × Custom TaqMan® Gene Expression Assay (Applied Biosystems) or equivalent primer/probe working solutions, where the forward primer sequence (5′-3′) is GCT CTACACCCACCTGCTGAA, the reverse primer sequence

(5′-3′) is ACAGCGACCACGCTCTTGA , and the probe sequence (5′-3′) is CCTGGAGGCACAGGG . The primer and probe working concentrations in the qPCR reaction are 900 nM and 250 nM, respectively.

11. 2× qPCR master mix containing *taq* polymerase, buffer, and dNTPs such as TaqMan® Fast Advanced Master Mix (Applied Biosystems).
12. PCR-quality gDNA prep from a transgenic *Sorghum* plant carrying the transgene to be quantified (*see* **Note 3**).

### 2.5 Determination of Transgene Copy Number

1. Accurately calibrated PIPETMAN with appropriate sterile filter tips.
2. Ice.
3. Fisher Vortex Genie 2 or equivalent.
4. A real-time qPCR system such as the Applied Biosystems ViiA7™ or equivalent.
5. Suitable plasticware for the qPCR system (e.g., Applied Biosystems MicroAmp® Fast Reaction Tube strips and caps), including PCR tubes, caps, and nanocentrifuges for spin-down of components and mixes.
6. Endogenous control gene *enol-2* (*Sorghum bicolor* L.) [12, 13] 20 × Custom TaqMan® Gene Expression Assay (Applied Biosystems) or equivalent primer/probe working solutions, where the forward primer sequence (5′-3′) is GTTCATGGGCAT CATTGCAA , the reverse primer sequence (5′-3′) is TGA CAAGTAAATCATCACCCACAA , and the probe sequence (5′-3′) is CCTCAGTCGATATCCA . The primer and probe working concentrations are 900 and 250 nM, respectively.
7. *bar* transgene Custom TaqMan® Gene Expression Assay (Applied Biosystems) or equivalent primer/probe working solutions, where the forward primer sequence (5′-3′) is GCT CTACACCCACCTGCTGAA , the reverse primer sequence (5′-3′) is ACAGCGACCACGCTCTTGA , and the probe sequence (5′-3′) CCTGGAGGCACAGGG . The primer and probe working concentrations are 900 and 250 nM, respectively.
8. 2× qPCR master mix containing *taq* polymerase, buffer, and dNTPs such as TaqMan® Fast Advanced Master Mix (Applied Biosystems).
9. PCR-quality gDNA prep from a single-copy hemizygous transgenic *Sorghum* plant carrying the transgene to be quantified in the unknown samples (*see* **Note 3**).

## 3 Methods

### 3.1 Determination of Transgene Copy Number and Zygosity

The ability to discern plants with one or two transgene insertions requires an assay that is capable of distinguishing twofold differences in the PCR [14, 15]; the same is true for zygosity analysis. There are two conditions required for valid qPCR copy number and zygosity assay results. First, successful comparative ($2^{-\Delta\Delta Ct}$) quantification requires identification of an endogenous *Sorghum* control gene that will amplify with an approximately equal efficiency as that of the transgene target. As the number of copies of an endogenous gene per genome is a constant, the endogenous control gene normalizes the multiplexed reaction by providing a reference Ct value that is compared to the transgene Ct, yielding ΔCt [ΔCt = Ct (transgene) – Ct (endogenous gene)]. This removes the need to precisely quantify template contributions to the individual qPCR reactions and corrects for any well-to-well variation created by unequal template contributions. Therefore, the selection of a suitable endogenous control gene is critical [5, 14–17]. Second, a "calibrator" sample (plant) must be identified that carries a defined number of T-DNA inserts [16, 18]. For the assay presented here, this sample must contain a single transgene insert per diploid *Sorghum* genome (and is thus hemizygous, single copy) of the same transgene sequence that will be quantified in the "unknown" transgenic plants (see: [2]). Once characterized and identified, seed should be collected and a large stock of gDNA from this sample plant should be collected and kept at −80 °C, as it can perform the role of **h**emizygous **s**ingle-**c**opy **r**eference **c**alibrator (HSCRC) in all subsequent experiments involving *Sorghum* and the identical transgene. This same procedure can be followed for multiple transgenes, i.e., additional primer/probe sets can be developed and validated for the assay for different transgenes. In the assay below, the *bar* transgene [19] is used; however, in principle and in practice [15–18], any transgene can be used.

The HSCRC allows for an accurate copy number determination in unknown samples because all two-copy or greater plants (regardless of zygosity) will have *a lower ΔCt value* than the HSCRC ΔCt, and the difference [ΔΔCt = ΔCt(unknown) – ΔCt (HSCRC)] allows for a copy number calculation using $2^{(-\Delta\Delta Ct)}$ [16]. It follows that all plants carrying three copies or more of the transgene will have even lower ΔCt values relative to the HSCRC, and thus the calculation of $2^{(-\Delta\Delta Ct)}$ will yield an estimate of the relative transgene copy number (*see* **Note 4**) for all unknown samples. Furthermore, this technique can be used to verify the zygosity of $T_1$ segregants produced from a single-copy hemizygous $T_0$ plant, as $T_1$ homozygotes will show a two-copy score, whereas one-copy hemizygotes will show a one-copy score and nulls a 0 score relative to the HSCRC (i.e., a 1 AA:2 Aa:1 aa segregation ratio for the transgene "A"). DNA samples from the $T_1$ homozygotes and hemizygotes can now themselves function as additional quantitative references for future qPCR assays, so long as the unknown DNA samples are prepared in the same fashion as the $T_1$ samples were prepared [14] (*see* **Note 5**). $T_1$ homozygotes can then be moved forward into a suitable breeding scheme.

### *3.1.1 Urea-Based High-Molecular-Weight gDNA Extraction (Modified from [11] for Southern Blotting and qPCR*

This procedure can be used to obtain high-quality, high molecular weight gDNA for Southern Blotting in Sorghum. It can also be used in the qPCR assay.

### *3.1.2 Tissue Collection and Processing (see **Note 1**)*

1. Collect 0.6 g fresh young leaves clipped 2 cm long into 15 mL tubes containing five 4.33 mm stainless steel beads.
2. Remove the caps from the centrifuge tubes and place them back on tube, but do not fasten all the way (i.e., cap loosely).
3. Freeze tissue at −80 °C as quickly as possible after harvest. Let freeze completely.
4. Once completely frozen, remove the tubes and quickly place them into the precooled freeze dryer and freeze-dry the samples following the freeze dryer instructions.

### *3.1.3 DNA Extraction*

1. Homogenize the tissue in 2- to 3-min intervals on a suitable bead mill; this may take up to ten cycles. Periodically check the tissue for the appropriate level of processing as completely powdered is the desired outcome (*see* **Note 1**).
2. Add 6 mL UEB to each tube and vortex vigorously until well-mixed.
3. Incubate the tubes in the 65 °C water bath for 15 min. Then, remove the tubes and let them cool to room temperature.
4. In a fume hood, carefully add 5 mL 1:1 phenol-chloroform from the premade stock to each tube. Cap tightly and mix vigorously inside the fume hood until well-mixed.
5. Spin the tubes in the centrifuge for 10 min at 1811 × *g*.
6. Remove the tubes to a fume hood and using a Pasteur pipette, carefully remove the aqueous phase above the interface into fresh 15 mL tubes containing 5 mL of 100% 2-propanol (IPA), and mix the tubes thoroughly.
7. Spin tubes in centrifuge for 15 min at 1811 × *g*.
8. Remove the tubes and pour off the alcohol into a waste beaker. In a laminar flow hood or on a benchtop, invert the tubes on paper towels (be careful not to lose the pellet) and let air dry for approximately 20 min (or until no alcohol remains).

9. Add 5 mL 5 × TE + acetate + RNase. The pellet must be fully dissolved before the next step. Therefore, the tubes should be incubated at 37 °C overnight.
10. Once the pellets are dissolved, carefully add 5 mL 1:1 phenol-chloroform to each tube from the premade stock. Cap tightly and mix vigorously inside the fume hood until well-mixed.
11. Spin tubes in centrifuge for 5 min 1811 × *g*.
12. Remove the tubes and using a Pasteur pipette; carefully remove the aqueous phase above the interface into fresh 15 mL tubes containing 5 mL of 100% 2-propanol (IPA). Be careful not to break the phase/suck up any organic phase liquid. Mix the tubes thoroughly but gently.
13. Spin the tubes in centrifuge for 10 min at 1811 × *g*.
14. Pour off the alcohol into waste beaker and add 5 mL of 70% ethanol to rinse pellet. Vortex to wash the pellet. Pour off the alcohol into a waste beaker. Respin with alcohol, if it is necessary to re-pellet DNA.
15. In a laminar flow hood, invert the tubes on paper towels (be careful not to lose the pellet!), and let dry for roughly 15 min (or until no alcohol remains).
16. Add 4 mL of TE buffer to each sample; the pellets need to be dissolved as in **step 2** above and can be placed in a warm water bath to aid in dissolving the DNA. Once dissolved, add 400μL of 2.5 M ammonium acetate to each sample. Then, add 9 mL of 95% ethanol, and mix thoroughly but gently by inversion approximately 10 × to precipitate the DNA.
17. Spin the tubes at 1811 × *g* for 5 min.
18. After the spin, pour off the alcohol into a waste beaker taking care not to lose the pellet. Add 5 ml of 70% ethanol to rinse the pellet as described in **step** 7 above. Respin with alcohol if necessary to re-pellet DNA. Pour off alcohol into waste beaker, invert the tubes on paper towels (taking care not to lose the pellet), and let dry for ~15 min (or until no alcohol remains).
19. Once completely dry, add 0.4 ml (400μL) TE. The pellet must completely dissolve into solution. If necessary, transfer the tubes to a 37 °C warm water bath as described above, and let sit until completely dissolved.
20. Once completely in solution, transfer the solutions to pre-labelled (sample ID #) sterile 1.5 mL microcentrifuge tubes. Label a second set of identical tubes. Spin the gDNA tubes in a microcentrifuge at 21,130 × *g* for 5 min. This should remove insoluble material from the prep (you should see a pellet in the tube). Pipette away the liquid into the second tube set. These can now be kept at 4 °C temporarily or at −20 °C for long-term storage.

### 3.2 Rapid gDNA Extraction Using the DNeasy Plant Mini Kit

The DNeasy Plant Mini Kit enables phenol- and chloroform-free recovery of gDNA of sufficient quality and quantity for qPCR. The method described below will yield enough gDNA prep for 50 qPCR assays.

1. Sample 20 4.0 mm leaf disks per plant into a 1.7 mL safe lock tube.
2. Leaving the caps ajar, cover the tubes with parafilm.
3. Poke a single hole in the parafilm for each tube, and freeze the tubes at −80 °C.
4. When the tissue is completely frozen, freeze-dry the material according to the freeze dryer manufacturer's instructions.
5. Once the material is freeze dried, follow the manufacturer's instructions for the kit, eluting in a total volume of 100–150μL AE buffer (10 mM Tris-Cl, 0.5 mM EDTA, pH 9.0) or water.

### 3.3 Serial Dilution of Transgenic gDNA for qPCR Efficiency Determination

The comparative ($2^{-\Delta\Delta Ct}$) quantification method requires that both the endogenous *Sorghum* control gene and target transgene amplify with approximately equal efficiencies. Thus, a 5-log dilution series, using gDNA from a transformed plant carrying the target transgene, is constructed, and duplexed assays are run in triplicate to ensure that the assay meets this condition. The dilution series is used to construct a standard curve, and data from this experiment yields information about the performance of the reaction, in particular the correlation coefficient ($R^2$). Ideally, $R^2 = 1.0$ (0.999 is the maximum value in practice); thus, the $R^2$ value reflects the linearity of the standard curve, and the closer the $R^2$ value is to 1.0, the more efficient the reaction. The $R^2$ values and efficiencies for the endogenous control (*enol-2*) and transgene (*bar*) reactions must be very similar, indicating that (1) there is no competition for reagents and both reactions are highly efficient (i.e., one reaction is not inhibiting another), and therefore (2) the assay is robust and can accurately determine copy number changes among samples having differing gDNA concentrations (Fig. 1).

#### 3.3.1 Preparing the Dilution Series

1. Using a NanoDrop 2000 spectrophotometer (Thermo Fisher Scientific) or fluorometer (such as a Qubit® 3.0, Life Technologies), quantify the concentration of a known transgenic gDNA sample according to the manufacturer's instruction. In this example, a sample having a concentration of 200 ng/μL (approximately 250,000 copies of *enol-2*) is used.
2. Label one sterile 0.2 mL PCR tube with the number "1." Mix the known sample by vortexing, and transfer 10μL of the known sample to tube 1.
3. Prepare four sterile 0.2 mL PCR tubes (numbered 2 through 5) with 18μL of sterile type I or II water.

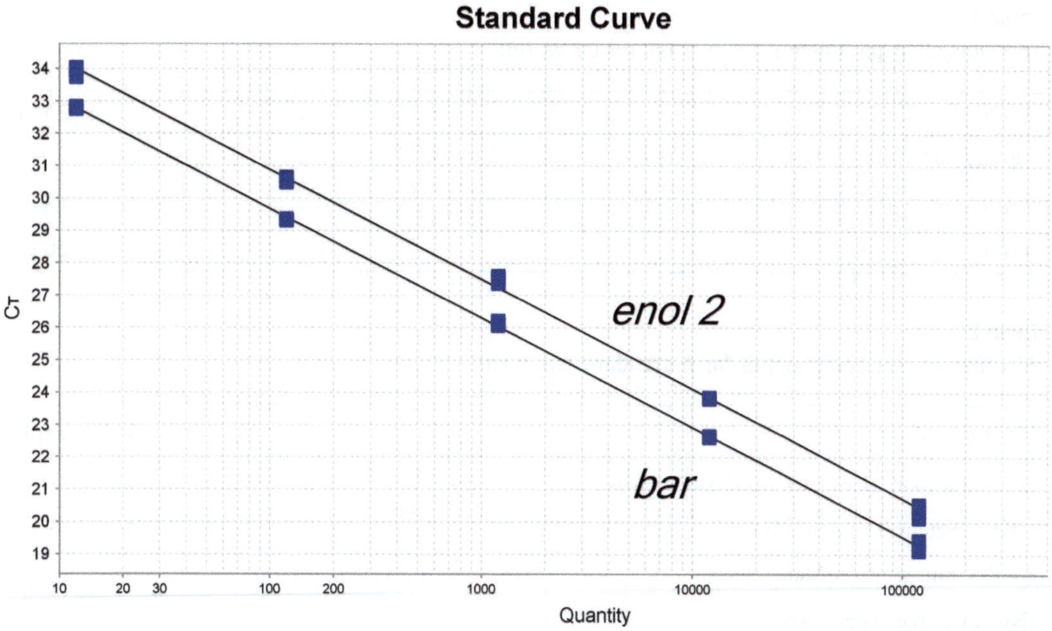


**Fig. 1** A TaqMan® standard curve assay for *bar* and *enol-2* primer/probe sets. gDNA from a transgenic *Sorghum bicolor* plant carrying four *bar* transgene inserts was serially diluted ten-fold and used in multiplexed TaqMan® copy number assays. A logarithmic curve was constructed from the data at each dilution point (represented as approximate total copies present per reaction). The upper curve is for *enol-2*, and has a slope of −3.388, an $R^2 = 0.998$, and an efficiency of 97.33. The lower curve is for *bar* and has a slope of −3.380, $R^2 = 1.0$, and an efficiency of 97.65. The $R^2$ and efficiency values demonstrate that the *bar* and *enol-2* primer and probe sets are suitable for a multiplexed comparative quantification ($2^{-\Delta\Delta Ct}$) assay

4. Mix tube 1 by vortexing, and transfer 2μL of the known sample to tube 2, making 20μL of solution having a 20 ng/μL concentration (approximately 25,000 copies of *enol-2*). In each reaction tube, 2μL of template gDNA is needed, and each reaction is performed in triplicate, so 6μL total is needed per dilution point. This method of preparation produces enough solution to perform the experiment three times; plus allow for minor pipetting errors. Be sure to mix this solution (tube 2) by vortexing prior to the next transfer.
5. Carry out the remaining dilutions in the same fashion, which should result in tube 3 having 2 ng/μL (2500 copies of *enol-2*), tube 4 having 0.2 ng/μL (250 copies of *enol-2*), and tube 5 having 0.02 ng/μL (25 copies of *enol-2*) (*see* **Note 6**).

#### *3.3.2 Carrying out the Standard Curve Experiment*

To set up the standard curve experiment, a reaction for each dilution point as well as a non-template control (NTC) must be performed in triplicate (i.e., three qPCR reactions per dilution point).

1. If necessary, choose the quantification method [e.g., standard curve rather than comparative Ct (ΔΔCt)] using the qPCR system's user interface.

**Table 1**
**Thermocycling parameters for a standard curve run**

| | |
|---|---|
| 95 °C | 2 min |
| *40 cycles of* | |
| 95 °C | 1 s |
| 60 °C | 20 s |

**Table 2**
**PCR master mix components for a standard curve run**

| Component | Volume (μL) |
|---|---|
| 20 × TaqMan® Fast Advanced Master Mix | 10 |
| *bar* transgene TaqMan® Assay | 1 |
| *enol-2* transgene TaqMan® Assay | 1 |
| Nuclease-free, Type I water | 6 |
| gDNA sample | 2 |
| | 20μL total |

2. Enter the tube IDs (i.e., sample names) and locations for the qPCR run using the qPCR system's user interface. Define the diluted samples as standards and the NTC as a negative control.
3. Depending upon the qPCR system's user interface and software, a threshold may be automatically calculated by the software or set by the user. In practice, it is best to let the software calculate the threshold whenever possible.
4. The thermocycling run should follow the manufacturer's instructions; for the 20 × TaqMan® Fast Advanced Master Mix, the following thermocycling parameters can be used (*see* Table 1):
5. Using the 20 × TaqMan® Fast Advanced Master Mix (Applied Biosystems), each PCR reaction is comprised of 20μL total. Given five dilution points and a no-template control, a master mix sufficient for 18 individual reactions should be prepared according to the following list (*see* Table 2):
6. Label the qPCR tubes if necessary, for tracking.
7. Gently vortex the master mix, and then pipette 18μL of master mix into each tube.
8. Carefully pipette 2μL of sterile, DNA-free, type I water into each NTC tube. Immediately cap these tubes (*see* **Note 7**).

9. Pipette 2μL of the diluted gDNA samples into the appropriate tubes, ensuring that the correct sample is placed in the correct tube and no cross-contamination occurs (*see* **Note** 7).
10. Carefully cap the tubes, gently vortex them, and then centrifuge the tubes to ensure each reaction solution is at the bottom of each tube.
11. Place the samples into the qPCR system and start the run.
12. Following the run, check that the NTC samples have Ct values of zero/absent or ≥ 35 cycles (*see* **Note** 7). Check the data to see that results of the experiment fit the parameters (i.e., linearity, efficiency, and $R^2$) laid out above (*see* Fig. 1).

### 3.4 Determination of Transgene Copy Number in Unknown Samples ($T_0$ Transgenics)

When using the Applied Biosystems ViiA7™ qPCR system, the transgene copy number is automatically calculated relative to the **h**emizygous **s**ingle-**c**opy **r**eference **c**alibrator (HSCRC) by the software at the end of the run (*see* Figs. 2 and 3).

1. Choose the quantification method [i.e., comparative Ct (ΔΔCt)] using the qPCR system's user interface if necessary.
2. Enter the tube IDs (i.e., sample names) and locations for the qPCR run using the qPCR system's user interface. Designate the NTC samples as the negative control and remaining samples (including the WT sample) as the unknown samples to be tested.

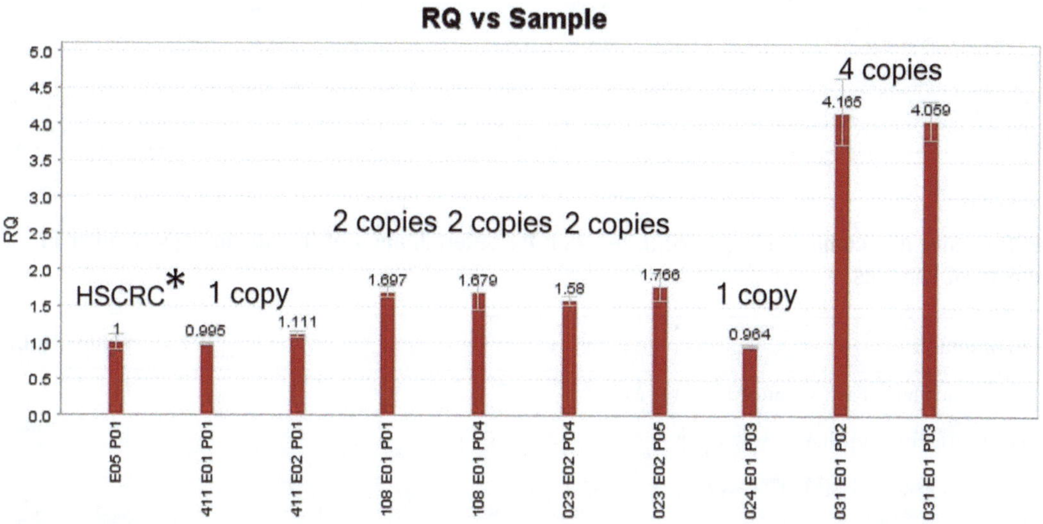


**Fig. 2** Copy number determination in primary transformants ($T_0$ generation). gDNA from $T_0$ events were used in multiplexed TaqMan® copy number assays using *enol-2* and *bar* primer/probe sets. The **h**emizygous **s**ingle-copy **r**eference **c**alibrator (HSCRC) E05 P01 (indicated by an asterisk) is the leftmost sample. Samples to the right of the HSCRC are from individual transgenic plants derived from different transgenic events. Transgenic events carrying one copy, two copies, and four copies of the *bar* transgene are shown

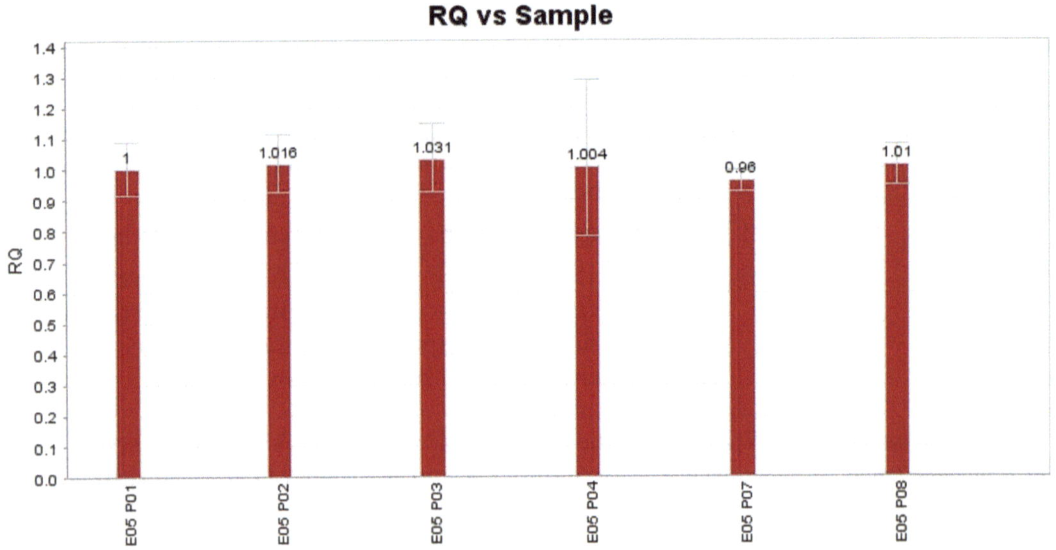


**Fig. 3** Copy number determinations in a single $T_0$ event. gDNA from individual plants regenerated from a single $T_0$ event was used in multiplexed TaqMan® copy number assays. Plant #1 (P01, indicated by an asterisk) was shown to carry a single transgene insert via Southern blot (data not shown) and is a **h**emizygous **s**ingle-**c**opy **r**eference **c**alibrator (HSCRC). Samples to the right of the HSCRC are individual transgenic plants regenerated in tissue culture from the same bialaphos-resistant callus as Plant #1

**Table 3**
**Thermocycling conditions for determination of transgene copy number in unknown samples**

| | |
|---|---|
| 95 °C | 2 min |
| *40 cycles of* | |
| 95 °C | 1 s |
| 60 °C | 20 s |

**Table 4**
**PCR master mix components per 20μL reaction for determination of transgene copy number in unknown samples**

| Component | Volume (μL) |
|---|---|
| 20 × TaqMan® Fast Advanced Master Mix | 10 |
| *bar* transgene TaqMan® Assay | 1 |
| *enol-2* transgene TaqMan® Assay | 1 |
| Nuclease-free, Type I water | 6 |
| gDNA sample | 2 |
| | 20μL total |

3. Depending upon the qPCR system's user interface and software, a threshold may be automatically calculated by the software or set by the user. In practice, it is best to let the software calculate the threshold whenever possible.
4. The thermocycling run should follow the manufacturer's instructions; for the 20 × TaqMan® Fast Advanced Master Mix, the following thermocycling parameters can be used (*see* Table 3):
5. Determine the total number of reactions to be run. Include both an NTC and wild-type (WT) sample as negative controls, and then prepare a master mix using the 20 × TaqMan® Fast Advanced Master Mix (Applied Biosystems) according to the following list (*see* Table 4):
6. Label the qPCR tubes if necessary, for tracking.
7. Gently vortex the master mix, and then pipette 18μL of master mix into each tube.
8. Carefully pipette 2μL of sterile, DNA-free, type I water into each NTC tube. Immediately cap these tubes. Repeat this step with the WT samples.
9. Pipette 2μL of the gDNA unknown samples into the appropriate tubes, ensuring that the correct sample is placed in the correct tube and no cross-contamination occurs.
10. Carefully cap the tubes, gently vortex them, and then centrifuge the tubes to ensure each reaction solution is at the bottom of each tube.
11. Place the samples into the qPCR system and start the run.
12. Following the run, check that the NTC samples have Ct values of zero/absent or ≥35 cycles (*see* **Note 7**).

### 3.5 Determination of Zygosity in $T_1$ Segregants

Once single-copy hemizygous $T_0$ plants are identified via the assay discussed above, these plants can be selfed to produce $T_1$ seed. Planting of the $T_1$ seed should produce plants segregating as homozygotes for the transgene, hemizygotes for the transgene, and nulls (no transgene) (*see* Fig. 4). Plants scored as homozygotes can be further confirmed as homozygotes by selfing and $T_2$ progeny analysis (either by standard PCR or phenotypic analysis). Homozygous lines should show no segregation for the transgene (s) and their associated phenotypes (e.g., all plants carrying a *bar* transgene will be resistant to glufosinate-ammonium herbicides).

## 4 Notes

1. Alternatively, collected plant tissues can be frozen at −80 °C and ground in the presence of liquid nitrogen (rather than

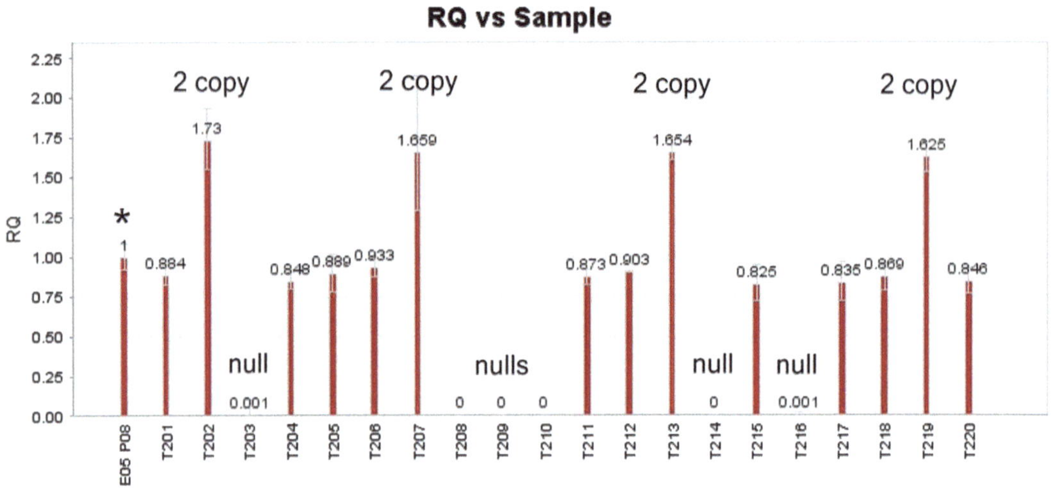


**Fig. 4** Identification of homozygotes, hemizygotes, and nulls in a segregating $T_1$ population. gDNA from $T_1$ plants was used in multiplexed TaqMan® copy number assays. Plant "08" is E05 P08 (event 5, plant 08, indicated by an asterisk), showed a single-copy score (data not shown) relative to a **h**emizygous **s**ingle-**c**opy **r**eference **c**alibrator (HSCRC, E05 P01, see Fig. 3). Plant "08" was allowed to self-fertilize and plants 01 through 20 (grown from $T_1$ seed from "08") are $T_1$ plants segregating for presence of the *bar* transgene in a 4 homozygotes: 10 hemizygotes: 6 nulls ratio. Homozygotes clearly score for two copies of the *bar* transgene relative to "08"

freeze-dried) using a mortar and pestle. Bead mills will be unnecessary for this procedure.

2. It is not necessary to use a Harris UniCore tool for this step, as tissue can be collected using scissors.
3. Although it is not necessary to have very high-quality gDNA for qPCR assays, high-throughput extractions can produce DNA of low-quality that contributes to standard deviations in Ct values. When standard deviations exceed 0.3, the assay accuracy declines and the ability to detect twofold differences is lacking.
4. As the number of transgene copies goes up, the Ct value between plants having a one-copy difference gets smaller, i.e., the Ct difference between an eight-copy plant and seven-copy is just 0.18 cycles [14]. Therefore, as copy number goes up, the likelihood of determining an exact copy number by qPCR goes down.
5. As discussed in Note 3, the gDNA preparation method can have a negative effect on assay accuracy. Therefore, gDNA samples should all be prepared in the same fashion, and preferably yield gDNA of good quality (i.e., having a $Abs_{260/280} = 1.8–2.0$ and $Abs_{260/230} > 2$) and be free of PCR inhibitors.
6. The mass of one copy of *enol-2* is approximately 0.8 pg.

7. If the NTC exhibits a Ct value less 35, it can indicate contamination of the master mix or one of its components (i.e., *bar* transgene TaqMan® assay). Cross-contamination of the master mix, its components, or the gDNA samples will reduce the accuracy of the assay results and should be avoided. If contamination is detected, the assay should be repeated with new reagents and samples if possible.

## References


1. Southern EM (1975) Detection of specific sequences among DNA fragments separated by gel electrophoresis. J Mol Biol 98 (3):503–517
2. Bhat S, Srinivasan S (2002) Molecular and genetic analyses of transgenic plants:: considerations and approaches. Plant Sci 163 (4):673–681
3. Kovalic D, Garnaat C, Guo L, Yan Y, Groat J, Silvanovich A, Ralston L, Huang M, Tian Q, Christian A (2012) The use of next generation sequencing and junction sequence analysis bioinformatics to achieve molecular characterization of crops improved through modern biotechnology. Plant Genome 5(3):149–163
4. Zastrow-Hayes GM, Lin H, Sigmund AL, Hoffman JL, Alarcon CM, Hayes KR, Richmond TA, Jeddeloh JA, May GD, Beatty MK (2015) Southern-by-sequencing: a robust screening approach for molecular characterization of genetically modified crops. Plant Genome 8:1. https://doi.org/10.3835/plantgenome2014.08.0037
5. Livak KJ, Schmittgen TD (2001) Analysis of relative gene expression data using real-time quantitative PCR and the 2− ΔΔCT method. Methods 25(4):402–408
6. Flavell RB (1994) Inactivation of gene expression in plants as a consequence of specific sequence duplication. Proc Natl Acad Sci 91 (9):3490–3496. https://doi.org/10.1073/pnas.91.9.3490
7. Iyer LM, Kumpatla SP, Chandrasekharan MB, Hall TC (2000) Transgene silencing in monocots. Plant Mol Biol 43(2):323–346. https://doi.org/10.1023/a:1006412318311
8. Kohli A, Twyman RM, Abranches R, Wegel E, Stoger E, Christou P (2003) Transgene integration, organization and interaction in plants. Plant Mol Biol 52(2):247–258. https://doi.org/10.1023/a:1023941407376
9. Kooter JM, Matzke MA, Meyer P (1999) Listening to the silent genes: transgene silencing, gene regulation and pathogen control. Trends Plant Sci 4(9):340–347. https://doi.org/10.1016/S1360-1385(99)01467-3
10. Svitashev SK, Pawlowski WP, Makarevitch I, Plank DW, Somers DA (2002) Complex transgene locus structures implicate multiple mechanisms for plant transgene rearrangement. Plant J 32(4):433–445. https://doi.org/10.1046/j.1365-313X.2002.01433.x
11. Chen J, Dellaporta S (1994) Urea-based plant DNA Miniprep. In: Freeling M, Walbot V (eds) The maize handbook. Springer New York, New York, NY, pp 526–527. https://doi.org/10.1007/978-1-4612-2694-9_85
12. Belide S, Vanhercke T, Petrie JR, Singh SP (2017) Robust genetic transformation of sorghum (*Sorghum bicolor* L.) using differentiating embryogenic callus induced from immature embryos. Plant Methods 13 (1):109. https://doi.org/10.1186/s13007-017-0260-9
13. Xue B, Guo J, Que Y, Fu Z, Wu L, Xu L (2014) Selection of suitable endogenous reference genes for relative copy number detection in sugarcane. Int J Mol Sci 15(5):8846–8862
14. Bubner B, Baldwin IT (2004) Use of real-time PCR for determining copy number and zygosity in transgenic plants. Plant Cell Rep 23 (5):263–271. https://doi.org/10.1007/s00299-004-0859-y
15. Li Z, Hansen JL, Liu Y, Zemetra RS, Berger PH (2004) Using real-time PCR to determine transgene copy number in wheat. Plant Mol Biol Report 22(2):179–188. https://doi.org/10.1007/bf02772725
16. Ingham DJ, Beer S, Money S, Hansen G (2001) Quantitative real-time PCR assay for determining transgene copy number in transformed plants. BioTechniques 31(1):132–140. https://doi.org/10.2144/01311rr04
17. Schmidt M, Parrott W (2001) Quantitative detection of transgenes in soybean [Glycine max (L.) Merrill] and peanut (Arachis hypogaea L.) by real-time polymerase chain

reaction. Plant Cell Rep 20(5):422–428. https://doi.org/10.1007/s002990100326

18. Fletcher SJ (2014) qPCR for quantification of transgene expression and determination of transgene copy number. In: Fleury D, Whitford R (eds) Crop breeding: methods and protocols. Springer New York, New York, NY, pp 213–237. https://doi.org/10.1007/978-1-4939-0446-4_17
19. Thompson CJ, Movva NR, Tizard R, Crameri R, Davies JE, Lauwereys M, Botterman J (1987) Characterization of the herbicide-resistance gene bar from *Streptomyces hygroscopicus*. EMBO J 6(9):2519–2523

# Chapter 14

# A Road Map for Undertaking Quantitative Proteomics in Plants: New Opportunities for Cereal Crops

**Ibrahim Khodabocus, Qiaomu Li, Devang Mehta, and R. Glen Uhrig**

## Abstract

Proteomics has been a key tool in the molecular biochemical characterization of plant systems since the early 2000s. Initially relying upon one-dimensional and two-dimensional gel electrophoresis to quantify changes in tens to hundreds of proteins, proteomics today has evolved to be completely gel free and capable of quantifying abundance and protein post-translational modification changes in thousands of proteins in a single mass spectrometry run. Here, we provide a framework for understanding the landscape of modern discovery-based proteomic technologies as well as a road map to deploy the latest workflows to analyse crop proteomes quantitatively. In particular, we focus on how quantitative proteomics can be applied to cereal crops, such as barley and wheat, and how proteomics is being applied today in light of advances in genome sequencing.

**Key words** Mass spectrometry, Quantitative proteomics, Data-dependent acquisition, Data-independent acquisition, *Arabidopsis thaliana*, Barley, Wheat

## 1 Introduction

Proteomics technology has evolved significantly over the past 20 years, making faster and deeper analyses of the proteome possible. Accompanying these technological advancements has been the development of quantitation workflows, which allow those acquiring proteomic data to quantify relative changes in protein abundance and/or modification between conditions or genotypes. Despite these advancements, the adoption of quantitative proteomics, especially in agriculturally relevant plants, has been slow compared to other fields of biology. Across molecular plant sciences, genomics and transcriptomics technologies have been the dominant choices for systems-level analysis and have been vital in our development of new crop varieties. However, quantitative changes in DNA or RNA are only proxy measurements for

Ibrahim Khodabocus and Qiaomu Li contributed equally to this work.

Andriy Bilichak and John D. Laurie (eds.), *Accelerated Breeding of Cereal Crops*, Springer Protocols Handbooks, https://doi.org/10.1007/978-1-0716-1526-3_14, © Springer Science+Business Media, LLC, part of Springer Nature 2022

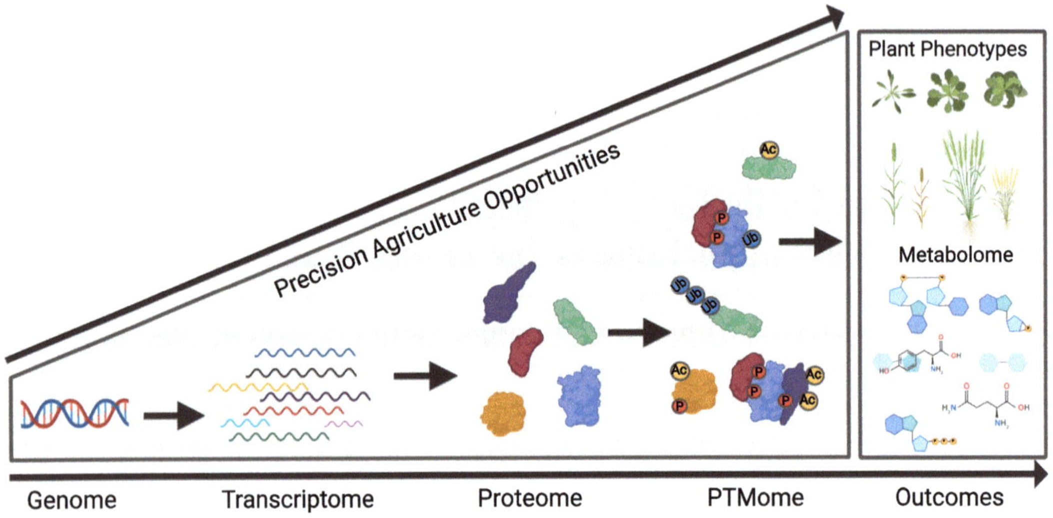


**Fig. 1** Complexity in the molecular landscape of plants. From genome to metabolome and phenotype, quantitative proteomics offers new opportunities for developing next-generation crops, by providing essential information related to the proteome, post-translational modifications (PTMs) and protein-protein interactions (adapted from Bludau and Aebersold 2020 [1])

functional biochemical outcomes, which can only be determined by assessing changes in the proteome (Fig. 1 [1]). Owing to the technological advancements in mass spectrometry over the past 20 years, proteomics is now a legitimate technology for uncovering new breeding and biotechnology targets across the crop landscape.

The proteome is immensely complex, comprising changes in protein levels as well as protein post-translational modifications (PTMs), which can augment protein function, localization and protein-protein interaction partners, leading to profound changes in phenotype. PTMs, such as phosphorylation, acetylation and ubiquitination, are some of the most well-understood PTMs in plants, each resulting in a complex array of outcomes. Correspondingly, global PTM analysis is a new frontier of opportunity for the development of next-generation crops. Given the importance of measuring protein-level changes and the impact of PTMs on protein structure-function relationships, it is imperative that crop scientists understand how proteomics can be deployed in plant systems.

High-throughput analysis of the proteome requires the use of a high-resolution mass spectrometer (MS) to analyse and sequence individual peptides from a complex protein sample. In the modern, widely-used, bottom-up MS approach, proteins are first digested into peptides through the use of proteases. Digested peptides are then chromatographically separated (usually by reverse-phase high-performance liquid chromatography; HPLC) for ionization and injection into a MS (e.g. Thermo Q Exactive or LUMOS mass spectrometer). Once injected into the MS, there are two key analysis steps, which is why the method is also termed tandem MS, or MS/MS. The first step is the mass analysis of intact peptide ions (also referred to as 'parent' ions). This step is called the MS1 analysis and produces a mass-to-charge (m/z) versus intensity profile (or mass spectrum) of the injected peptide pool. Depending on the instrument and method used, a subpopulation of these parent ions are then selected for further MS2 analysis. For MS2 analysis, parent ions are fragmented to produce several 'daughter' ions, producing a unique mass spectrum for each parent ion. As daughter ions are products of the fragmentation of the parent peptide, their MS2 analysis allows the mass spectrometer to deduce the amino acid sequence of the parent peptide. MS2 analysis is also able to identify the specific locations of PTMs on peptides. Quantification of the amounts of each peptide and protein present in a sample is made possible by measuring the intensity and number of spectra at both levels of analysis.

In this chapter, we present a methodological road map for the deployment of quantitative proteomics in plants (Fig. 2 [2]). We describe the necessary information and considerations for undertaking plant proteomic experimentation, while describing the latest approaches and technologies. In particular, we explore each of the key steps involved in setting up a successful proteomics experiment: protein extraction and digestion (Sect. 1), quantification (Sect. 2), increasing the depth of analysis and PTMs (Sect. 3) and data analysis resources for quantification (Sect. 4). Finally, we outline the latest proteomic technologies (Sect. 5) and how proteomics can be applied to accelerate cereal crop breeding (Sect. 6).

## 2 Protein Extraction and Digestion

To achieve desirable proteomics results through mass spectrometry, it is critical to design an optimal sample preparation workflow based on research objectives and selected crop species. This starts with an effective protein extraction approach, which will assist downstream mass spectrometry analyses and elucidate the desired protein-level phenomenon [3], and requires tailoring extraction procedures, in particular extraction buffers, based upon experimental goals and crop tissues/organs. Although studies examining protein extraction methods in proteomics have been undertaken in *Arabidopsis thaliana* (*Arabidopsis*) [4], comparatively fewer studies have been performed in crops [5, 6]. Here, we discuss both classic and, more recently devised, protein extraction options and discuss their feasibility for implementation in cereal crops.

Non-denaturing protein extraction methods typically use non-ionic detergents, such as Triton X-100 (Sigma-Aldrich), Tween-20 (Sigma-Aldrich) and NP-40 (Thermo Scientific),

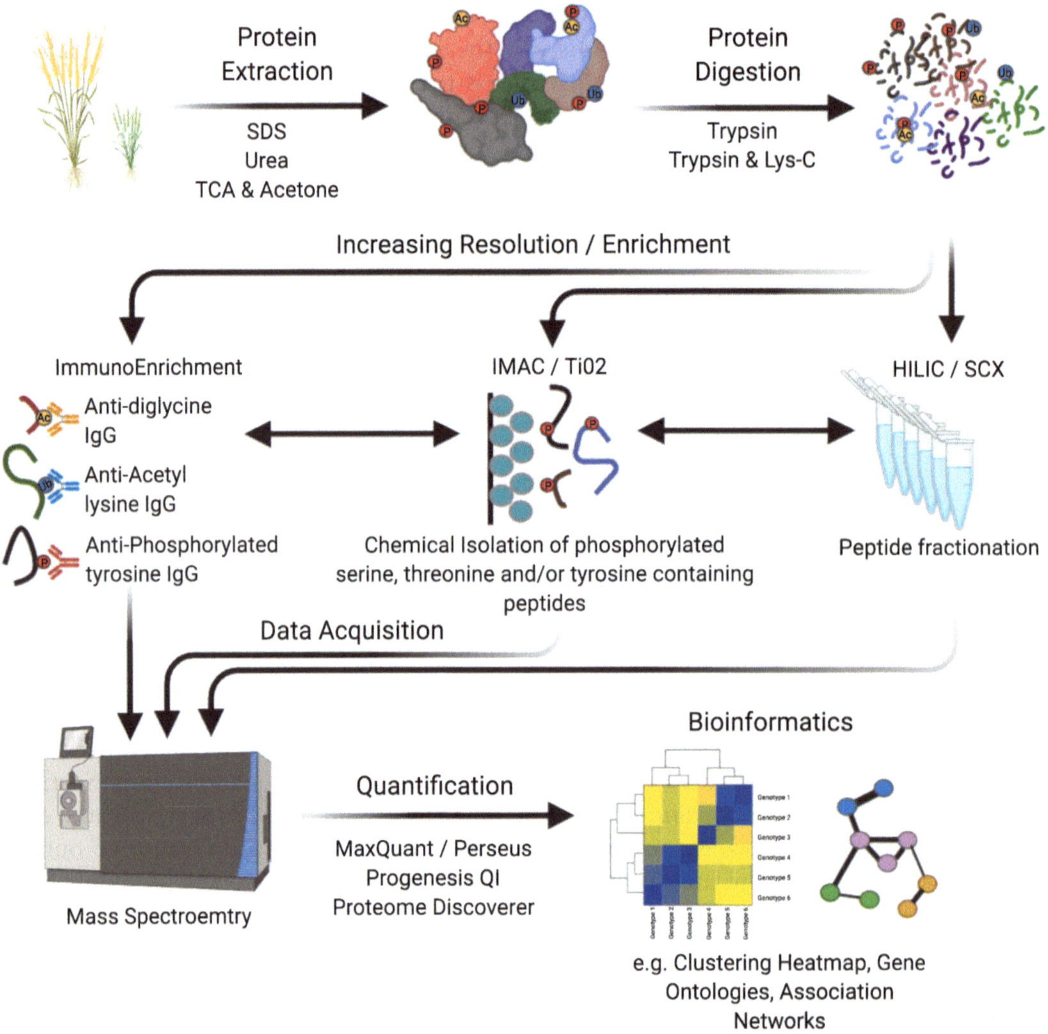


**Fig. 2** Visual representation of a quantitative proteomic workflow. The generalized workflow depicts the key steps and considerations for executing the latest quantitative proteomic workflows (adapted from Altelaar et al., 2012 [2])

which can solubilize membrane proteins while preserving native protein-protein interactions [7, 8]. This type of extraction buffer is typically used for proteomics studies aimed at identifying protein-protein interactions. For example, NP-40-containing extraction buffers have been used to elucidate numerous interactors in *Arabidopsis* [9–12]. For systemic analyses of protein abundance and/or PTM changes, denaturing extraction buffers are more commonly used. The most common denaturants are as follows: urea and/or thiourea and sodium dodecyl sulfonate (SDS). The presence of urea enables effective protein denaturation, resulting in high protein yields. However, the addition of urea can also result in artefactual protein modifications [13]. Furthermore, as high concentrations of urea can impede protein digestion, it is usually advisable to dilute urea to a concentration below 2 M prior to digestion.

SDS is considered one of the strongest surfactants, as it is able to break down both membrane-localized proteins and cytosolic proteins alike, enhancing overall extraction efficiency. However, unlike urea, SDS should be completely removed prior to digestion in order to avoid the unintended denaturation of trypsin [4, 8]. To eliminate SDS before digestion, trichloroacetic acid (TCA)/acetone can be employed, which precipitates proteins away from the SDS-containing solution [5, 14]. However, the resultant protein pellet can be difficult to solubilize, potentially incurring protein losses in samples [5]. More recently, several novel sample preparation approaches have been developed. Filter-aided sample preparation (FASP) takes advantage of centrifugal regenerated cellulose filters in order to eliminate SDS prior to mass spectrometry, in turn maximizing amounts of extracted protein [15]. Numerous studies have successfully applied FASP in crops, including the following: rice [16], barley [17], maize [18] and wheat [19], amongst others. Most recently, a new technique called Sample Preparation by Easy Extraction and Digestion (SPEED) was developed as a purely chemical method using trifluoroacetic acid (TFA) to lyse cells [20], thus easily retaining original protein or protein PTM profiles of a given sample while minimizing sample processing time. However, the use of TFA requires important considerations for user safety.

With a large number of protein lysis choices available for extracting proteins for proteomic analysis, it is important to select a lysis method that efficiently solubilizes proteins with the least number of manipulation steps [21, 22]. This also requires considerations for sample type (e.g. tissue) and research objectives (e.g. global proteome vs. protein-protein interactome), as different types of crop tissues and organs can vary in cell wall thickness and secondary compounds. For example, the midrib in leaves of maize (*Zea mays* L.) has a thick cell wall as well as lignin deposition (phenolics-based), giving rise to difficulties in protein extraction. Here, Wang et al. found TCA/acetone extraction/precipitation produced the highest quantities of extractable protein, suggesting that this method is suitable for recalcitrant plant tissues [5]. Alternatively, analysis of maize seedlings and seeds employed FASP, also implying the effectiveness of SDS in key crop tissues [23, 24]. Lastly, PTMs, such as phosphorylation, acetylation and ubiquitination, are relatively stable when utilizing denaturing extraction conditions. In fact, it may be the preferred approach as it reduces the need for expensive enzyme inhibitors (e.g. phosphatase or histone deacetylase inhibitors), which makes sample processing more cost effective.

## 3 Quantification Technologies and Approaches

A wide range of protein quantification methods based on mass spectrometry have been published. These can broadly be grouped into two strategies: label-free quantitation and chemical label-based quantitation. Both types of methods are established in both *Arabidopsis* and crops, with different approaches having trade-offs ranging from differences in precision to cost of implementation. In this section, we discuss several quantitative approaches.

### 3.1 Label-Free Quantitation Approaches

Label-free quantitation (LFQ) is the most widely applied quantification approach. It is fairly affordable from a reagent standpoint but requires individual mass spectrometry injections for each sample, increasing analysis time. Moreover, it can provide deeper proteome coverage due to the lack of sample multiplexing. In general, quantification in LFQ methods is conducted by measuring ion intensities of peptides (at MS1 quantification level) or by counting obtained spectra for every protein (at the MS2 quantification level) [25, 26]. In LFQ, protein-level quantification requires precise liquid chromatography to ensure the alignment of peptide elution peaks between samples during downstream data processing. To ensure this, internal retention time standards can be utilized [27–30]. Since LFQ compares MS1 and/or MS2 data across individual mass spectrometry runs, reproducible chromatography and sample preparation are critical. In *Arabidopsis*, this technology has been successfully used to quantitatively investigate Golgi-localized proteins [31], to shed light on roles of phosphorylation and acetylation throughout the diel cycle [32] as well as to reveal different signal transducers (biomarkers) in response to local wounding and virus infection [33]. Similarly, LFQ has also been widely employed to uncover changes in the protein abundance of crops. Some examples of LFQ in crop plants include the following: understanding leaf greening proteins in maize seedlings [34], generating a phosphoproteome landscape for maize primary roots [35], elucidating the roles of autophagic recycling in maize [36] and uncovering distinct early developmental patterns of barley grains [37]. Additionally, LFQ has been used to assess crop fitness. In rice, it was applied to elucidate changes in protein profiles in response to drought stress [38], heat stress [39] and biotic stress, including viruses and flg22 challenge [40]. In maize, LFQ has been harnessed to gain insights into response mechanisms following exposure to drought stress [24, 41, 42], to heat stress [43] and to fungi infection [44].

### 3.2 Label-Based Quantitation Approaches

Unlike LFQ, label-based strategies were designed to boost quantitative precision. The tandem mass tags (TMT; Thermo) method uses a sample specific isobaric tag to modify all peptides in that specific sample to allow multiple samples to be mixed together for a single mass spectrometry injection. Here, the same peptide for each sample will be analysed simultaneously, with each differentially labelled peptide species quantified upon fragmentation. Most recently, a 16plex TMT kit has become available, offering opportunities to maximize sample comparison. However, TMT-labelling is limited to the number of samples it can label (e.g. 16plex TMT = 16 samples) and can be cost prohibitive depending on the labelling kit utilized and the available budget for experimentation. Like TMT, isobaric tag for relative and absolute quantitation (iTRAQ; Sigma) is also a chemical labelling approach. Similar to TMT, a labelled peptide will present as a single peak in MS1, avoiding overlapping peak issues, that will be fragmented and quantified in MS2. There are at least eight iTRAQ reagents available, thereby making it possible to conduct comparative analysis of several samples via a single MS scan [45].

Both TMT and iTRAQ are popular choices for use in plant proteomic analyses. In particular, Mergner et al. applied the TMT labelling approach to quantify the proteomes and phosphoproteomes of 30 tissues from *Arabidopsis* [46]. In crop plants, Lin et al. used TMT to identify interacting proteins of OsFD2 in rice protoplasts [47], while Qin et al. took advantage of this method to understand how rapeseed (*Brassica napus*), a significant global oil crop, establishes root responses to nitrogen starvation conditions [48]. The iTRAQ technique has been employed to determine differentially expressed proteins involved in heat stress signalling in rice grains [49] and to quantitatively identify proteins related to immune responses in maize following exposure to fungi [44].

Compared to the peptide labelling options of TMT and iTRAQ, the stable-isotope dimethyl labelling approach represents a cost-effective alternative. Although it has not been extensively applied in crops, it has been shown to be a proven alternative in *Arabidopsis* [50–52] and other plants [53]. Dimethyl peptide labelling is based on the dimethylation of the primary amine of peptide lysine residues after tryptic digestion, using different isotopic forms of formaldehyde [53–55]. Such modification leads to a mass shift, enabling the quantification based on the mass difference of the dimethyl tag in each peptide following MS analysis [56]. However, the added mass of 2 or 4 Da can result in an increase in sample complexity at the MS1-level. However, this increase in complexity can be solved through offline peptide fractionation [55]. Due to the combination of cost-effectiveness and quantitative precision through peptide labelling, dimethyl labelling may represent a balanced approach for label-based quantitation in crop systems.

Lastly, in vivo labelling approaches also exist but are difficult to employ in plants and crops. These approaches involve labelling proteins as they are made within an organism. Two of these in vivo labelling approaches include SILAC (stable-isotope labelling by/with amino acids in cell culture; [55, 57]) and heavy nitrogen labelling ($N^{15}$ labelling; [58]). Both of these technologies have been implemented in either plant cell culture [6, 59] or seedlings [57, 58]. However, both SILAC and $N^{15}$ labelling have technical limitations largely due to the autotrophic nature of plants and inconsistency in labelled amino acid uptake. Although $N^{15}$ labelling may have potential for greater use in plants, LFQ, TMT, iTRAQ and dimethyl labelling remain the dominant quantification approaches in the field.

## 4 Increasing Proteome Resolution and Protein Post-Translational Modifications

The plant cellular proteome is enormously complex, encompassing proteins present in a very wide range of concentrations. In plants, chloroplastic (e.g. RuBisCo), storage, ribosomal and cytoskeletal proteins are some of the most abundant [60, 61]. When undertaking mass spectrometry analysis of samples containing these types of proteins (e.g. leaf tissues), these high-abundant species can contribute to ion suppression, whereby peptides from low abundant proteins are obstructed during MS1 selection, leading to reduced detection [62]. To overcome this, protein (e.g. size exclusion chromatography) or peptide (strong cation exchange; SCX or hydrophilic interaction liquid chromatography; HILIC) fractionation can be employed to reduce sample complexity and the corresponding number of ion suppressing species in any given peptide fraction (Fig. 2). Similarly, PTM protein variants are also low abundant species and therefore have been historically difficult to identify by MS analyses [63]. To overcome this, numerous PTM-specific enrichment strategies have been developed, including both chemical and immunological technologies (Fig. 2). These technologies function to isolate PTM modified proteins or peptides from the broader total proteome or peptide fraction. Here, we describe a selection of approaches to increase the detection of low-abundance proteins as to enrich phosphorylated, acetylated and ubiquitylated protein/peptide species.

### 4.1 Increasing Depth of Analysis through Offline Peptide Fraction

Peptide fractionation prior to MS/MS analysis is a well-established, long-standing approach to increase proteome coverage and increase the detection of low-abundance proteins. Strong cation exchange (SCX), as a type of ion-exchange chromatography, can separate peptides based on differences in the electric charge. SCX is deployed using a salt gradient or a salt plug, eluting peptides with the least positive net charge first, thus fractionating across a gradient. However, this approach requires multiple desalting steps, rendering it lengthy and risks peptide losses with each step [64, 65]. Alternatively, pH gradients that enable the elution of peptides according to isoelectric point have also been used to elute SCX gradients [64, 65]. Both approaches to SCX fractionation have been successfully applied in plant proteome studies. For example, a pH-based SCX gradient was used to undertake comparative proteomic analysis of maize leaves responding to photoperiod [66]. Alternatively, salt-based SCX gradients have been employed to determine the molecular basis of a salt-tolerant rice line [67], to elucidate the phosphoproteome of maize across its development [68] and to investigate the roles of protein kinase in response to high-temperature condition in tomato [69].

Alternatively, hydrophilic interaction chromatography (HILIC) has also shown immense success in fractionating peptides. Here, peptides are fractionated based on their polarity [70, 71]. Consequently, HILIC is compatible with a wide range of buffer conditions. It is equipped with a hydrophobic mobile phase and a hydrophilic stationary phase,which is able to strongly interact with the hydrophilic regions of peptides [72]. The mobile phase then aims to elute these polar analytes in a gradient manner, with higher polarity peptides possessing a longer retention time. Peptides are then eluted by reducing the concentration of organic solvent and increasing the concentration of compounds such as TFA [73]. The HILIC technique has not only been a key offline peptide fractionation tool but has also been utilized in the investigation of the glycoproteome of wheat flour albumins [74] and to explore phosphoproteome in rice pistils [75].

### 4.2 PTM Enrichment Technologies and Approaches

Given the importance of PTMs in regulating the cellular environment, huge efforts have been made to efficiently detect and quantify PTMs [76–78]; in particular, protein phosphorylation [32, 46, 52], ubiquitination [79] and acetylation [32, 50]. Phosphorylation is the most abundant PTMs [80], with intimate association with almost all aspects of plant growth and stress responses. To characterize phosphorylation events on proteins, chemical-based enrichment technologies, such as immobilized metal affinity chromatography (IMAC) and titanium dioxide (Ti02), have been the most robust. IMAC exploits the negatively charged phosphate on phosphopeptides to selectively bind them from complex peptide mixtures. This enrichment workflow has been applied to multiple plant proteomic studies, ranging from identification of novel phosphoproteins to mapping phosphorylation sites [81–83]. Similar to IMAC, TiO2 is also an ionic interaction-based method for enriching phosphorylated peptides [70]. In crops, these technologies have been utilized to varying degrees, but in cereals, TiO2 enrichment has revealed changes in grain development under well-watered and water-deficient conditions [84], elucidated phosphoproteome in maize leaves upon drought conditions [85] as well as characterized phosphoproteome changes in response to cadmium stress in rice seedlings [86].

Both IMAC and TiO2 techniques are widely used in crop phosphoprotein studies and have often been combined with SCX or HILIC peptide fractionation in the same workflow to increase depth of analysis [75, 87, 88]. In phosphoproteomics, chemical enrichment techniques have gradually surpassed antibody-based affinity enrichment methods as they are typically more robust. Antibody-based methods rely on recognizing a specific modification site to pull down either proteins or peptides. Where antibody-based enrichment and phosphoproteomics maintain comparable efficacy is in the enrichment of phosphorylated tyrosine proteins and/or peptides [89, 90]. However, unlike phosphorylation, antibody-based enrichment methods are still routinely employed to enrich ubiquitylated and/or acetylated proteins and/or peptides. In plants, pan-acetyl-lysine antibodies have been developed to undertake maize acetylome profiling [32, 44, 50], while various anti-K-ε-GG-specific antibodies have been used to investigate the ubiquitinome of *Arabidopsis* [91], peach [92], rice [93] and wheat [94].

## 5 Approaches to Protein Quantification

Once proteomic data has been acquired by mass spectrometry, processing this data is arguably one of the greatest challenges. Typically, in plant and crop research, mass spectrometry data is acquired in data-dependent acquisition (DDA) mode, more colloquially known as 'shotgun' proteomics. Several software packages exist to analyse data acquired through DDA-driven methods. Here, we highlight three of the most commonly deployed packages: MaxQuant, Progenesis QI and Proteome Discoverer. MaxQuant is a popular, free accessible program utilized by plant proteomics laboratories around the world, while Progenesis QI (Nonlinear Dynamics) and Proteome Discoverer (Thermo) are proprietary commercial analysis packages. Each of these software packages have different capabilities, pricing and popularity within the proteomics community. MaxQuant allows its users to conduct a wide range of quantification analyses including, but not limited to, LFQ, dimethyl labelling, TMT and PTM analyses, while Progenesis QI has focused primarily on LFQ approaches. Similar to MaxQuant, Proteome Discoverer possesses LFQ, dimethyl labelling, TMT and PTM analysis capabilities. Within this section, we attempt to cover some of the features associated with each of these three software packages.

When analysing proteomic data using any of these analysis packages, a few generic parameters should be considered.

1. Fixed modifications generally include carbamidomethylation due to cysteine alkylation by iodoacetamide.
2. Common variable modifications include the oxidation of methionine and n-terminal acetylation; however, variable modifications may change based on prior knowledge. For example,

when searching your phosphoproteomic data, phosphorylated serine, threonine and/or tyrosine should be set as variable modifications.

3. The availability of a reliable proteome for the organism of interest greatly increases the power of analyses and should be used in conjunction with either concatenated decoy sequences or by selecting corresponding decoy options within each software package. Correspondingly, the inclusion of decoy sequences is important in determining false discovery rate (FDR). Typically, a 1% threshold for protein and peptide FDR is utilized.

### 5.1 MaxQuant

Likely owing to its free availability, MaxQuant is the most cited proteomics software for DDA-driven proteomics. MaxQuant was initially released in 2008 with its basal protocol version [95, 96]. MaxQuant can be freely downloaded for academic use at http://www.maxquant.org [97, 98] and is available for both Windows and Linux machines. MaxQuant is frequently updated and modified to include new, more user-friendly functionality, with later versions including a built-in peptide search engine, called Andromeda [99], eliminating the need for a third-party search engine to conduct peptides searches, such as MASCOT (Matrix Science). MaxQuant can perform multiple types of quantification including LFQ, TMT and SILAC; dimethyl labelling; and many more quantification techniques [100]. In addition to analysing changes in protein abundance, MaxQuant is also equipped to identify PTMs, being able to provide probabilistic site identifications that allow for parsing of more reliable site identifications from less reliable sites. Typically, a threshold of 0.75–0.80 is utilized as a threshold for PTM site reliability [32, 50, 52].

### 5.2 Progenesis QI

Progenesis QI is a proprietary software package available to perform quantitative proteomic data analysis. Progenesis QI facilitates LFQ quantification [32, 52] by utilizing a highly interactive graphical user interface to allow users to observe their omics data in 2D or 3D. Moreover, Progenesis QI allows the researcher to utilize a variety of databases for sample analysis [101]. Unfortunately, Progenesis QI is exclusively geared towards LFQ analysis pipelines and is unable to support isotope-based (e.g. SILAC)/label-based (e.g. TMT) data at this time. Progenesis QI also requires a third-party search engine subscription (e.g. Mascot).

### 5.3 Thermo Proteome Discoverer

Thermo Proteome Discoverer is a proprietary product that allows users to analyse from TMT, SILAC and dimethyl labelling experimentation but has a more limited capacity for LFQ analyses [102, 103]. PTM identification, quantification and site reliability analyses are also possible within the Proteome Discoverer workspace, along with the flexibility to build customizable data analysis workflows as needed. This differs from both MaxQuant and Progenesis QI, which deploy predetermined analysis pipeline structures, preventing the construction of customizable workflows for data processing. Proteome Discoverer also comes equipped with its own search engine (SEQUEST), eliminating the need for a third-party search engine.

## 6 Emerging Quantitative Proteomic Technologies

In recent history, plant proteomics has utilized DDA-driven LC-MS/MS methods to conduct plant proteome research [104]. DDA-driven methodologies set thresholds for deciding which parent ions (MS1) to fragment, leading to potential bias towards higher abundant proteins and technical variability with regard to the detection and quantification of lower abundant proteins between MS runs [105, 106]. This often leaves gaps in acquired data for proteins across samples, requiring the application of data imputation methods to replace these missing values. Alternatively, data-independent acquisition (DIA) methodologies fragment all peptides for a given MS1 m/z window, significantly reducing reliance on data imputation and providing a more precise picture of the captured proteome [107]. DIA protocols notably reduce data variability and the under-representation of low-abundance peptides [105]. DIA workflows are already being implemented in multiple plant species including *Arabidopsis thaliana* [108], tomato [109] and wheat [110].

### 6.1 Data-Independent Acquisition (DIA)

The strength of DIA is that all peptides are selected for fragmentation based on m/z windows. The nature of DIA analysis ensures that data reproducibility is sustained throughout multiple sample analyses [104, 110, 111]. This ensures that DIA can produce an exhaustive record of the fragment-ion signals across an entire m/z range, irrespective of peptide intensity; thus, unlike DDA, DIA has increased reproducibility between MS runs. The data acquired through DIA also require specialized software tools for downstream analyses and quantification. Open source options for analysing DIA data include OpenSWATH software (http://openswath.org/en/latest/; [111]), DIA-Umpire (https://diaumpire.nesvilab.org/; [112]) and ISOQuant [113], while proprietary options include Spectronaut (https://biognosys.com/; [114]). Each of these options allow users to compile custom spectral libraries derived from in-depth DDA analyses to be used in DIA analyses. The major caveat of DIA analyses is that the research group needs to create a spectral library from pooled samples for each experiment, covering the various tissue types and/or conditions prior to being able to run an effective DIA workflow. Most recently, plant proteomics researchers have begun to assemble tissue-specific libraries that can be used by other groups in their own DIA experiments [16]. Spectral libraries need to be as exhaustive as possible prior to conducting any DIA analyses in order to maximize its ability to quantify proteome changes. DIA may represent the future of plant-based proteomics if comprehensive spectral libraries become established for each plant species of interest [107].

### 6.2 Direct Data-Independent Acquisition (directDIA)

Both DDA and DIA have caveats with respect to data acquisition. DDA data acquired from identical samples can have upwards of 30% variance and bias for high-abundant proteins [105], which can be detrimental for plant proteomics, while DIA requires a spectral library assembly to undertake quantification analyses [107, 114]. In early 2020, a new approach called directDIA was developed that combined beneficial elements of both DDA and DIA while eliminating some of the major drawbacks of both approaches [115, 116]. Unlike DIA, directDIA utilizes previous and future analyses to continuously build a dynamic spectral library, which eliminates the immediate need for the development of an exhaustive spectral library (Bekker-Jensen et al. 2020). As directDIA represents the newest iteration of systems-level proteomic workflows, its disadvantages remain to be resolved, while its upside seems very promising in advancing new opportunities for analysing the proteome of plants.

## 7 Application of Quantitative Proteomics to Cereal Systems

Quantitative proteomics has increased our understanding of protein-level abundance changes and where PTMs impact plant processes using the model plant *Arabidopsis*. With increased accessibility to sequenced genomes, the techniques and protocols that have been pioneered in *Arabidopsis* can now provide an invaluable insight into cereal crops, such as wheat and barley. Furthermore, when combined with other validation technologies, including targeted proteomics and metabolomics, new breeding and biotechnology workflows can be developed that can help significantly accelerate the development of new crop varieties (Fig. 3).

In agriculture, gel-based proteomics (e.g. two-dimensional electrophoresis mass spectrometry; 2DE-MS) has been the norm for quantitative proteomics. However, in the past 5 years, multiple gel-free quantitative proteomics techniques and technologies have begun to be employed, already demonstrating promise in revealing new information for breeding. In tomato, DIA proteomics has been used to analyse the pathogenesis response progression [117] as well as understand the fundamental mechanisms of fruit development [118], while in rapeseed, iTRAQ-based DDA proteomics has elucidated stress-induced proteins [119]. In rice, insight into

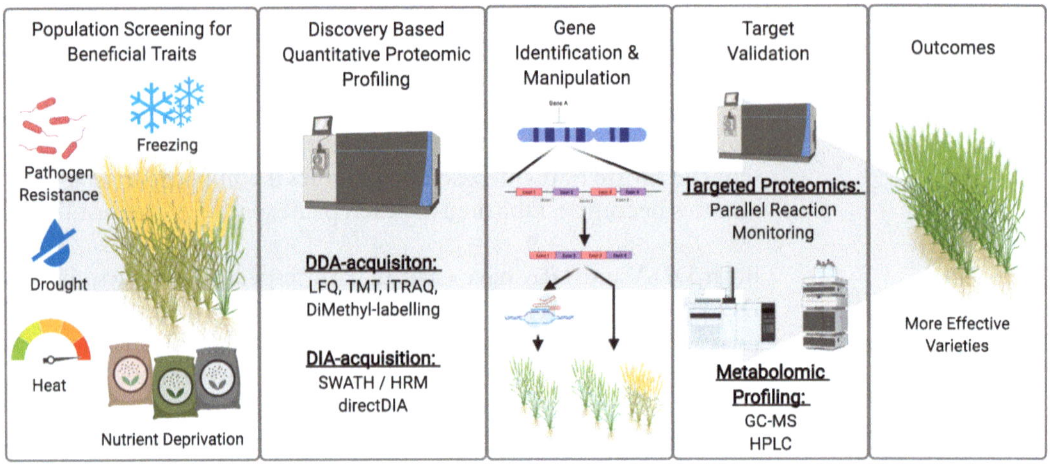


**Fig. 3** Integrating quantitative proteomics into the development of next generation agricultural crops

the evolutionary arms race between rice and the rice pathogen *Magnaporthe oryzae* has been explored using LC-MS/MS proteomics and iTRAQ-based quantitative proteomics [120, 121], while in potato, the impacts of *Phytophthora infestans* was investigated using TMT-based proteomics [122], amongst others. Lastly, grain crops have also recently seen a surge in the use of advanced quantitative proteomics, utilizing DIA proteomics better to detect mycotoxins [123].

### 7.1 Proteomics in Wheat

Wheat (*Triticum spp.*) is one of the three major food crops in the world [124]. The wheat genome is approximately 17 gigabases in size, due primarily to the several polyploidy events that occurred several thousands of years ago. From a proteomics perspective, studies employing 2DE-MS were the first attempts at revealing systemic changes in the wheat proteome. These studies have resolved the initial proteomes of leaves, flowers and amyloplasts in addition to identifying proteins involved in early and late grain development. By today's standards, 2DE-MS now represents a classic proteomic technique, where proteins are gel separated by isoelectric focus point and size, followed by spot picking and MS analysis. Although gel-based proteomics is still used [124, 125], more advanced proteomic approaches are now being deployed.

Recently, Bromilow et al. (2016) employed a combined protocol of DDA and DIA proteomics to increase the coverage of the wheat proteome by as much as 85% [110]. Other studies using label and label-free quantitative proteomics have analysed the effectiveness of biostimulants and fertilizers in wheat [126], to contrast genotypes under drought [127] and salt [120] stress, to assess genotypes for gluten allergy factors [128] and to assess carbohydrate metabolism [129]. Advanced quantitative proteomics has also been used to quantify changes in the wheat acetylome

[130]. Examples such as these suggest that advanced quantitative techniques, such as DDA and DIA proteomics, can be immensely useful in uncovering new breeding targets for biotic (e.g. *Fusarium sp.* resistance; [131]) or abiotic (e.g. drought) stress as well as for increasing the yield and digestibility of wheat.

Systematic proteomic investigations into wheat have largely lagged behind other species due to the size and complexity of the wheat genome. With major strides being made in creating a more complete genome [132], coupled with the proteomic technologies available today, the time for increased deployment of proteomics in wheat has now arrived.

### 7.2 Proteomics in Barley

Similar to wheat, agricultural researchers have predominantly used 2DE-MS approaches to understand changes in the barley proteome. However, more recently, advanced DDA and DIA approaches have started to come into the picture. In particular, DIA proteomics was used to assess the predominance of the four families of gluten proteins within barley [133] and was able to show a lowered presence of gluten family proteins in the triple-null barley line, compared to the wild-type cultivar [134]. This use of proteomics could have major benefits for the breeding or biotechnological engineering of high gluten crops, such as wheat or barley. DIA proteomics has also been deployed in barley to understand how the powdery mildew pathogen impacts the seed proteome [135]. In addition to instances of DIA workflow implementation, DDA approaches have also been employed to assess a number of agronomic traits in barley [136], such as abiotic [137–139] and biotic [140] stress as well as development [141–143]. As barley has a lower-complexity genome, advanced quantitative proteomics has been more widely adopted in the field. Given this, it would be interesting to begin to use these proteomic approaches to examine multiple barley genotypes en masse for the identification of new breeding targets.

## 8 Considerations for Cereal Agriculture into the Future

Although *Arabidopsis thaliana* continues to be the main model organism utilized in plant biology research, the decreased cost of genome sequencing, coupled with advances in proteomics such as directDIA acquisition methodologies, holds enormous potential for undertaking quantitative proteomics in crop plants, including cereals. Obstacles still exist with crops like wheat due to their ploidy, large genome size and genetic redundancy. However, diploid grains, such as barley and millet, offer immediate opportunities to harness the power of today's quantitative proteomics workflows. One key application for quantitative proteomics lies in the identification of new stress response proteins for crop breeding or biotechnology, as well as an understanding of how metabolism is augmented under particular stresses to inform stress management practices in the field. Unlike earlier gel-based proteomic analyses, the ability of today's mass spectrometers and data processing workflows to quantify multiple thousands of proteins in a single 1.5–2 h gradient permits systems-level quantification of the most abundant up- and downregulated proteins. Moving forward, modern quantitative proteomics offers a dual advantage to agricultural researchers: first, as a discovery tool to identify new stress-responsive, or biosynthetic proteins, and second, as a diagnostic tool to monitor specific protein biomarkers to accompany conventional phenotyping and genotyping during breeding trials.

## References


1. Bludau I, Aebersold R (2020) Proteomic and interactomic insights into the molecular basis of cell functional diversity. Nat Rev Mol Cell Biol 21(6):327–340. https://doi.org/10.1038/s41580-020-0231-2
2. Altelaar AF, Munoz J, Heck AJ (2013) Next-generation proteomics: towards an integrative view of proteome dynamics. Nat Rev Genet 14(1):35–48. https://doi.org/10.1038/nrg3356
3. Niu L, Yuan H, Gong F, Wu X, Wang W (2018) Protein extraction methods shape much of the extracted proteomes. Front Plant Sci 9:802. https://doi.org/10.3389/fpls.2018.00802
4. Krahmer J, Hindle MM, Martin SF, Le Bihan T, Millar AJ (2015) Sample preparation for phosphoproteomic analysis of circadian time series in Arabidopsis thaliana. Methods Enzymol 551:405–431. https://doi.org/10.1016/bs.mie.2014.10.022
5. Wang N, Wu X, Ku L, Chen Y, Wang W (2016) Evaluation of three protein-extraction methods for proteome analysis of maize leaf midrib, a compound tissue rich in sclerenchyma cells. Front Plant Sci 7:856. https://doi.org/10.3389/fpls.2016.00856
6. Luís IM, Alexandre BM, Oliveira MM, Abreu IA (2016) Selection of an appropriate protein extraction method to study the Phosphoproteome of maize photosynthetic tissue. PLoS One 11(10):e0164387. https://doi.org/10.1371/journal.pone.0164387
7. Dhabaria A, Cifani P, Reed C, Steen H, Kentsis A (2015) A high-efficiency cellular extraction system for biological proteomics. J Proteome Res 14(8):3403–3408. https://doi.org/10.1021/acs.jproteome.5b00547
8. Moore SM, Hess SM, Jorgenson JW (2016) Extraction, enrichment, solubilization, and digestion techniques for membrane proteomics. J Proteome Res 15(4):1243–1252. https://doi.org/10.1021/acs.jproteome.5b01122
9. Nishimura N, Sarkeshik A, Nito K, Park S-Y, Wang A, Carvalho PC, Lee S, Caddell DF, Cutler SR, Chory J, Yates JR, Schroeder JI (2010) PYR/PYL/RCAR family members are major in-vivo ABI1 protein phosphatase 2C-interacting proteins in Arabidopsis. Plant J 61(2):290–299. https://doi.org/10.1111/j.1365-313X.2009.04054.x
10. Uhrig RG, Labandera AM, Tang LY, Sieben NA, Goudreault M, Yeung E, Gingras AC, Samuel MA, Moorhead GB (2017) Activation of mitochondrial protein phosphatase SLP2 by MIA40 regulates seed germination. Plant Physiol 173(2):956–969. https://doi.org/10.1104/pp.16.01641
11. Van Leene J, Eeckhout D, Cannoot B, De Winne N, Persiau G, Van De Slijke E, Vercruysse L, Dedecker M, Verkest A, Vandepoele K, Martens L, Witters E, Gevaert K, De Jaeger G (2015) An improved toolbox to unravel the plant cellular machinery by tandem affinity purification of Arabidopsis protein complexes. Nat Protoc 10(1):169–187. https://doi.org/10.1038/nprot.2014.199
12. Van Leene J, Han C, Gadeyne A, Eeckhout D, Matthijs C, Cannoot B, De Winne N, Persiau G, Van De Slijke E, Van de Cotte B, Stes E, Van Bel M, Storme V, Impens F, Gevaert K, Vandepoele K, De Smet I, De Jaeger G (2019) Capturing the phosphorylation and protein interaction landscape of the

plant TOR kinase. Nat Plants 5(3):316–327. https://doi.org/10.1038/s41477-019-0378-z

13. Dapic I, Uwugiaren N, Jansen PJ, Corthals GL (2017) Fast and simple protocols for mass spectrometry-based proteomics of small fresh frozen uterine tissue sections. Anal Chem 89 (20):10769–10775. https://doi.org/10.1021/acs.analchem.7b01937
14. Koontz L (2014) TCA precipitation. Methods Enzymol 541:3–10. https://doi.org/10.1016/B978-0-12-420119-4.00001-X
15. Wiśniewski JR, Zougman A, Nagaraj N, Mann M (2009) Universal sample preparation method for proteome analysis. Nat Methods 6 (5):359–362. https://doi.org/10.1038/nmeth.1322
16. Zhang H, Liu P, Guo T, Zhao H, Bensaddek D, Aebersold R, Xiong L (2019) Arabidopsis proteome and the mass spectral assay library. Sci Data 6(1):278. https://doi.org/10.1038/s41597-019-0294-0
17. Wang W-Q, Jensen ON, Møller IM, Hebelstrup KH, Rogowska-Wrzesinska A (2018) Evaluation of sample preparation methods for mass spectrometry-based proteomic analysis of barley leaves. Plant Methods 14:72. https://doi.org/10.1186/s13007-018-0341-4
18. Hu X, Wu L, Zhao F, Zhang D, Li N, Zhu G, Li C, Wang W (2015) Phosphoproteomic analysis of the response of maize leaves to drought, heat and their combination stress. Front Plant Sci 6:298. https://doi.org/10.3389/fpls.2015.00298
19. Li L, Xu Y, Ren Y, Guo Z, Li J, Tong Y, Lin T, Cui D (2019) Comparative proteomic analysis provides insights into the regulatory mechanisms of wheat primary root growth. Sci Rep 9(1):11741. https://doi.org/10.1038/s41598-019-47926-7
20. Doellinger J, Schneider A, Hoeller M, Lasch P (2020) Sample preparation by easy extraction and digestion (SPEED) - a universal, rapid, and detergent-free protocol for proteomics based on acid extraction. Mol Cell Proteomics 19(1):209–222. https://doi.org/10.1074/mcp.TIR119.001616
21. Agrawal GK, Rakwal R (2006) Rice proteomics: a cornerstone for cereal food crop proteomes. Mass Spectrom Rev 25(1):1–53. https://doi.org/10.1002/mas.20056
22. Wu X, Gong F, Wang W (2014) Protein extraction from plant tissues for 2DE and its application in proteomic analysis. Proteomics 14(6):645–658. https://doi.org/10.1002/pmic.201300239
23. Li C, Yue Y, Chen H, Qi W, Song R (2018) The ZmbZIP22 transcription factor regulates 27-kD γ-Zein gene transcription during maize endosperm development. Plant Cell 30(10):2402–2424. https://doi.org/10.1105/tpc.18.00422
24. Liu S, Zenda T, Dong A, Yang Y, Liu X, Wang Y, Li J, Tao Y, Duan H (2019) Comparative proteomic and Morpho-physiological analyses of maize wild-type Vp16 and mutant vp16 germinating seed responses to PEG-induced drought stress. Int J Mol Sci 20:22. https://doi.org/10.3390/ijms20225586
25. Lai X, Wang L, Witzmann FA (2013) Issues and applications in label-free quantitative mass spectrometry. Int J Proteomics 2013:756039. https://doi.org/10.1155/2013/756039
26. Schulze WX, Usadel B (2010) Quantitation in mass-spectrometry-based proteomics. Annu Rev Plant Biol 61:491–516. https://doi.org/10.1146/annurev-arplant-042809-112132
27. Escher C, Reiter L, MacLean B, Ossola R, Herzog F, Chilton J, MacCoss MJ, Rinner O (2012) Using iRT, a normalized retention time for more targeted measurement of peptides. Proteomics 12(8):1111–1121. https://doi.org/10.1002/pmic.201100463
28. Li Y, Li L (2020) Retention time shift analysis and correction in chemical isotope labeling liquid chromatography/mass spectrometry for metabolome analysis. Rapid Commun Mass Spectrom 34(Suppl 1):e8643. https://doi.org/10.1002/rcm.8643
29. Parker SJ, Rost H, Rosenberger G, Collins BC, Malmström L, Amodei D, Venkatraman V, Raedschelders K, Van Eyk JE, Aebersold R (2015) Identification of a set of conserved eukaryotic internal retention time standards for data-independent acquisition mass spectrometry. Mol Cell Proteomics 14(10):2800–2813. https://doi.org/10.1074/mcp.O114.042267
30. Tabata T, Sato T, Kuromitsu J, Oda Y (2007) Pseudo internal standard approach for label-free quantitative proteomics. Anal Chem 79 (22):8440–8445. https://doi.org/10.1021/ac701628m
31. Nikolovski N, Shliaha PV, Gatto L, Dupree P, Lilley KS (2014) Label-free protein quantification for plant Golgi protein localization and abundance. Plant Physiol 166(2):1033–1043. https://doi.org/10.1104/pp.114.245589
32. Uhrig RG, Schläpfer P, Roschitzki B, Hirsch-Hoffmann M, Gruissem W (2019) Diurnal

changes in concerted plant protein phosphorylation and acetylation in Arabidopsis organs and seedlings. Plant J 99(1):176–194. https://doi.org/10.1111/tpj.14315

33. Niehl A, Zhang ZJ, Kuiper M, Peck SC, Heinlein M (2013) Label-free quantitative proteomic analysis of systemic responses to local wounding and virus infection in Arabidopsis thaliana. J Proteome Res 12(6):2491–2503. https://doi.org/10.1021/pr3010698

34. Shen Z, Li P, Ni R-J, Ritchie M, Yang C-P, Liu G-F, Ma W, Liu G-J, Ma L, Li S-J, Wei Z-G, Wang H-X, Wang B-C (2009) Label-free quantitative proteomics analysis of etiolated maize seedling leaves during greening. Mol Cell Proteomics 8(11):2443–2460. https://doi.org/10.1074/mcp.M900187-MCP200

35. Marcon C, Malik WA, Walley JW, Shen Z, Paschold A, Smith LG, Piepho H-P, Briggs SP, Hochholdinger F (2015) A high-resolution tissue-specific proteome and phosphoproteome atlas of maize primary roots reveals functional gradients along the root axes. Plant Physiol 168(1):233–246. https://doi.org/10.1104/pp.15.00138

36. McLoughlin F, Augustine RC, Marshall RS, Li F, Kirkpatrick LD, Otegui MS, Vierstra RD (2018) Maize multi-omics reveal roles for autophagic recycling in proteome remodelling and lipid turnover. Nat Plants 4 (12):1056–1070. https://doi.org/10.1038/s41477-018-0299-2

37. Kaspar-Schoenefeld S, Merx K, Jozefowicz AM, Hartmann A, Seiffert U, Weschke W, Matros A, Mock H-P (2016) Label-free proteome profiling reveals developmental-dependent patterns in young barley grains. J Proteome 143:106–121. https://doi.org/10.1016/j.jprot.2016.04.007

38. Hamzelou S, Pascovici D, Kamath KS, Amirkhani A, McKay M, Mirzaei M, Atwell BJ, Haynes PA (2020) Proteomic responses to drought vary widely among eight diverse genotypes of Rice (Oryza sativa). Int J Mol Sci 21:1. https://doi.org/10.3390/ijms21010363

39. Timabud T, Yin X, Pongdontri P, Komatsu S (2016) Gel-free/label-free proteomic analysis of developing rice grains under heat stress. J Proteome 133:1–19. https://doi.org/10.1016/j.jprot.2015.12.003

40. Meng Q, Gupta R, Min CW, Kim ST (2018) Label-free quantitative proteome data associated with MSP1 and flg22 induced signaling in rice leaves. Data Brief 20:204–209. https://doi.org/10.1016/j.dib.2018.07.063

41. Wang X, Zenda T, Liu S, Liu G, Jin H, Dai L, Dong A, Yang Y, Duan H (2019) Comparative proteomics and physiological analyses reveal important maize filling-kernel drought-responsive genes and metabolic pathways. Int J Mol Sci 20:15. https://doi.org/10.3390/ijms20153743

42. Zenda T, Liu S, Wang X, Jin H, Liu G, Duan H (2018) Comparative proteomic and physiological analyses of two divergent maize inbred lines provide more insights into drought-stress tolerance mechanisms. Int J Mol Sci 19:10. https://doi.org/10.3390/ijms19103225

43. Zhao F, Zhang D, Zhao Y, Wang W, Yang H, Tai F, Li C, Hu X (2016) The difference of physiological and proteomic changes in maize leaves adaptation to drought, heat, and combined both stresses. Front Plant Sci 7:1471. https://doi.org/10.3389/fpls.2016.01471

44. Walley JW, Shen Z, McReynolds MR, Schmelz EA, Briggs SP (2018) Fungal-induced protein hyperacetylation in maize identified by acetylome profiling. Proc Natl Acad Sci U S A 115(1):210–215. https://doi.org/10.1073/pnas.1717519115

45. Wiese S, Reidegeld KA, Meyer HE, Warscheid B (2007) Protein labeling by iTRAQ: a new tool for quantitative mass spectrometry in proteome research. Proteomics 7 (3):340–350. https://doi.org/10.1002/pmic.200600422

46. Mergner J, Frejno M, List M, Papacek M, Chen X, Chaudhary A, Samaras P, Richter S, Shikata H, Messerer M, Lang D, Altmann S, Cyprys P, Zolg DP, Mathieson T, Bantscheff M, Hazarika RR, Schmidt T, Dawid C, Dunkel A, Hofmann T, Sprunck S, Falter-Braun P, Johannes F, Mayer KFX, Jürgens G, Wilhelm M, Baumbach J, Grill E, Schneitz K, Schwechheimer C, Kuster B (2020) Mass-spectrometry-based draft of the Arabidopsis proteome. Nature 579 (7799):409–414. https://doi.org/10.1038/s41586-020-2094-2

47. Lin Q, Zhou Z, Luo W, Fang M, Li M, Li H (2017) Screening of proximal and interacting proteins in Rice protoplasts by proximity-dependent Biotinylation. Front Plant Sci 8:749. https://doi.org/10.3389/fpls.2017.00749

48. Qin L, Walk TC, Han P, Chen L, Zhang S, Li Y, Hu X, Xie L, Yang Y, Liu J, Lu X, Yu C, Tian J, Shaff JE, Kochian LV, Liao X, Liao H (2019) Adaption of roots to nitrogen deficiency revealed by 3D quantification and proteomic analysis. Plant Physiol 179

(1):329–347. https://doi.org/10.1104/pp.18.00716
49. Zhang H-Y, Lei G, Zhou H-W, He C, Liao J-L, Huang Y-J (2017) Quantitative iTRAQ-based proteomic analysis of rice grains to assess high night temperature stress. Proteomics 17:5. https://doi.org/10.1002/pmic.201600365
50. Hartl M, Füßl M, Boersema PJ, Jost J, Kramer K, Bakirbas A, Sindlinger J, Plöchinger M, Leister D, Uhrig G, GBG M, Cox J, Salvucci ME, Schwarzer D, Mann M, Finkemeier I (2017) Lysine acetylome profiling uncovers novel histone deacetylase substrate proteins in Arabidopsis. Mol Syst Biol 13(10):949. https://doi.org/10.15252/msb.20177819
51. Liu S, Yu F, Yang Z, Wang T, Xiong H, Chang C, Yu W, Li N (2018) Establishment of dimethyl Labeling-based quantitative Acetylproteomics in Arabidopsis. Mol Cell Proteomics 17(5):1010–1027. https://doi.org/10.1074/mcp.RA117.000530
52. Uhrig RG, Echevarría-Zomeño S, Schlapfer P, Grossmann J, Roschitzki B, Koerber N, Fiorani F, Gruissem W (2020) Diurnal dynamics of the Arabidopsis rosette proteome and Phosphoproteome. https://doi.org/10.1101/2020.09.11.293779
53. Hsu C-C, Zhu Y, Arrington JV, Paez JS, Wang P, Zhu P, Chen IH, Zhu J-K, Tao WA (2018) Universal plant Phosphoproteomics workflow and its application to tomato Signaling in response to cold stress. Mol Cell Proteomics 17(10):2068–2080. https://doi.org/10.1074/mcp.TIR118.000702
54. Boersema PJ, Raijmakers R, Lemeer S, Mohammed S, Heck AJR (2009) Multiplex peptide stable isotope dimethyl labeling for quantitative proteomics. Nat Protoc 4 (4):484–494. https://doi.org/10.1038/nprot.2009.21
55. Shih Y-C, Hsiao J-T, Sheu F (2019) Feasibility of utilizing stable-isotope dimethyl Labeling in liquid chromatography⁻Tandem mass spectrometry-based determination for food allergens-case of kiwifruit. Molecules 24:10. https://doi.org/10.3390/molecules24101920
56. Hennrich ML, Mohammed S, Altelaar AFM, Heck AJR (2010) Dimethyl isotope labeling assisted de novo peptide sequencing. J Am Soc Mass Spectrom 21(12):1957–1965. https://doi.org/10.1016/j.jasms.2010.08.007
57. Lewandowska D, ten Have S, Hodge K, Tillemans V, Lamond AI, Brown JWS (2013) Plant SILAC: stable-isotope labelling with amino acids of arabidopsis seedlings for quantitative proteomics. PLoS One 8(8): e72207. https://doi.org/10.1371/journal.pone.0072207
58. Kline KG, Sussman MR (2010) Protein quantitation using isotope-assisted mass spectrometry. Annu Rev Biophys 39:291–308. https://doi.org/10.1146/annurev.biophys.093008.131339
59. Schütz W, Hausmann N, Krug K, Hampp R, Macek B (2011) Extending SILAC to proteomics of plant cell lines. Plant Cell 23 (5):1701–1705. https://doi.org/10.1105/tpc.110.082016
60. Chen S, Harmon AC (2006) Advances in plant proteomics. Proteomics 6 (20):5504–5516. https://doi.org/10.1002/pmic.200600143
61. Fukao Y, Yoshida M, Kurata R, Kobayashi M, Nakanishi M, Fujiwara M, Nakajima K, Ferjani A (2013) Peptide separation methodologies for in-depth proteomics in Arabidopsis. Plant Cell Physiol 54(5):808–815. https://doi.org/10.1093/pcp/pct033
62. Fonslow BR, Carvalho PC, Academia K, Freeby S, Xu T, Nakorchevsky A, Paulus A, Yates JR 3rd (2011) Improvements in proteomic metrics of low abundance proteins through proteome equalization using ProteoMiner prior to MudPIT. J Proteome Res 10 (8):3690–3700. https://doi.org/10.1021/pr200304u
63. Thingholm TE, Jørgensen TJD, Jensen ON, Larsen MR (2006) Highly selective enrichment of phosphorylated peptides using titanium dioxide. Nat Protoc 1(4):1929–1935. https://doi.org/10.1038/nprot.2006.185
64. Batth TS, Francavilla C, Olsen JV (2014) Off-line high-pH reversed-phase fractionation for in-depth phosphoproteomics. J Proteome Res 13(12):6176–6186. https://doi.org/10.1021/pr500893m
65. Ly L, Wasinger VC (2011) Protein and peptide fractionation, enrichment and depletion: tools for the complex proteome. Proteomics 11(4):513–534. https://doi.org/10.1002/pmic.201000394
66. Wu L, Tian L, Wang S, Zhang J, Liu P, Tian Z, Zhang H, Liu H, Chen Y (2016) Comparative proteomic analysis of the response of maize (Zea mays L.) leaves to long photoperiod condition. Front Plant Sci 7:752. https://doi.org/10.3389/fpls.2016.00752
67. Peng P, Gao Y, Li Z, Yu Y, Qin H, Guo Y, Huang R, Wang J (2019) Proteomic analysis of a Rice mutant sd58 possessing a novel d1 allele of heterotrimeric G protein alpha

subunit (RGA1) in salt stress with a focus on ROS scavenging. Int J Mol Sci 20:1. https://doi.org/10.3390/ijms20010167

68. Walley JW, Sartor RC, Shen Z, Schmitz RJ, Wu KJ, Urich MA, Nery JR, Smith LG, Schnable JC, Ecker JR, Briggs SP (2016) Integration of omic networks in a developmental atlas of maize. Science 353 (6301):814–818. https://doi.org/10.1126/science.aag1125
69. Ding H, He J, Wu Y, Wu X, Ge C, Wang Y, Zhong S, Peiter E, Liang J, Xu W (2018) The tomato mitogen-activated protein kinase SlMPK1 is as a negative regulator of the high-temperature stress response. Plant Physiol 177(2):633–651. https://doi.org/10.1104/pp.18.00067
70. Dudley E, Bond AE (2014) Phosphoproteomic techniques and applications. Adv Protein Chem Struct Biol 95:25–69. https://doi.org/10.1016/B978-0-12-800453-1.00002-6
71. Hernandez-Hernandez O, Lebron-Aguilar R, Quintanilla-Lopez JE, Sanz ML, Moreno FJ (2010) Development of a new method using HILIC-tandem mass spectrometry for the characterization of O-sialoglycopeptides from proteolytically digested caseinomacropeptide. Proteomics 10(20):3699–3711. https://doi.org/10.1002/pmic.201000156
72. Boersema PJ, Mohammed S, Heck AJR (2008) Hydrophilic interaction liquid chromatography (HILIC) in proteomics. Anal Bioanal Chem 391(1):151–159. https://doi.org/10.1007/s00216-008-1865-7
73. Bensaddek D, Nicolas A, Lamond AI (2015) Evaluating the use of HILIC in large-scale, multi dimensional proteomics: horses for courses? Int J Mass Spectrom 391:105–114. https://doi.org/10.1016/j.ijms.2015.07.029
74. Dedvisitsakul P, Jacobsen S, Svensson B, Bunkenborg J, Finnie C, Hägglund P (2014) Glycopeptide enrichment using a combination of ZIC-HILIC and cotton wool for exploring the glycoproteome of wheat flour albumins. J Proteome Res 13 (5):2696–2703. https://doi.org/10.1021/pr401282r
75. Wang K, Zhao Y, Li M, Gao F, Yang M-K, Wang X, Li S, Yang P (2014) Analysis of phosphoproteome in rice pistil. Proteomics 14(20):2319–2334. https://doi.org/10.1002/pmic.201400004
76. Heazlewood JL, Durek P, Hummel J, Selbig J, Weckwerth W, Walther D, Schulze WX (2008) PhosPhAt: a database of phosphorylation sites in Arabidopsis thaliana and a plant-specific phosphorylation site predictor. Nucleic Acids Res 36(Database issue):D1015–D1021. https://doi.org/10.1093/nar/gkm812
77. Willems P, Horne A, Van Parys T, Goormachtig S, De Smet I, Botzki A, Van Breusegem F, Gevaert K (2019) The plant PTM viewer, a central resource for exploring plant protein modifications. Plant J 99 (4):752–762. https://doi.org/10.1111/tpj.14345
78. Yao Q, Ge H, Wu S, Zhang N, Chen W, Xu C, Gao J, Thelen JJ, Xu D (2014) $P^3DB$ 3.0: from plant phosphorylation sites to protein networks. Nucleic Acids Res 42(Database issue):D1206–D1213. https://doi.org/10.1093/nar/gkt1135
79. Romero-Barrios N, Monachello D, Dolde U, Wong A, San Clemente H, Cayrel A, Johnson A, Lurin C, Vert G (2020) Advanced Cataloging of Lysine-63 Polyubiquitin networks by genomic, Interactome, and sensor-based proteomic analyses. Plant Cell 32 (1):123–138. https://doi.org/10.1105/tpc.19.00568
80. Khoury GA, Baliban RC, Floudas CA (2011) Proteome-wide post-translational modification statistics: frequency analysis and curation of the swiss-prot database. Sci Rep 1. https://doi.org/10.1038/srep00090
81. Wang R-J, Peng J, Li QX, Peng Y-L (2017) Phosphorylation-mediated regulatory networks in mycelia of Pyricularia oryzae revealed by Phosphoproteomic analyses. Mol Cell Proteomics 16(9):1669–1682. https://doi.org/10.1074/mcp.M116.066670
82. Zhao Y, Jensen ON (2009) Modification-specific proteomics: strategies for characterization of post-translational modifications using enrichment techniques. Proteomics 9 (20):4632–4641. https://doi.org/10.1002/pmic.200900398
83. Mehta D, Ghahremani M, Pérez-Fernández M, Tan M, Schläpfer P, Plaxton WC, Uhrig RG (2020) Phosphate and phosphite differentially impact the proteome and phosphoproteome of Arabidopsis suspension cell cultures. bioRXiv. https://doi.org/10.1101/2020.05.29.124040
84. Zhang M, Ma C-Y, Lv D-W, Zhen S-M, Li X-H, Yan Y-M (2014) Comparative phosphoproteome analysis of the developing grains in bread wheat (Triticum aestivum L.) under well-watered and water-deficit conditions. J Proteome Res 13(10):4281–4297. https://doi.org/10.1021/pr500400t
85. Vu LD, Stes E, Van Bel M, Nelissen H, Maddelein D, Inzé D, Coppens F, Martens L, Gevaert K, De Smet I (2016)

Up-to-date workflow for plant (Phospho)-proteomics identifies differential drought-responsive phosphorylation events in maize leaves. J Proteome Res 15(12):4304–4317. https://doi.org/10.1021/acs.jproteome.6b00348

86. Zhong M, Li S, Huang F, Qiu J, Zhang J, Sheng Z, Tang S, Wei X, Hu P (2017) The Phosphoproteomic response of Rice seedlings to cadmium stress. Int J Mol Sci 18:10. https://doi.org/10.3390/ijms18102055
87. Chaudhary R, Atamian HS, Shen Z, Briggs SP, Kaloshian I (2015) Potato aphid salivary proteome: enhanced salivation using resorcinol and identification of aphid phosphoproteins. J Proteome Res 14(4):1762–1778. https://doi.org/10.1021/pr501128k
88. Endler A, Reiland S, Gerrits B, Schmidt UG, Baginsky S, Martinoia E (2009) In vivo phosphorylation sites of barley tonoplast proteins identified by a phosphoproteomic approach. Proteomics 9(2):310–321. https://doi.org/10.1002/pmic.200800323
89. Bergström Lind S, Molin M, Savitski MM, Emilsson L, Aström J, Hedberg L, Adams C, Nielsen ML, Engström A, Elfineh L, Andersson E, Zubarev RA, Pettersson U (2008) Immunoaffinity enrichments followed by mass spectrometric detection for studying global protein tyrosine phosphorylation. J Proteome Res 7(7):2897–2910. https://doi.org/10.1021/pr8000546
90. Fíla J, Honys D (2012) Enrichment techniques employed in phosphoproteomics. Amino Acids 43(3):1025–1047. https://doi.org/10.1007/s00726-011-1111-z
91. Johnson A, Vert G (2016) Unraveling K63 Polyubiquitination networks by sensor-based proteomics. Plant Physiol 171 (3):1808–1820. https://doi.org/10.1104/pp.16.00619
92. Song Y, Shi X, Zou Y, Guo J, Huo N, Chen S, Zhao C, Li H, Wu G, Peng Y (2020) Proteome-wide identification and functional analysis of ubiquitinated proteins in peach leaves. Sci Rep 10(1):2447. https://doi.org/10.1038/s41598-020-59342-3
93. Xie X, Kang H, Liu W, Wang G-L (2015) Comprehensive profiling of the rice ubiquitome reveals the significance of lysine ubiquitination in young leaves. J Proteome Res 14 (5):2017–2025. https://doi.org/10.1021/pr5009724
94. Zhang N, Zhang L, Shi C, Tian Q, Lv G, Wang Y, Cui D, Chen F (2017) Comprehensive profiling of lysine ubiquitome reveals diverse functions of lysine ubiquitination in common wheat. Sci Rep 7(1):13601. https://doi.org/10.1038/s41598-017-13992-y
95. Gonzalez-Galarza FF, Lawless C, Hubbard SJ, Fan J, Bessant C, Hermjakob H, Jones AR (2012) A critical appraisal of techniques, software packages, and standards for quantitative proteomic analysis. OMICS 16 (9):431–442. https://doi.org/10.1089/omi.2012.0022
96. Tyanova S, Temu T, Cox J (2016) The MaxQuant computational platform for mass spectrometry-based shotgun proteomics. Nat Protoc 11(12):2301–2319. https://doi.org/10.1038/nprot.2016.136
97. Cox J, Matic I, Hilger M, Nagaraj N, Selbach M, Olsen JV, Mann M (2009) A practical guide to the MaxQuant computational platform for SILAC-based quantitative proteomics. Nat Protoc 4(5):698–705. https://doi.org/10.1038/nprot.2009.36
98. Tyanova S, Mann M, Cox J (2014) MaxQuant for in-depth analysis of large SILAC datasets. In: Warscheid B (ed) Stable isotope Labeling by amino acids in cell culture (SILAC): methods and protocols. Springer New York, New York, NY, pp 351–364. https://doi.org/10.1007/978-1-4939-1142-4_24
99. Cox J, Neuhauser N, Michalski A, Scheltema RA, Olsen JV, Mann M (2011) Andromeda: a peptide search engine integrated into the MaxQuant environment. J Proteome Res 10 (4):1794–1805. https://doi.org/10.1021/pr101065j
100. Cox J, Mann M (2008) MaxQuant enables high peptide identification rates, individualized p.p.b.-range mass accuracies and proteome-wide protein quantification. Nat Biotechnol 26(12):1367–1372. https://doi.org/10.1038/nbt.1511
101. Zhang J, Yang W, Li S, Yao S, Qi P, Yang Z, Feng Z, Hou J, Cai L, Yang M, Wu W, Guo D-A (2016) An intelligentized strategy for endogenous small molecules characterization and quality evaluation of earthworm from two geographic origins by ultra-high performance HILIC/QTOF MS(E) and Progenesis QI. Anal Bioanal Chem 408 (14):3881–3890. https://doi.org/10.1007/s00216-016-9482-3
102. Colaert N, Barsnes H, Vaudel M, Helsens K, Timmerman E, Sickmann A, Gevaert K, Martens L (2011) Thermo-msf-parser: an open source Java library to parse and visualize Thermo proteome discoverer msf files. J Proteome Res 10(8):3840–3843. https://doi.org/10.1021/pr2005154

103. Veit J, Sachsenberg T, Chernev A, Aicheler F, Urlaub H, Kohlbacher O (2016) LFQProfiler and RNP(xl): open-source tools for label-free quantification and protein-RNA cross-linking integrated into proteome discoverer. J Proteome Res 15(9):3441–3448. https://doi.org/10.1021/acs.jproteome.6b00407
104. Doerr A (2015) DIA mass spectrometry. Nat Methods 12(1):35–35. https://doi.org/10.1038/nmeth.3234
105. Egertson JD, Kuehn A, Merrihew GE, Bateman NW, MacLean BX, Ting YS, Canterbury JD, Marsh DM, Kellmann M, Zabrouskov V, Wu CC, MacCoss MJ (2013) Multiplexed MS/MS for improved data-independent acquisition. Nat Methods 10(8):744–746. https://doi.org/10.1038/nmeth.2528
106. Egertson JD, MacLean B, Johnson R, Xuan Y, MacCoss MJ (2015) Multiplexed peptide analysis using data-independent acquisition and skyline. Nat Protoc 10 (6):887–903. https://doi.org/10.1038/nprot.2015.055
107. Deutsch EW, Perez-Riverol Y, Chalkley RJ, Wilhelm M, Tate S, Sachsenberg T, Walzer M, Käll L, Delanghe B, Böcker S, Schymanski EL, Wilmes P, Dorfer V, Kuster B, Volders P-J, Jehmlich N, Vissers JPC, Wolan DW, Wang AY, Mendoza L, Shofstahl J, Dowsey AW, Griss J, Salek RM, Neumann S, Binz P-A, Lam H, Vizcaíno JA, Bandeira N, Röst H (2018) Expanding the use of spectral libraries in proteomics. J Proteome Res 17(12):4051–4060. https://doi.org/10.1021/acs.jproteome.8b00485
108. Zhang X, Yin F, Xiao S, Jiang C, Yu T, Chen L, Ke X, Zhong Q, Cheng Z, Li W (2019) Proteomic analysis of the rice (Oryza officinalis) provides clues on molecular tagging of proteins for brown planthopper resistance. BMC Plant Biol 19(1):30. https://doi.org/10.1186/s12870-018-1622-9
109. He Y, Yu S, Liu S, Tian H, Yu C, Tan W, Zhang J, Li Z, Jiang F, Duan L (2020) Data-independent acquisition proteomics unravels the effects of iron ions on Coronatine synthesis in pseudomonas syringae pv. Tomato DC3000. Front Microbiol 11:1362. https://doi.org/10.3389/fmicb.2020.01362
110. Bromilow SNL, Gethings LA, Langridge JI, Shewry PR, Buckley M, Bromley MJ, Mills ENC (2016) Comprehensive proteomic profiling of wheat gluten using a combination of data-independent and data-dependent acquisition. Front Plant Sci 7:2020. https://doi.org/10.3389/fpls.2016.02020
111. Röst HL, Rosenberger G, Navarro P, Gillet L, Miladinović SM, Schubert OT, Wolski W, Collins BC, Malmström J, Malmström L, Aebersold R (2014) OpenSWATH enables automated, targeted analysis of data-independent acquisition MS data. Nat Biotechnol 32(3):219–223. https://doi.org/10.1038/nbt.2841
112. Tsou C-C, Avtonomov D, Larsen B, Tucholska M, Choi H, Gingras A-C, Nesvizhskii AI (2015) DIA-umpire: comprehensive computational framework for data-independent acquisition proteomics. Nat Methods 12(3):258–264, 257 p following 264. https://doi.org/10.1038/nmeth.3255
113. Distler U, Kuharev J, Navarro P, Levin Y, Schild H, Tenzer S (2014) Drift time-specific collision energies enable deep-coverage data-independent acquisition proteomics. Nat Methods 11(2):167–170. https://doi.org/10.1038/nmeth.2767
114. Müller F, Kolbowski L, Bernhardt OM, Reiter L, Rappsilber J (2019) Data-independent acquisition improves quantitative cross-linking mass spectrometry. Mol Cell Proteomics 18(4):786–795. https://doi.org/10.1074/mcp.TIR118.001276
115. Bekker-Jensen DB, Bernhardt OM, Hogrebe A, Martinez-Val A, Verbeke L, Gandhi T, Kelstrup CD, Reiter L, Olsen JV (2020) Rapid and site-specific deep phosphoproteome profiling by data-independent acquisition without the need for spectral libraries. Nat Commun 11(1):787. https://doi.org/10.1038/s41467-020-14609-1
116. Guan S, Taylor PP, Han Z, Moran MF, Ma B (2020) Data dependent-independent acquisition (DDIA) proteomics. J Proteome Res 19 (8):3230–3237. https://doi.org/10.1021/acs.jproteome.0c00186
117. Fan K-T, Wang K-H, Chang W-H, Yang J-C, Yeh C-F, Cheng K-T, Hung S-C, Chen Y-R (2019) Application of data-independent acquisition approach to study the proteome change from early to later phases of tomato pathogenesis responses. Int J Mol Sci 20:4. https://doi.org/10.3390/ijms20040863
118. Martin LBB, Sherwood RW, Nicklay JJ, Yang Y, Muratore-Schroeder TL, Anderson ET, Thannhauser TW, Rose JKC, Zhang S (2016) Application of wide selected-ion monitoring data-independent acquisition to identify tomato fruit proteins regulated by the CUTIN DEFICIENT2 transcription factor. Proteomics 16(15–16):2081–2094. https://doi.org/10.1002/pmic.201500450

119. Farooq MA, Zhang K, Islam F, Wang J, Athar HUR, Nawaz A, Ullah Zafar Z, Xu J, Zhou W (2018) Physiological and iTRAQ-based quantitative proteomics analysis of methyl Jasmonate-induced tolerance in Brassica napus under arsenic stress. Proteomics 18 (10):e1700290. https://doi.org/10.1002/pmic.201700290
120. Ma Z, Wang L, Zhao M, Gu S, Wang C, Zhao J, Tang Z, Gao H, Zhang L, Fu L, Yin Y, He N, Zheng W, Xu Z (2020) iTRAQ proteomics reveals the regulatory response to Magnaporthe oryzae in durable resistant vs. susceptible rice genotypes. PLoS One 15(1):e0227470. https://doi.org/10.1371/journal.pone.0227470
121. Wang P, Yao S, Kosami K-I, Guo T, Li J, Zhang Y, Fukao Y, Kaneko-Kawano T, Zhang H, She Y-M, Wang P, Xing W, Hanada K, Liu R, Kawano Y (2020) Identification of endogenous small peptides involved in rice immunity through transcriptomics- and proteomics-based screening. Plant Biotechnol J 18(2):415–428. https://doi.org/10.1111/pbi.13208
122. Xiao C, Gao J, Zhang Y, Wang Z, Zhang D, Chen Q, Ye X, Xu Y, Yang G, Yan L, Cheng Q, Chen J, Shen Y (2019) Quantitative proteomics of potato leaves infected with Phytophthora infestans provides insights into coordinated and altered protein expression during early and late disease stages. Int J Mol Sci 20:1. https://doi.org/10.3390/ijms20010136
123. Renaud JB, Sumarah MW (2016) Data independent acquisition-digital archiving mass spectrometry: application to single kernel mycotoxin analysis of Fusarium graminearum infected maize. Anal Bioanal Chem 408 (12):3083–3091. https://doi.org/10.1007/s00216-016-9391-5
124. Ghatak A, Chaturvedi P, Weckwerth W (2017) Cereal crop proteomics: systemic analysis of crop drought stress responses towards marker-assisted selection breeding. Front Plant Sci 8:757. https://doi.org/10.3389/fpls.2017.00757
125. Kosová K, Vítámvás P, Klíma M, Prášil IT (2019) Breeding drought-resistant crops: G×E interactions, proteomics and pQTLS. J Exp Bot 70(10):2605–2608. https://doi.org/10.1093/jxb/erz116
126. Pichereaux C, Laurent E-A, Gargaros A, Viudes S, Durieu C, Lamaze T, Grieu P, Burlet-Schiltz O (2019) Analysis of durum wheat proteome changes under marine and fungal biostimulant treatments using large-scale quantitative proteomics: a useful dataset of durum wheat proteins. J Proteome 200:28–39. https://doi.org/10.1016/j.jprot.2019.03.003
127. Mehri N, Fotovat R, Mirzaei M, Fard EM, Parsamatin P, Hasan MT, Wu Y, Ghaffari MR, Salekdeh GH (2020) Proteomic analysis of wheat contrasting genotypes reveals the interplay between primary metabolic and regulatory pathways in anthers under drought stress. J Proteome 226:103895. https://doi.org/10.1016/j.jprot.2020.103895
128. Afzal M, Pfannstiel J, Zimmermann J, Bischoff SC, Würschum T, Longin CFH (2020) High-resolution proteomics reveals differences in the proteome of spelt and bread wheat flour representing targets for research on wheat sensitivities. Sci Rep 10(1):14677. https://doi.org/10.1038/s41598-020-71712-5
129. Tahir A, Kang J, Choulet F, Ravel C, Romeuf I, Rasouli F, Nosheen A, Branlard G (2020) Deciphering carbohydrate metabolism during wheat grain development via integrated transcriptome and proteome dynamics. Mol Biol Rep 47(7):5439–5449. https://doi.org/10.1007/s11033-020-05634-w
130. Guo W, Han L, Li X, Wang H, Mu P, Lin Q, Liu Q, Zhang Y (2020) Proteome and lysine acetylome analysis reveals insights into the molecular mechanism of seed germination in wheat. Sci Rep 10(1):13454. https://doi.org/10.1038/s41598-020-70230-8
131. Varzakas T (2016) Quality and safety aspects of cereals (wheat) and their products. Crit Rev Food Sci Nutr 56(15):2495–2510. https://doi.org/10.1080/10408398.2013.866070
132. Maccaferri M, Harris NS, Twardziok SO, Pasam RK, Gundlach H, Spannagl M, Ormanbekova D, Lux T, Prade VM, Milner SG, Himmelbach A, Mascher M, Bagnaresi P, Faccioli P, Cozzi P, Lauria M, Lazzari B, Stella A, Manconi A, Gnocchi M, Moscatelli M, Avni R, Deek J, Biyiklioglu S, Frascaroli E, Corneti S, Salvi S, Sonnante G, Desiderio F, Marè C, Crosatti C, Mica E, Özkan H, Kilian B, De Vita P, Marone D, Joukhadar R, Mazzucotelli E, Nigro D, Gadaleta A, Chao S, Faris JD, Melo ATO, Pumphrey M, Pecchioni N, Milanesi L, Wiebe K, Ens J, MacLachlan RP, Clarke JM, Sharpe AG, Koh CS, Liang KYH, Taylor GJ, Knox R, Budak H, Mastrangelo AM, Xu SS, Stein N, Hale I, Distelfeld A, Hayden MJ, Tuberosa R, Walkowiak S, Mayer KFX, Ceriotti A, Pozniak CJ, Cattivelli L (2019) Durum wheat genome highlights past domestication signatures and future improvement

targets. Nat Genet 51(5):885–895. https://doi.org/10.1038/s41588-019-0381-3

133. Bose U, Broadbent JA, Byrne K, Blundell MJ, Howitt CA, Colgrave ML (2020) Proteome analysis of Hordein-null barley lines reveals storage protein synthesis and compensation mechanisms. J Agric Food Chem 68 (20):5763–5775. https://doi.org/10.1021/acs.jafc.0c01410
134. Colgrave ML, Byrne K, Blundell M, Heidelberger S, Lane CS, Tanner GJ, Howitt CA (2016) Comparing multiple reaction monitoring and sequential window Acquisition of all Theoretical Mass Spectra for the relative quantification of barley gluten in selectively bred barley lines. Anal Chem 88 (18):9127–9135. https://doi.org/10.1021/acs.analchem.6b02108
135. Kerr ED, Phung TK, Caboche CH, Fox GP, Platz GJ, Schulz BL (2019) The intrinsic and regulated proteomes of barley seeds in response to fungal infection. Anal Biochem 580:30–35. https://doi.org/10.1016/j.ab.2019.06.004
136. Kosová K, Vítámvás P, Prášil IT (2014) Proteomics of stress responses in wheat and barley-search for potential protein markers of stress tolerance. Front Plant Sci 5:711. https://doi.org/10.3389/fpls.2014.00711
137. Chmielewska K, Rodziewicz P, Swarcewicz B, Sawikowska A, Krajewski P, Marczak Ł, Ciesiołka D, Kuczyńska A, Mikołajczak K, Ogrodowicz P, Krystkowiak K, Surma M, Adamski T, Bednarek P, Stobiecki M (2016) Analysis of drought-induced proteomic and Metabolomic changes in barley (Hordeum vulgare L.) leaves and roots unravels some aspects of biochemical mechanisms involved in drought tolerance. Front Plant Sci 7:1108. https://doi.org/10.3389/fpls.2016.01108
138. Shen Q, Fu L, Dai F, Jiang L, Zhang G, Wu D (2016) Multi-omics analysis reveals molecular mechanisms of shoot adaption to salt stress in Tibetan wild barley. BMC Genomics 17 (1):889. https://doi.org/10.1186/s12864-016-3242-9
139. Wang Y, Sang Z, Xu S, Xu Q, Zeng X, Jabu D, Yuan H (2020) Comparative proteomics analysis of Tibetan hull-less barley under osmotic stress via data-independent acquisition mass spectrometry. Gigascience 9:3. https://doi.org/10.1093/gigascience/giaa019
140. Eggert K, Pawelzik E (2011) Proteome analysis of fusarium head blight in grains of naked barley (Hordeum vulgare subsp. nudum). Proteomics 11(5):972–985. https://doi.org/10.1002/pmic.201000322
141. Guo B, Luan H, Lin S, Lv C, Zhang X, Xu R (2016) Comparative proteomic analysis of two barley cultivars (Hordeum vulgare L.) with contrasting grain protein content. Front. Plant Sci 7:542. https://doi.org/10.3389/fpls.2016.00542
142. Lewandowska D, Zhang R, Colas I, Uzrek N, Waugh R (2019) Application of a sensitive and reproducible label-free proteomic approach to explore the proteome of individual meiotic-phase barley anthers. Front Plant Sci 10:393. https://doi.org/10.3389/fpls.2019.00393
143. Mahalingam R (2017) Shotgun proteomics of the barley seed proteome. BMC Genomics 18(1):44. https://doi.org/10.1186/s12864-016-3408-5

# Chapter 15

# DNA Affinity Purification Sequencing (DAP-Seq) for Mapping Genome-Wide Transcription Factor Binding Sites in Plants

**Miaomiao Li and Shao-Shan Carol Huang**


## Abstract

DNA affinity purification sequencing (DAP-seq) couples in vitro expression of transcription factor (TF) and binding to a genomic DNA library with next-generation sequencing, providing a robust method to identify genome-wide transcription factor binding sites (TFBS) on endogenous genomic DNA. Here we describe a detailed protocol of DAP-seq, including DNA library construction, cell-free protein expression, DNA affinity purification, and DNA sequencing. DAP-seq is an efficient and high-throughput approach to creating genome-wide binding site maps. Moreover, the genomic DNA library retains the chemical modifications on DNA such as cytosine methylation, allowing the effect of these modifications on TF binding to be analyzed. This method has been applied to the model plant *Arabidopsis* as well as cereal crops such as rice and maize.


**Key words** DAP-seq, Transcription factor, Transcription factor binding site, Transcription factor-DNA interaction, Next-generation sequencing

## 1 Introduction

Transcriptional regulation plays an important role in controlling gene expression in plant development and responses to environmental stimuli [1]. Transcription factors (TF) regulate the expression of their target genes by recognizing DNA sequences known as transcription factor binding sites (TFBS) located at proximal or distal regions of the genes. TF binding activities are also influenced by DNA features such as DNA secondary structures and modifications [2]. Systematically identifying genome-wide TFBS is required for comprehensive understanding of the functions and regulatory mechanisms of TFs [3].

Chromatin immunoprecipitation sequencing (ChIP-seq) is commonly used to identify in vivo binding sites of TFs. However, the requirement for antibodies or tagged transgenic lines hinders the wide application of this method in plants [4]. Systematic evolution of ligands by exponential enrichment (SELEX) and

Andriy Bilichak and John D. Laurie (eds.), *Accelerated Breeding of Cereal Crops*, Springer Protocols Handbooks,
https://doi.org/10.1007/978-1-0716-1526-3_15, © Springer Science+Business Media, LLC, part of Springer Nature 2022

protein-binding microarrays (PBM) are two main in vitro approaches to characterize DNA-binding motifs of TFs by screening synthesized DNA oligos [5, 6]. The accuracy of these two methods can be limited by the lack of genome context of the synthetic oligos. The DAP-seq technique, combining affinity purification of genomic DNA fragments and next-generation sequencing, provides a high-throughput platform for mapping the genome-wide DNA binding sites of single TF or TF families [7]. In DAP-seq, the TF of interest is fused with an affinity tag, such as HaloTag, and expressed in vitro. Genomic DNA is extracted from plant tissues and fragmented by sonication, and Illumina-compatible sequencing adapters are ligated to the DNA fragments to create a genomic DNA library. After incubating the Halo-TF with the DNA library, the TF/DNA complex is purified by HaloTag ligand-conjugated magnetic beads, unbound DNA is washed away, and TF-bound DNA is eluted. Eluted DNA is PCR amplified to construct library for sequencing. DAP-seq is able to identify TFBS at high accuracy because the genomic DNA in the binding reaction contains both the binding site motifs and the endogenous genomic context of the binding sites, such as DNA structure and modifications. DAP-seq can also be modified to examine the effect of DNA modifications on TF binding. For example, ampDAP-seq with an amplified DNA library was used to reveal binding events sensitive to DNA methylation [7, 8]. Originally developed in the model plant *Arabidopsis thaliana*, DAP-seq has been applied to study TFs in crops, such as rice and maize [9, 10], and has the potential to be used for mapping cis-regulatory elements for large number of TFs in the genomes of many plant species.

## 2 Materials

### 2.1 DNA Library Preparation

1. Mortar and pestle.
2. Liquid nitrogen.
3. 15 mL Falcon tube.
4. Genomic DNA (gDNA) extraction buffer: 1 M Tris, 0.5 M NaCl, 50 mM EDTA, adjust pH to 8.0 with HCl.
5. Phenol:Chloroform:Isoamyl Alcohol 25:24:1 (Sigma-Aldrich).
6. Chloroform (Sigma-Aldrich).
7. Isopropanol (Sigma-Aldrich).
8. Ethanol, 100% and 70% (vol/vol).
9. DNA low-bind microcentrifuge tubes (Eppendorf).
10. 3 M NaOAc, pH 5.2 (Thermo Scientific, cat. no. PI17888).
11. Elution Buffer (Qiagen, cat. no. 19086).

12. Sonicator: Covaris S220 Series.
13. microTUBE holder (Covaris, cat. no. 500114).
14. microTUBE AFA Fiber Pre-Slit Snap-Cap (Covaris, cat. no. 520045).
15. End-It DNA End-Repair Kit (Epicentre, cat. no. ER81050).
16. 1 mM dATP.
17. Klenow (3′–5'exo-; NEB, cat. no. M0212).
18. T4 DNA Ligase (Promega, cat. no. M1804).
19. Y-adapter (Illumina-based) sequences:
    Adapter A: 5'-ACACTCTTTCCCTACACGACGCTCTTCC GATCT-3′.

    Adapter B: 5'-Phos-GATCGGAAGAGCACACGTCTGAACTC CAGTCAC-3′ ("Phos" represents a 5′-phosphate group).

#### 2.2 Vector Construction and Protein Expression

1. pIX-HALO vector (ABRC, stock no. CD3–1742).
2. TNT SP6 Coupled Wheat Germ Extract System (Promega, L4130) or TNT SP6 Coupled Reticulocyte Lysate System (Promega, L4600).
3. Recombinant RNasin Ribonuclease Inhibitor (Promega, cat. no. N2515).
4. Anti-HaloTag Monoclonal Antibody (Promega, cat. no. G9211).

#### 2.3 Affinity Purification

1. Magne HaloTag Beads (Promega, cat. no. G7282).
2. Magnetic rack: DynaMag™-2 Magnet (Life Technologies, cat. no. 12321D) or DynaMag™-96 Side Magnet (Life Technologies, cat. no. 12331D).
3. PBS buffer: Add 8 g of NaCl, 0.2 g of KCl, 1.44 g $Na_2HPO_4$, 0.24 g of $KH_2PO_4$, and 800 mL Nanopure water. Adjust the pH to 7.4 with HCl and add water into 1 L. Sterilize by autoclaving.
4. PBS + NP-40: Add 200 μL of 25% (vol/vol) NP-40 into 1 L PBS buffer, and slowly shake to mix well (0.05% NP-40 in PBS). Store the solution at room temperature less than 1 month or 4 °C less than 6 months.

#### 2.4 PCR Enrichment

1. 10 mM dNTPs (BioPioneer, cat. no. MDM-4).
2. Phusion High-Fidelity DNA Polymerase (NEB, cat. no. M0530).
3. Primer A (dual index i5):
    5'- AATGATACGGCGACCACCGAGATCTACACNNNNN NNNACACTCTTTCCCTAC-3'.

Primer B (dual index i7): 5'-CAAGCAGAAGACGGCATAC GAGATNNNNNNNNGTGACTGGAGTTCAG-3'.

The "NNNNNNNN" represents the 8-mer index sequence for multiplexing.

### 2.5 DNA Quantification and Size Selection

1. Gel Extraction Kit (Qiagen, cat. no. 28704).
2. Qubit dsDNA High-Sensitivity Assay Kit (Life Technologies, cat. no. Q32851).
3. Kapa Library Quantification Kit for Illumina platforms (Kapa, cat. no. KK4824).

## 3 Methods

### 3.1 DNA Library Construction

1. Extract genomic DNA from leaves of 3-week-old *Arabidopsis* plants using phenol-chloroform method as follows:
   (a) Freeze tissue in liquid nitrogen and grind in a ceramic mortar. Quickly transfer ground plant material to a 15 mL Falcon tube. Remember to keep the sample frozen when grinding.
   (b) Add 1.5–2× volume of gDNA extraction buffer containing 100 μg/mL RNase A to the 15 mL tube, and vortex until well mixed at room temperature for 3 min.
   (c) In a fume hood, add equal volume of phenol/chloroform to each tube and vortex.
   (d) Centrifuge at 4000 × g for 20 min. In a fume hood, take the aqueous layer (top layer) from each tube, and transfer to a new tube.
   (e) In a fume hood, add equal volume of chloroform and vortex. Centrifuge 4000 × g for 20 min. Transfer the aqueous layer from each tube to new tubes. Repeat this step twice.
   (f) Add 0.1× volume of 3 M NaOAc and 0.7× volume of 100% isopropanol to each tube, and vortex thoroughly.
   (g) Centrifuge at 4 °C at full speed for 25 min to precipitate the DNA. Invert and discard the supernatant.
   (h) Add 1 mL 70% ethanol to each tube. Centrifuge at 4 °C full speed for 10 min to wash and precipitate the DNA. Invert and discard the supernatant.
   (i) Pipette off the remaining ethanol from the tube after quick spin. Do not disturb the pellet.
   (j) Put tubes in a 37 °C oven, or dry bath for 10–20 min until all residual ethanol is evaporated.

(k) Add the desired volume of the elution buffer to each tube. Check the concentration using the Qubit HS Assay Kit. For *Arabidopsis*, 2–5 μg DNA is required for one library preparation reaction. The amount should be adjusted for other plant species.

2. DNA fragmentation. Sonicate 5 μg DNA in 130 μL Elution Buffer in a Covaris microTUBE using Covaris S2 ultrasonicator with "200 bp target peak size" protocol setup (Peak Power 175.0, Duty Factor 10.0, Cycles/Burst 200, Duration 180 s, sample temperature 4–8 °C).
3. Clean up the sample. Transfer sonicated sample from microTUBE into a 1.5 mL DNA low-bind microcentrifuge tube, and add 0.1× volume of 3 M NaOAc and 2× volume of cold 100% ethanol. Mix the sample and keep it on ice or in −20 °C for 15 min to 24 h. Centrifuge at maximum speed (20,000 × g) for 20 min at 4 °C. Discard the supernatant carefully without disturbing the DNA pellet. A quick spin and pipetting off the remaining ethanol are recommended. Dry the pellet at 37 °C for 10–20 min.
4. Dissolve the DNA pellet in 34 μL elution buffer at 37 °C for 5–10 min.
5. Set up the end-repair reaction using the End-Repair Kit. Mix 5 μL of 10× End-It buffer, 5 μL of 10 mM dNTP mix, 5 μL of 10 mM ATP, 1 μL of End-It enzyme mix, and 34 μL of sample, and add water to 50 μL. Incubate the mixture at room temperature for 45 min.
6. Clean up the sample as in **step 3**.
7. Dissolve the DNA pellet with 32 μL of elution buffer at 37 °C for 5–10 min.
8. Set up the A-tailing reaction. Mix 5 μL of 10x NEBuffer 2, 10 μL of 1 mM dATP, and 3 μL of Klenow Fragment (3′–5′ exo-; 5 U/μL); add water to 50 μL. Incubate the mixture at 37 °C for 30 min.
9. Clean up the sample as in **step 3**.
10. Dissolve the DNA pellet in 32 μL elution buffer at 37 °C for 5–10 min.
11. Set up the adapter ligation reaction.

(a) Annealing Y adapters:
Mix 30 μL 100 μM adapter A, 30 μL100 μM adapter B, and 40 μL $ddH_2O$ to make 100 μL 30 μM Y adapters in PCR tubes. Incubate the mixture at 96 °C for 2 min. Remove the tubes from the heat source, and cool to room temperature (around 22 °C) on the benchtop for

2 h. Annealed Y adapters can be stored at −20 °C for less than 6 months.

(b) Ligating Y adapters:
Mix 5 μL of 10x T4 DNA Ligase buffer, 10 μL of 30 μM Y adapter, and 5 μL of T4 DNA Ligase (3 U/μL), and add water to 50 μL. Incubate the mixture at room temperature for 3 h. Samples can be stored at 4 °C overnight or at −20 °C for a week.

12. Clean up the sample. Repeat **step 3**.
13. Dissolve the DNA pellet with 31 μL of elution buffer at 37 °C for 5–10 min. Take 1 μL sample for measuring the concentration with Qubit dsDNA HS Assay Kit. Total library amount will be 2–5 μg from the starting 5 μg genomic DNA (*see* **Note 1**) (Fig. 1a).

### 3.2 In Vitro Protein Expression and Purification

1. Set up the in vitro protein expression reaction. Mix 25 μL of TNT wheat germ extracts, 5 μL of 10x buffer, 1 μL of Lys, 1 μL of SP6 polymerase, 1 μL of RNasin (20 U/μL), and 1000 ng of pIX-HALO-TF plasmid, and add water to 50 μL. The HaloTag expressed by pIX-HALO empty vector should be included as a negative control for the following DAP-seq and data analysis (*see* **Note 2**). Incubate the mixture at 30 °C for 2 h. Set aside 5 or 10 μL products for testing protein expression by Western blot (*see* **Note 3**).
2. Prepare the HaloTag ligand-coupled magnetic beads (Halo beads). Wash the HaloTag ligand-coupled magnetic beads with PBS + NP-40 on a magnetic rack for three to five times. One DAP reaction requires 10 μL Halo beads (*see* **Note 4**).
3. Transfer 10 μL Halo beads into each 0.2 mL PCR tube or a well in a 96-well plates. Position the tube or plate on a magnetic rack, and pipette off the supernatant after the beads are settled on the tube wall. Add 35 μL PBS + NP-40 and 45 μL protein expression products from **step 1** to each tube. Seal the tubes or 96-well plates carefully, and rotate it gently at room temperature for 1 h.
4. Quickly spin the tubes or plates and put them on a magnetic rack. Remove the supernatant, and wash off the unbounded proteins using 85 μL of PBS + NP-40 for three times (Fig. 1b).

### 3.3 DNA Affinity Purification

1. Resuspend the Halo bead-protein complex in 40 μL PBS + NP-40 and add 40–100 ng DNA library. Fill the final volume to 80 μL with water. Seal the tube or plate tightly and incubate at room temperature with rotation for 1–2 h.
2. Wash off the unbounded DNA for five times with 85 μL PBS + NP-40 on a magnetic rack. For the last wash, transfer the buffer and beads to a new plate (*see* **Note 5**).

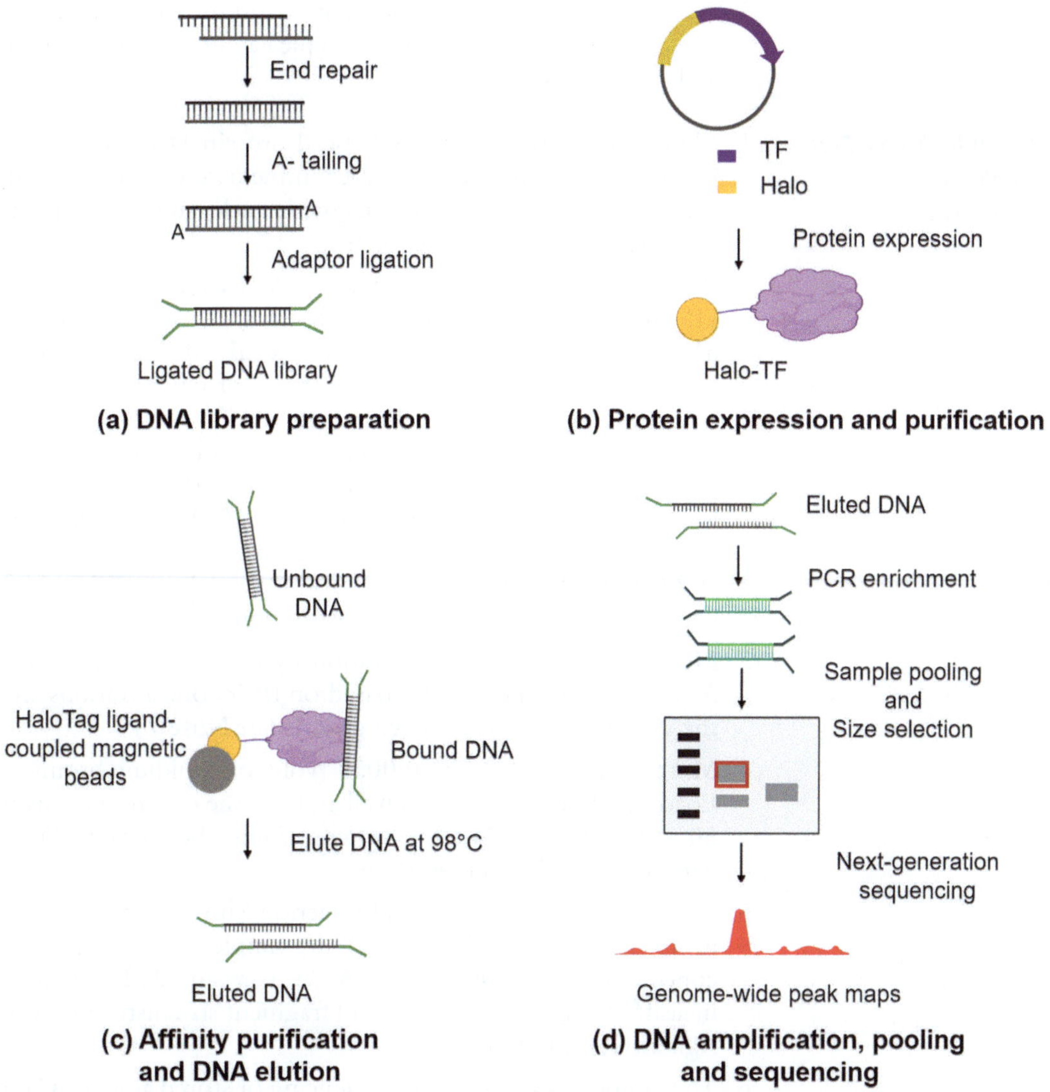


**Fig. 1** Workflow of DAP-seq. (**a**) DNA library preparation (Sect. 3.1). Genomic DNA is sonicated into ~200 bp size fragments, and after A-tailing the end-repaired fragments are ligated with Illumina-based sequencing adapters. (**b**) Protein expression and purification (Sect. 3.2). Transcription factor coding sequence is fused to an N-terminal HaloTag in an expression vector. Halo-TF fusion proteins are synthesized using a cell-free, in vitro protein expression system. (**c**) Affinity purification and DNA elution (Sect. 3.3). Halo-TF proteins bound to the HaloTag ligand-coupled magnetic beads are incubated with the adapter-ligated DNA library. Unbound DNA is washed away, and bound DNA is eluted by boiling the beads at 98 °C and placed on the ice immediately to recover the DNA. (**d**) DNA amplification, pooling, and sequencing (Sect. 3.4). Amplify the eluted DNA by PCR with Illumina-compatible sequencing primers. The indexed DNA products are pooled and purified by gel size selection for sequencing. The binding events can be viewed and analyzed in a genome viewer such as JBrowse

3. Place the samples on a magnet. Pipette off the supernatant and add 30 μL elution buffer. This sample can be stored overnight at 4 °C (Fig. 1c).

### 3.4 DNA Elution, PCR Enrichment, and Pooling

1. Elute DNA by boiling the Halo bead-protein-DNA complex at 98 °C for 10 min, and put on ice immediately for 5 min. Put the samples on a precooled magnetic rack, and take out the supernatant for PCR enrichment.
2. Prepare the PCR reactions as follows: Mix 10 μL of 5x Phusion HF Buffer, 2.5 μL of 10 mM dNTPs, 1 μL of Primer A (25 μM), 1 μL of Primer B (25 μM), 1 μL of Phusion DNA Polymerase (2000 U/mL), and 25 μL of eluted DNA, and add water to 50 μL. The Primer A and Primer B sequences contain unique indexes for each sample to be pooled in one sequencing run (*see* **Note 6**). Amplify the DNA with the following PCR conditions: 98 °C for 2 min, 15–19 cycles of 98 °C for 15 s, 60 °C for 30 s, and 72 °C for 1–2 min, final extension at 72 °C for 10 min. The PCR products can be stored at 4 °C overnight or at −20 °C for at least 6 months.
3. Measure the sample concentration by the Qubit dsDNA HS Assay Kit. Pool the samples based on their concentrations and the ratio of planned sequencing reads (*see* **Note** 7).
4. Make 1% agarose gel with 0.005% (vol/vol) ethidium bromide. Load 35 μL samples in each well and run the electrophoresis at 50 V for 40 min. Cut the gel containing the smeared DNA band between 200 and 400 bp.
5. Extract the DNA using a Gel Extraction Kit with a final elution in 33 μL elution buffer. Take 1 μL sample to measure DNA concentration in a Qubit dsDNA HS Assay and 2 μL sample to measure DNA concentration and fragment size distribution by Agilent TapeStation.
6. The samples can be sent for sequencing or stored at −20 °C for at least 6 months (Fig. 1d).

### 3.5 Auxiliary Protocol for ampDAP-Seq

1. Amplify the DNA to remove secondary DNA modifications, such as methylation, and create an ampDAP library. Set up the PCR reaction as follows: Mix 10 μL of 5x Phusion HF Buffer, 2.5 μL of 10 mM dNTPs, 1 μL of Primer A (25 μM), 1 μL of Primer B (25 μM), 1 μL of Phusion DNA Polymerase (2000 U/mL), and 5–15 ng of ligated DNA library, and add water to 50 μL. Amplify the DNA with the following PCR conditions: 95 °C for 2 min, 9–11 cycles of 98 °C for 15 s, 60 °C for 30 s, and 72 °C for 1–2 min, final extension at 72 °C for 10 min. The Primer A and Primer B used here should be the same primers used in the PCR enrichment in Sect. 3.4.

2. Transfer the PCR products into a 1.5 mL tube. Add 0.1x volume of 3 M NaOAc and 2x volume of cold 100% (v/v) ethanol to the samples, and mix thoroughly.
3. Clean up the samples. Refer to Sect. 3.1, **step 3**.
4. Add 31 μL of elution buffer. This is the ampDAP library. Use 1 μL of sample to measure the DNA concentration by the Qubit dsDNA HS Assay Kit. AmpDNA libraries can be stored at 4 °C for overnight or at −20 °C for 1 month.
5. 40–100 ng of ampDAP library can be used in one DAP experiment with the same protocol of protein expression and DNA binding, beginning at Sect. 3.3 DNA Affinity Purification (*see* **Note 8**).

### 3.6 DNA Sequencing and Data Analysis

1. Sequence the DAP products by next-generation sequencing on the Illumina platform. The read configuration and the requirement of read number depend on the genome size and complexity (*see* **Note 9**).
2. Align FASTQ files of sequencing reads to a reference genome using a standard short read mapping software, such as Bowtie2 [11]. Trimming of reads and removal of mapped reads based on mapping quality and overlap with repeats may be necessary depending on data quality and the reference genome.
3. Peak calling with software, such as GEM [12] or MACS2 [13], with the mapped read files will identify peaks that indicate TFBS. A negative control sample (DAP reaction with the empty Halo vector) can be used for background subtraction in the peak calling program.
4. Examine the mapped reads and peaks in a genome browser, such as JBrowse [14] or Integrative Genomics Viewer [15], and compare the reads in peak regions with those in the negative control sample. Successful DAP-seq experiments will produce peaks with substantial enrichment over the negative control and typically yield >5% of reads in peaks.
5. For identifying binding sequences of the TF, sequences underlying the peaks could be extracted and used in motif discovery programs, such as MEME [16], to discover enriched motifs.

## 4 Notes

1. A DNA library preparation starting from 5 μg of genomic DNA will yield a 2–5 μg DNA library. This also applies to other plant genomes such as rice and maize. DNA amount and quality are the key factors for a successful DAP-seq experiment. We recommend using 40–100 ng of DNA library for each DAP experiment in *Arabidopsis*. For experiments in rice and maize,

library amounts of 200–500 ng and 400–1000 ng, respectively, have been reported in the literature [9, 10]. The quality of genomic DNA libraries should be checked before starting the DAP protocol. On a 1% agarose gel, the size of the adapter-ligated DNA library should be about 100 bp larger than the sonicated DNA.

2. A sample with an empty pIX-HALO vector must be included as a negative control to normalize the background when comparing with experimental samples. For positive control we suggest performing DAP with the AtTGA5 vector (ABRC, stock no. HALO_TF-22_H03), which works robustly in *Arabidopsis* DAP-seq reactions.
3. DAP-seq experiments were successfully carried out with plasmid amounts ranging from 400 ng to 2500 ng. Performing Western blot using an aliquot of the protein expression reaction can verify the protein expression.
4. To avoid unevenly distributing the beads, make sure to thoroughly mix the beads by pipetting up and down before aliquoting.
5. This step washes away excess DNA that is not bound to proteins. Performing fewer washes may result in high background. After the last wash, transferring the beads to a new tube is crucial to eliminate the background from DNA bound to the tube wall.
6. Do not pool samples with the same index. You will not be able to demultiplex them in your sequencing data. Use a unique index for each sample in the same sequencing pool. Be sure to record properly the index used for each sample, as this will be necessary for pooling and eventual demultiplexing. The number of samples that can be pooled together will depend on the species and sequencing output.
7. Optionally, before pooling, the quality and amount of each sample can be checked by Agilent TapeStation and Kapa Library Quantification Kit, respectively.
8. Nine to eleven cycles of PCR of the ampDAP DNA library provide a good balance between yield and amplification bias. A greater number of cycles will result in higher amplification bias that could interfere following peak analysis.
9. The target number of sequencing reads will depend on the size and complexity of the experimented genome and repeat content. For *Arabidopsis* samples (genome size of 125 Mb), we typically pool around 40 samples in an Illumina NextSeq 2500 Mid Output flow cell that produces 120 million reads, aiming for 2–4 million reads per DAP-seq sample. Extrapolating from these numbers, maize samples (genome size of 2.5 Gb) should

be pooled aiming for at least 20 million reads per sample. Therefore, sequencing pools for rice and maize should contain fewer samples, as they will require greater sequencing depth based on their genome size and complexity.

## Acknowledgments

This work was supported by NSF Plant Genome Research Program grant to SCH (IOS-1916804).

### References

1. Alexandre CM, Urton JR, Jean-Baptiste K, Huddleston J, Dorrity MW, Cuperus JT, Sullivan AM, Bemm F, Jolic D, Arsovski AA, Thompson A, Nemhauser JL, Fields S, Weigel D, Bubb KL, Queitsch C (2017) Complex relationships between chromatin accessibility, sequence divergence, and gene expression in *Arabidopsis* thaliana. Mol Biol Evol 35(4):837–854
2. Inukai S, Kock KH, Bulyk ML (2017) Transcription factor–DNA binding: beyond binding site motifs. Curr Opin Genet Dev 43:110–119
3. Huang S-sC, Ecker JR (2018) Piecing together cis-regulatory networks: insights from epigenomics studies in plants. Wiley Interdiscip Rev Syst Biol Med 10(3):e1411
4. Chen X, Bhadauria V, Ma B (2018) ChIP-Seq: a powerful tool for studying protein-DNA interactions in plants. Curr Issues Mol Biol 27:171–180
5. Jolma A, Kivioja T, Toivonen J, Cheng L, Wei G, Enge M, Taipale M, Vaquerizas JM, Yan J, Sillanpää MJ, Bonke M, Palin K, Talukder S, Hughes TR, Luscombe NM, Ukkonen E, Taipale J (2010) Multiplexed massively parallel SELEX for characterization of human transcription factor binding specificities. Genome Res 20(6):861–873
6. Berger MF, Bulyk ML (2009) Universal protein-binding microarrays for the comprehensive characterization of the DNA-binding specificities of transcription factors. Nat Protoc 4(3):393–411
7. O'Malley RC, Huang SC, Song L, Lewsey MG, Bartlett A, Nery JR, Galli M, Gallavotti A, Ecker JR (2016) Cistrome and epicistrome features shape the regulatory DNA landscape. Cell 165(5):1280–1292
8. Bartlett A, O'Malley RC, Huang SC, Galli M, Nery JR, Gallavotti A, Ecker JR (2017) Mapping genome-wide transcription-factor binding sites using DAP-seq. Nat Protoc 12 (8):1659–1672
9. Cerise M, Giaume F, Galli M, Khahani B, Lucas J, Podico F, Tavakol E, Parcy F, Gallavotti A, Brambilla V, Fornara F (2020) OsFD4 promotes the rice floral transition via florigen activation complex formation in the shoot apical meristem. New Phytol. https://doi.org/10.1111/nph.16834
10. Galli M, Khakhar A, Lu Z, Chen Z, Sen S, Joshi T, Nemhauser JL, Schmitz RJ, Gallavotti A (2018) The DNA binding landscape of the maize AUXIN RESPONSE FACTOR family. Nat Commun 9(1):4526
11. Langmead B, Salzberg SL (2012) Fast gapped-read alignment with bowtie 2. Nat Methods 9 (4):357–359
12. Guo Y, Mahony S, Gifford DK (2012) High resolution genome wide binding event finding and motif discovery reveals transcription factor spatial binding constraints. PLoS Comput Biol 8(8):e1002638
13. Zhang Y, Liu T, Meyer CA, Eeckhoute J, Johnson DS, Bernstein BE, Nusbaum C, Myers RM, Brown M, Li W, Liu XS (2008) Model-based analysis of ChIP-Seq (MACS). Genome Biol 9(9):R137
14. Buels R, Yao E, Diesh CM, Hayes RD, Munoz-Torres M, Helt G, Goodstein DM, Elsik CG, Lewis SE, Stein L, Holmes IH (2016) JBrowse: a dynamic web platform for genome visualization and analysis. Genome Biol 17:66
15. Robinson JT, Thorvaldsdóttir H, Winckler W, Guttman M, Lander ES, Getz G, Mesirov JP (2011) Integrative genomics viewer. Nat Biotechnol 29(1):24–26
16. Machanick P, Bailey TL (2011) MEME-ChIP: motif analysis of large DNA datasets. Bioinformatics 27(12):1696–1697

Chapter 16

# Basics of Sensor-Based Phenotyping in Wheat

**Emily Thoday-Kennedy, Nathan Good, and Surya Kant**

## Abstract

Phenotyping is a crucial building block in crop breeding research, integrating our understanding of crop development, growth and yield under different environments, with the genetic inheritance of key traits. Acquiring crop trait data traditionally involves time-consuming, laborious and destructive sampling techniques, limiting the unification of phenomics data with high-throughput genomics data. Advanced phenotyping platforms developed in the past two decades, deployed both in controlled environments and field, have revolutionized data capture, analysis and output. The combination of high-throughput, accurate sensor technology, with advanced image analysis pipelines, computer learning techniques and data storage systems, have contributed to remote, precise and non-destructive phenotyping. Visible, multispectral, hyperspectral, fluorescence and laser sensors and cameras can non-invasively monitor biochemical, water content, biomass, morphological, phenological, yield- and root-related traits. Here, we describe developments of sensor-based phenotyping for cereals, especially wheat. By examining traits amenable to efficient phenotyping, and the usage of various high-throughput platforms in crop research and breeding, the key benefits of sensor-based phenotyping are demonstrated.

**Key words** Plant phenotyping, Phenotyping platforms, RGB imaging, Multispectral imaging, Hyperspectral imaging, Laser imaging, Sensors, Cereals, Crop breeding

## 1 Introduction

While it's been over 100 years since the introduction of the term phenotyping [1], radical changes to the practice of plant phenotyping have occurred over the last two decades. In this time, the sequencing and annotation of plant genomes from model *Arabidopsis* to complex cereals, including rice (*Oryza sativa*), maize (*Zea mays*), barley (*Hordeum vulgare*) and most recently wheat (*Triticum aestivum*), have revolutionized the crop breeding landscape. As global food demands continue to increase with a growing world population, yield increases achieved using traditional methods have become insufficient to meet current and future food demands, particularly in major cereal crops [2–4]. This has led to the amplified development of high-throughput phenotyping platforms

Andriy Bilichak and John D. Laurie (eds.), *Accelerated Breeding of Cereal Crops*, Springer Protocols Handbooks,
https://doi.org/10.1007/978-1-0716-1526-3_16, © Springer Science+Business Media, LLC, part of Springer Nature 2022

(HTPPs) utilizing a wide range of sensor-based technologies and encouraging enhanced data collection to assist in faster breeding of improved crop varieties.

A plant's phenotype, its observed morphological, phenological, physiological, agronomical and biochemical traits, is produced in response to environmental conditions during growth and its inheritable genetic composition. Plants are able to alter phenotypically under complex environments to ensure survival, especially under limiting conditions, with this adaptive capacity known as phenotypic plasticity [5]. The relationship between genotype, phenotype and environment can be simply conceptualized by the following:

$$\text{Genotype (G)} \times \text{environment (E)} \rightarrow \text{phenotype (P)}$$

Phenotyping traditionally occurs on a whole plant or plot basis, at a single or multiple defined time points, combining all defined traits into one phenotype. On the other hand, phenotyping can also relate to the observation or measurement of an individual trait or group of related traits, which can be in constant flux [6]. A plant phenome can be defined at all levels of biological organization, from individual cells to individual plant organs, to whole individual plants to plots and paddocks, and across different time points [6–8]. It is the challenge of modern phenotyping to capture these dynamic changes in effective, accurate and efficient platforms. Plant phenotyping encompasses the assessment of both basic and complex traits relative to growth, development, phenology, physiology, architecture, agronomy, resistance/tolerance and biochemistry. Comprehensive characterizations of basic traits, relative to dynamic spatial and temporal variability, can be integrated to further understand complex, economically important traits, including stress tolerance, yield and nutritional quality [9–11]. The assessment of phenotype-environment dynamics therefore plays a significant role in the advancement of crop breeding and management programs, through the power to correlate gene function, plant performance and environmental responses [12].

The popularization of marker-assisted selection (MAS) and genomic selection (GS) in recent decades, which can achieve genetic gains by improving selection intensity and accuracy, has increased focus on high-quality, precise and defined phenotyping techniques [13–16]. While these new breeding approaches rely heavily on genotyping, phenotypic data is essential as the basis for the development and training of statistical models used to predict phenotypic performance [15–17]. Therefore, the use of quality phenotypic and genotypic data, in conjunction with statistical tools, will allow candidate genes to be assessed against gene × variety, and gene × environment interactions [18]. Accurate, high-throughput phenomic techniques, which can screen large germplasm populations commonly utilized in MAS or GS across multiple environments and provide high-quality quantitative data on dynamic phenotypic changes, will provide significant benefits to the plant breeding discipline [19, 20]. Advancements in phenotyping techniques consequently offer researchers and breeders the tools to increase the rate of genetic gain and ultimately increase the arrival of environmentally adapted, high-yielding varieties [21].

## 2 High-Throughput Phenotyping Platforms

Developing precise, accurate and integrative HTPPs capable of phenotyping many plant traits has proven challenging in the past, due to restrictive and specialized manual protocols, lack of multidisciplinary approaches, and time and resource exhaustive methods [17]. These traditional phenotyping techniques are often resource-consuming and frequently use destructively harvested plant material for analysis. In order to address these shortcomings, next-generation HTPPs, utilizing sensors, automation, computer learning and data integration, are increasing the precision, accuracy and throughput of data collection while reducing costs and resource usage. These non-destructive, precise, data-driven HTPPs have been developed and harnessed to automate or semi-automate phenotyping of simple traits while exploring the vast array of complex or laborious traits [17]. Next-generation phenotyping methods are therefore assisting in easing the 'phenotyping bottleneck' that has developed with the rapid expansion of high-throughput genetic sequencing [22]. Swift strides have been made in the rapidly evolving field of non-invasive, high-throughput phenotyping in past two decades [12]. As many next-generation HTPPs are able to more accurately measure trait values and improved differentiation between genotypes, this is assisting in refining the understanding of relationships between growth conditions and plant development [17].

HTPPs utilizing modern technologies can collect accurate observations in open field, controlled field and controlled environment settings, which, when combined with cost effective data acquisition and analysis systems, can provide physiological, morphological, agronomical and biochemical information with massive datasets [20]. Technologies, such as handheld sensors and cameras, automated glasshouses, ground-based vehicles, unmanned aerial vehicles (UAVs) and satellites, allow us not only to access data on key traits efficiently and effectively but also to access novel traits previously unexplored due to method or resource difficulties [6, 20, 23, 24]. Much of the revolution in HTPPs has occurred due to developments in various imaging or sensing techniques, utilizing sensors and software pipelines, which can record and analyse data on plant structure, phenology, biomass production, plant health and biochemical parameters. Based on these available technologies, next-generation phenotyping can be classified into

**Table 1**
**Imaging techniques and their applications in plant phenotyping**

| Imaging technique | Imaging environment | Applications |
|---|---|---|
| Visible (RGB) | Controlled environment; field | Biomass production, senescence, leaf area, growth dynamics, shoot architecture, canopy coverage, emergence, height, spike identification, spike traits, phenology, yield prediction, grain quality, root architecture |
| Multispectral | Controlled environment; field | Biochemical content, nutrient status, plant water status, plant health, canopy coverage, senescence, emergence, yield prediction, grain quality |
| Hyperspectral | Controlled environment; field | Biochemical content, photosynthetic capacity, nutrient status, plant water status, plant health, canopy coverage, senescence, yield prediction, grain quality |
| Fluorescence | Controlled environment; field | Photosynthetic capacity, plant health, root architecture |
| Thermal | Controlled environment; field | Canopy temperature, plant water status, plant health |
| Laser | Controlled environment; field | Biomass production, shoot architecture, leaf area and angle, growth dynamics, canopy coverage, height, spike identification |
| X-ray CT | Controlled environment | Root architecture |
| MRI | Controlled environment | Root architecture, water content |

roughly five sensing classes (Table 1), visible (red-green-blue; RGB), spectroscopy (multispectral and hyperspectral), thermal, fluorescence and topographical/tomographical (LiDAR, MRI, X-ray CT) [23, 25], with many HTTPs integrating multiple strategies. RGB sensing can be used to capture size, morphology, architecture, growth dynamics and colour of plants and canopies. Multispectral or hyperspectral sensing is used to phenotype an ever-increasing range of traits, including plant or specific organ health in response to stress, growth and development, biochemical (especially water and pigments) composition, architecture and seed traits. Thermal sensing captures plant temperature data, while fluorescence sensing measures photosynthetic status, both of which can be used to monitor changes in plant health due to abiotic and biotic stresses. A range of other technologies can be group as topographical or tomographical sensors, including LiDAR (light detection and ranging) used to estimate canopy architecture; height and biomass; MRI (nuclear magnetic resonance imaging), which can be used to acquire information on plant, root and seed structures, as well as water content; and X-ray CT (X-ray computed tomography), which can generate volumetric data on plant and root structures. Overall, these techniques can be applied using a wide range of evolving HTPPs to phenotype for an ever-expanding range of traits in major crops, including cereals. In the following sections, we will use wheat as an example crop to demonstrate the new developments in phenotyping for key traits of interest in wheat.

## 3 Traits of Interest in Wheat Phenotyping

### 3.1 Biochemical Contents

Non-invasive HTPPs often focus on revealing and understanding variations in morphological and agronomical traits, but developments in spectroscopy-based phenotyping approaches have introduced new techniques to quantify biochemical traits non-destructively. Biochemical traits, such as the leaf or plant content of chlorophylls, carotenoids, other pigment contents, water-soluble carbohydrates (WSC), starch, proteins and other macronutrients, can be used to monitor crop health particularly under stress [26, 27]. Multispectral and hyperspectral sensors capture electromagnetic radiation reflected based from plants across the visible (400–700 nm), near-infrared (NIR; 700–1000 nm) and short-wave infrared (SWIR; 1100–2500 nm) regions of the spectrum, allowing them to record information on plant physiological and biochemical statuses [28]. Non-destructive measurements of biochemical compounds allow for further study of plant responses during development and under abiotic and biotic stresses, enhancing our understanding of these processes and efficiently distinguishing desirable germplasm. Data and image-based methods are also capable of capturing internal changes in composition well before physical manifestations are visible to the human eye. Sensors used can range from handheld devices utilized for single-leaf measurements to ground vehicle-mounted (specialized buggies or farm equipment) devices, aerial- (especially UAV) mounted sensors or satellites. Variations in spectral data at different wavelengths can infer variance in biochemical and structural components of plants or leaves and be used, along with statistical approaches such as partial least-squares regression approaches, to calculate indices estimating these traits. Foliar spectral signatures therefore have the potential to rapidly assess multiple biochemical traits in a single measurement [26, 29]. Reflectance spectroscopy (sensor and image-based) has successfully been used to distinguish variance in water-soluble carbohydrates [30], anthocyanin [31], carotenoids [32], chlorophyll [33], sugars [34] and macronutrient status [35, 36], using specific spectral wavelengths and indices.

#### 3.1.1 Chlorophyll Content

Leaf chlorophyll content (chlorophyll *a*, chlorophyll *b* and total chlorophyll) is an important agronomical trait due to its close relationship with photosynthetic capacity, nitrogen (N) status and plant health. Through the tracking of changes in chlorophyll content improved management of fertilizer application and stress relief treatments can occur throughout the season. Initial non-destructive phenotyping techniques, which are still used, involve handheld sensors such as SPAD-502 (Minolta corporation, Ltd., Osaka, Japan), which measures chlorophyll content by proxy and has demonstrated strong correlations between chlorophyll content, N status and yield [37–40]. Subsequently, a considerable number of indices have been developed based on narrow-band multispectral/hyperspectral imaging to infer chlorophyll content, with different indices developed to reliably track chlorophyll content in different plant organs at different developmental stages [29, 33, 41, 42].

Monitoring of photosynthetic capacity can give insight into plant health and development. Models based on hyperspectral data in conjunction with handheld equipment, such as infrared gas analyser (IRGA), have shown that the remote sensing of key photosynthetic traits, including electron transport rate and Rubisco activity, is possible [43]. Chlorophyll fluorescence imaging can be used as a non-invasive method to dissect photosynthesis efficiency, especially photosynthetic and non-photosynthetic quenching, which is sensitive to changes in development and stress levels. This technique has been effectively used to screen for stress tolerance, especially fungal infections, in a range of crops including wheat [29, 44–48].

#### 3.1.2 Nutrient Status

Nitrogen is one of the most important plant macronutrients, with low concentrations limiting plant growth and productivity; therefore, significant amounts of fertilizer application can be required to maximize agricultural outputs. Monitoring both N uptake efficiency and N utilization efficiency can lead to increased breeding for overall nitrogen use efficiency (NUE), a complex multigenic trait [49]. Understanding the dynamics of N status can optimize the application timing of fertilizers and assist in predicted grain yield and quality. Due to the high percentage of N in chlorophyll pigments, a simple way of phenotyping for plant N status is measuring chlorophyll content [33, 38, 42, 50]. Other multispectral sensors, which can be either handheld or mounted to ground vehicles or UAVs, use reflectance readings in the visible and NIR regions, to calculate spectral indices sensitive to variations in N. These sensors, such as Crop Circle (Holland Scientific, Inc., Lincoln, NE), GreenSeeker (NTech Industries, Inc., Ukiah, CA) and a range of UAV-mounted multispectral sensors, can capture specific spectra used to calculate indices including normalized difference vegetation index (NDVI), which is highly correlated to N status [38, 51–58]. Data captured using hyperspectral sensors, both handheld and vehicle based, has also been mined to produce and deploy various indices monitoring N and other nutrient content, sometimes in combination with other technology including

LiDAR [33, 36, 42, 50, 54, 59–62]. However, it should be noted that many indices actually do not calculate N content but are instead proxies for N content, allowing for difficulties in removing background/soil noise and saturation of N content after specific developmental stages [55, 63, 64]. Hyperspectral signature data has also been used to identify spectral regions or signatures specific to N or other nutrient excitation, allowing for differentiation of germplasm under different stress conditions over a range of developmental stages [35, 65–67].

### 3.2 Plant Water Relations

Maintaining plant water potential and content, which depends on water fluxes across the soil-plant-atmosphere continuum, is crucial to plant health. Changes in soil water content, water transport pathways and evapotranspiration affect physiological equilibriums within plants, impacting on growth, development, morphology and ultimately yield [68]. While water limitation conditions have a varied range of effects on plant growth and composition, only some can be phenotyped, often by proxy, using HTPPs [68, 69], while others can be quantified using laboratory methods [68–70].

#### 3.2.1 Leaf Water Content and Potential

Water absorbed and transported throughout plants is moved by negative pressure (tension) generated by evapotranspiration in a process commonly known as the cohesion-tension mechanism. As transpiration occurs, the water potential or tension drops, pulling water through the plant from the soil; although when soil moisture levels reduce, transpiration rates can exceed water absorption; thus, greater tension is required to draw water, utilizing more energy resources and resulting in water deficit at the leaves. Leaf water potential is therefore an indicator of plant water status, directly relating to plant water fluxes through the xylem and corresponding tension [71]. Leaf water potential changes in diurnal patterns relating to daily water usage. The changes in these patterns, especially the predawn water potential, midday water potential and the bulk turgor loss potential (wilting point), are important criteria for water use efficiency (WUE) and stress tolerance phenotyping [70, 72]. Relative leaf or canopy water content is another measure of plant water status, dependent on the balance between water supply and evapotranspiration [73]. Water content is a measurement of water status at the time of measurement relative to the maximum water holding capacity at full turgidity. Maintenance of leaf turgidity is vital in protecting and preserving photosynthesis, cell expansion and stomatal conductance. Water potential, content and overall status have become key traits for phenotyping under various watering regimes, enhancing selection of WUE varieties [74–76].

Traditionally, plant water status traits, including leaf water potential and relative water content, have been measured using destructive, time-consuming and laborious methods [70, 71,

73]. Subsequently, non-destructive, high-throughput techniques have been developed to remotely phenotype plant water traits. Leaf patch clamp pressure probes are able to observe plant water relations in real time, therefore monitoring temporal and spatial changes in water fluxes [77]. Probes are magnetically attached to individual representative leaves, with one side containing the sensor, measuring patch pressure, which is directly inversely proportional to leaf water potential. By remotely phenotyping water potential, differences in germplasm performance under drought conditions can be screened over time [78–80].

Spectral data from multispectral and hyperspectral sensors has been used to isolate wavelength regions found to be associated with water content and plant water status [72]. Regression techniques have been used to identify narrow bands in the NIR and SWIR regions associated with water content based on excitation of molecular bonds, which can enable monitoring of plant water content and overall status [65, 74, 81]. NIR spectroscopy can be used to monitor changes in spectral signatures and overall NIR intensity, as well as specific wavebands, which are associated with changes in water potential or water content under stress conditions [48, 82, 83]. Based on specific water-associated wavelengths, various spectral indices have been developed to remotely phenotype plant or leaf water content [75, 84, 85], leaf water potential [86, 87] or overall water stress status [56, 76, 88, 89].

#### *3.2.2 Stomatal Conductance*

Stomatal conductance is directly responsible for modifying photosynthesis and overall plant water relations, impacting stress responses, WUE, biomass production and therefore yield [90, 91]. Stomatal conductance is highly responsive and adaptive to changes in internal water potential, diurnal patterns, external temperature, vapour pressure deficit, wind and carbon dioxide concentrations [90–92]. The degree of stomata aperture directly correlates to the rate of gas exchange, regulating transpiration, photosynthesis and internal leaf temperature. Traditionally, diversity in stomatal conductance has been non-destructively phenotyped using handheld sensors, such as a porometer or IRGA system, which measure either the rate of carbon dioxide entering or water leaving the stomata. Due to the close relationship between stomatal conductance, photosynthesis and osmotic stress tolerance mechanism, stomatal conductance has been used as a quick proxy for osmotic adjustment and therefore photosynthetic maintenance under stress conditions [92–95].

Thermal sensing using long-wave infrared spectra to measure leaf temperature has been showed to have a strong curvilinear relationship to stomatal conductance [96]. Leaf temperature increases as stomatal conductance decreases, due to reduced capacity for evaporative cooling effects from the transpiration of water.

Leaf temperature captured using either handheld or mounted infrared temperature sensors or thermal cameras can therefore be utilized as a proxy to stomatal conductance for phenotyping germplasm for osmotic stress tolerance or WUE [92, 96, 97].

#### 3.2.3 Canopy Temperature

Canopy temperature, the temperature of leaf or canopy surfaces, provides insights into plant systems involved in WUE, including water status, water potential, evapotranspiration, stomatal conductance and yield. Canopy temperature is directly affected by changes in stomatal conductance, evapotranspiration and water potential, as differences in water fluxes causing reduced evaporative leaf cooling can increase leaf temperatures [98]. While water conservation strategies to reduce evapotranspiration benefit plants, this has to be balanced by the side effects of increased leaf temperatures, which can result in damage to critical biochemical pathways [99, 100]. Under favourable moisture conditions, genotypic variations in canopy temperature are often low; however, when exposed to unfavourable abiotic or biotic stress conditions, significant difference in canopy temperature can be seen [98]. Canopy or leaf temperature can therefore be used as a proxy measurement, through regression relationships, to phenotype for leaf water potential and stomatal conductance without interfering with plant processes, as well as being a non-destructive diagnostic tool for stress tolerance [86, 101–104]. Consequently, canopy temperature has become integrated in cereal breeding programs due to its strong relationships with key economical traits [97, 102, 105–108]. Measurements can be taken using a handheld or HTPP-mounted radiometric infrared thermometer (IRT) sensor or camera, including those mounted on UAVs, helicopters and satellites [102, 109, 110]. While canopy temperature measurements are non-destructive, for quantitative data usage, calibration and data transformation is needed to account for specific environmental and biophysical conditions, including background interference from soil and air temperatures, as well as other temporal environmental variations [101–103, 109].

### 3.3 Biomass and Morphological Traits

Maintaining biomass production is one of the most crucial elements to plant development; therefore, traits relating to physiology, morphology, growth and phenology are key descriptors of plant survival and the building blocks of plant phenotyping. Parameters, such as biomass production, leaf area, height and architecture, impact on nearly all other defining crop traits, especially yield and quality. RGB cameras have been key tools in quantifying biomass production, tracking colour changes and following morphology alterations. Developments in other sensor technologies, including multispectral imaging and LiDAR, have revealed new high-throughput phenotyping methodologies to quantify growth-related traits [23]. Revolutions in image analysis pipelines, in particular image segmentation, image classification and deep learning methods, are accompanying these sensor developments [111–117]. Phenotyping for biomass, architecture and phenology changes is therefore key to understanding crop behaviour under stress conditions.

#### 3.3.1 In Situ Biomass Production

Above-ground shoot biomass, defined as the plant fresh or dry weight, has been traditionally measured by destructive harvests. The in situ measurement of biomass production allows for the capture of data across multiple time points, resulting in reduced time and resource usage, and the understanding of growth dynamics. RGB cameras accompanied by image analysis pipelines, based on image segregation and classification, which calculate plant pixel area or estimated shoot biomass, have shown to achieve very strong correlations with manual fresh and dry biomass [48, 114, 118–122]. Most research into biomass production has occurred using glasshouse-based sensors, which have the ability to accurately quantify biomass variations between genotypes, especially under stress treatments, providing insights in genetic differences controlling various stress response strategies [83, 106, 114, 119–121, 123–126]. Through non-destructive RGB phenotyping methods, changes in pixel area can be mapped over time, allowing for differences in growth between genotypes and stress treatments to be charted in growth curves [33, 48, 83, 106, 119–121, 123, 124]. Remote image capture over many consecutive time points allows for the capture of small fluctuations in growth especially in response to stress application, as well as the differentiation between growth stage dynamics, often illustrating the prevalence of different stress tolerance mechanisms [119, 121, 122, 127, 128]. Non-destructive growth curves can also be used to quantify early vigour in young seedlings, a trait linked to improved development, stress tolerance and ultimately yield [129].

Sensors deployed in the field have tended to use different strategies to capture biomass accumulation under different stress environments. The monitoring and prediction of biomass accumulation, expressed as canopy coverage, biovolume or vegetation fraction, has relied on the relationship between biomass and canopy greenness for calculation, through the deployment of various broadband indices [58, 108, 130, 131]. Unfortunately, many of these indices saturate, struggling to capture accurate changes in biomass at high leaf area index values or at peak biomass production stages [132]. However, the use of regression methods to select narrow band spectral regions from hyperspectral sensor information has significantly improved the accuracy of vegetation indices across all growth stages [67, 133]. Crop structural information captured using aerial RGB, multispectral and hyperspectral sensors can be integrated with vegetation indices to significantly improve the correlation between manually harvested and predicted biomass traits [118, 134–136]. LiDAR technology, either terrestrial or aerial, which captures 3D image data, has also been demonstrated to accurately capture data for non-destructive biomass estimation [61, 137–139]. These technologies, used over a range of environments demonstrate that biomass production, crop coverage, etc., can be phenotyped remotely, thus allowing for finer resolution data collection during growth seasons and resulting in further understanding of biomass fluctuations under different stress, environmental or agronomical conditions.

#### 3.3.2 Leaf Area and Senescence

Leaf traits are key to understanding plant growth and development, as green plant surfaces determine light interception rates and photosynthetic activity, ultimately determining growth and yield. Leaf area or leaf area index (leaf area per unit ground area) has been extensively researched due to being important in determining biochemical, photosynthetic and biomass traits. Leaf area has been measured using electronic leaf area meters, but modern advances in sensors and image analysis software have introduced new methods of determining leaf traits. Using laser-based systems [121, 140], or RGB image classification and segmentation analysis pipelines [83, 141, 142], leaf area can be remotely measured in both controlled environment and field settings. These methods have shown that non-destructive leaf area predictions have strong correlations with destructively measured leaf area and biomass, as well as being a method to differentiate genotypes and stress treatments [121]. Leaf area index, which is important in determining photosynthetic area and therefore impacting on productivity, tends to be phenotyped remotely in field experiments. Using indices calculated from digital sensor data [133, 139, 143, 144] or laser-based sensors [140, 145], remote sensing of leaf area index can provide valuable insight in future crop performance.

The pattern of senescence in leaves strongly impacts on yield and yield quality parameters. Delayed leaf senescence, or stay-green traits, maintains photosynthetic activity and nutrient remobilization processes for longer periods, which can increase yield under certain conditions. In situ phenotyping of canopy greenness, onset of senescence and rate of senescence are therefore key to understanding their relationship to yield potential and quality. Canopy greenness and its decay can be measured in-field through vegetation indices, such as NDVI, crop senescence index, green chlorophyll index and total green area, calculated from multispectral or hyperspectral data [78, 108, 130, 146–148]. Senescence can also be scored utilizing high-resolution RGB images, and image segmentation and colour classification algorithms, which track senescence rates across different genotypes and treatments [83, 123, 124, 149, 150].

### 3.3.3 Crop Development and Morphology

Understanding differences in crop development and morphology gives insight into the effects of environmental conditions, abiotic and biotic stresses and genotypic variations on crop growth. Developments in sensors, software algorithms and computer learning techniques have led to the deployment of various remote phenotyping techniques developed to assess a multitude of crop morphology and development traits. Overall plant architecture is a key determinate of plant physiology, effecting yield potential, disease resistance and stress tolerance. High-resolution RGB images and LiDAR data, coupled with 3D reconstruction and object segmentation, classification and mapping pipelines, can also be used to phenotype leaf traits, including leaf length, leaf elongation rate, leaf number [129] and leaf angle [83, 140–142]. These pipelines can also be used to identify key architectural traits, including minimum enclosed area, convex hull, plant skeleton, tiller number and stem angle, as well visually representing overall plant shape [83, 119, 129, 138, 141, 142, 151].

Plant emergence count or density gives information on the genotypic vigour and assists with final yield estimates. While different techniques have been developed to estimate emergence counts in broadleaf species [152, 153], the thin upright nature of cereal seedlings has hindered development. High-resolution RGB images taken using angled cameras mounted on aerial or ground-based vehicles, combined with object classification pipelines, have shown strong potential in remotely estimating emergence density [154–156]. Multispectral imaging, which relies on differentiating wheat seedlings from background based on vegetation indices, has also been shown as a useful tool for predicting emergence density [157]. However, both of these methods are prone to saturation at high sowing densities and emergence numbers and are sensitive to growth stage, with imaging needing to occur no later than 1–2 leaves stage.

Plant height is a distinguishing factor between genotypes, phenological stages, agronomical treatments and stress conditions, as well as being a proxy for biomass estimation, canopy structure and yield potentials. Plant height or caliper length is one of the many parameters that can be calculated during image analysis of RGB images from glasshouse-based HTPPs, which shows high correlation with manual measurements [83, 119, 129, 141]. Structure-from-motion data captured by UAV or vehicle-based HTPPs can be extracted to remotely quantify plant height or crop height models, by calculating the difference between the digital surface model captured during the growth period and bare soil or the digital terrain model, which is captured before emergence [158]. Plant height captured using this method has shown strong correlations with manually measured height in the field [118, 127, 133, 134, 136, 159]. Crop height models can also be derived using the same method from data captured using laser-based HTPPs [24, 127,

137, 159]. However, it should be noted that crop height models can be systematically lower than the ground truth data, which tends to measure the highest point per plot, with models tending to measure plot averages [118]. Height data remotely captured over a time series can be utilized to track plant growth [129, 135, 158], score for lodging [107, 160] and differentiate stress treatments [83].

Quantifying grain head (panicle, ear or spike) numbers, scoring for heading dates and analysing head traits give insight into phenological processes and assist in projecting yield estimates. Using computer vision and deep learning algorithms remote head identification and tracking, it is now possible to further remotely understand grain head traits, using laboratory, glasshouse or field HTPPs. Analysing RGB images of individual severed heads using computer learning software and image analysis pipelines, grain heads can be measured for traits including length, volume, awn characteristics, grain number and distribution [115, 161–163]. Computer learning techniques have been developed to identify and count spikes as well as monitor their growth and development over time [164]. In situ field head detection and quantification pipelines built on deep learning methods have been developed using RGB [116, 165–172], multispectral [173] and LiDAR [24, 171] imagery data taken from a variety of angles. Regular in situ monitoring of detected heads also allows for quantification and dating of heading, flowering and maturity traits [166, 168, 170, 174].

### 3.4 Yield and Grain Quality Traits

The ultimate aim of phenotyping is to the assist breeding of new crop varieties with improved grain yield and quality under the increasingly complex growing conditions. Yield is dependent on a substantial number of variables, including biomass production, photosynthetic capacity, WUE and nutrient acquisition and remobilization, which interact both with each other and the surrounding environment across time. Predicting yield potential before harvest is therefore complicated and growth condition dependent. Correlations have been demonstrated between harvested yield and yield predicted using spectral indices, especially those relating to biomass, photosynthesis and water or nutrient use efficiency, although these are often depending on experimental conditions and growth stage [42, 74, 76, 88, 108, 162, 175, 176].

#### 3.4.1 Grain Quality

Grain end-use quality traits are key to successful wheat breeding efforts, but laboratory quality assays can be destructive and expensive. Flour traits, including nutrient content and end-use qualities, can be confidently predicted using NIR and SWIR hyperspectral sensors, including mapping their heritability with breeding populations [177]. NIR sensors are able to predict protein content from the single whole seed to milled flour, reducing the need for laborious wet chemistry [177–179]. Utilizing image analysis pipelines, automated HTPPs using RGB, hyperspectral and NIR sensors have been developed, which can classify individual grains based on size, weight, colour, protein content and overall classification class [177, 179–181]. Seed phenomic HTPPs can also classify insect and fungal damage by mapping differences in hyperspectral signatures and colourmetrics [181–183].

### 3.5 Root Phenotyping

Root systems are the most vital, but complex, of plant organs, controlling water and nutrient acquisition, and therefore linked to all important above-ground agronomical traits. Identifying root traits linked to increased capacity for water and nutrient mining, especially under limiting conditions, and associated with improved stress tolerance and yield potential is a key phenotyping priority. Root systems are underground, highly dynamic and considerably variable under different growth conditions, making it difficult to obtain data in situ and relevant to breeding efforts. Various growth methods, imaging devices and image analysis programs have been deployed to better understand key root traits under both controlled and field environments.

#### 3.5.1 Root Architecture

Cereals have two types of roots: seminal roots, which come directly from the embryo, and nodal roots, which emerge from lower tiller nodules, both of which later produce hair roots. The angle and number of seminal roots have been used as proxy traits in wheat breeding, as favourable root systems are narrow, compact and reach lower depths [184, 185]. Under controlled environment conditions, these traits can be phenotyped in young plants utilizing methods, including growth pouches [186–190], semi-hydroponic systems [191], clear pots, cylinders or soil tubes [186, 189, 192, 193], agar chambers [189, 190], soil baskets [189, 193] and rhizotrons [194–197], from which roots can be non-destructively viewed and imaged. Image analysis packages have been specifically designed, which can remotely identify roots, map key architectural traits and model 2D, 3D and 4D root structures using RGB or hyperspectral images. However, it should be noted that root phenotyping outcomes from controlled environment studies have often not been found to be correlated to those found in field [189]. Field-based methods to determine key root traits include 'shovelomics' or root crown phenotyping [198, 199], soil cores [189, 200] and excavating whole root systems [201, 202] to phenotype root architecture grown in situ. While the excavation methods to access root samples are laborious, these systems also rely on high-throughput imaging processing techniques to quantify root traits. Aside from RGB imaging and analysis tools, other sensors have been developed to assist in phenotyping key root traits. Fluorescence imagery can be used to count and measure roots, especially fine roots, in soil cores [200]. Electrical resistance tomography, which measures soil water content, can be a non-destructive proxy for soil water use and therefore root density across the growing season [203]. X-ray CT scanning can be used to quantify root system complexity and map 3D structure without destructive harvesting, although highly specialized equipment in needed [190, 204–206]. Similar to X-ray CT, MRI scanning is also used to non-destructively image root systems and remotely monitor key root traits [206, 207]. In situ, both controlled environment and field, root phenotyping has demonstrated dynamics of root growth over growing seasons, under different environment and management conditions and between genotypes, providing insight into ideal root architecture.

## 4 Conclusions

Here, we have described the fundamentals of plant phenotyping, focusing on cereals as well as giving insight into the range of sensor- and image-based technologies developed and deployed in cereal breeding and research. Over the past two decades, sensors and cameras, and accompanying HTPPs, have evolved considerably with the rapid advances in technology and image analysis pipelines. These next-generation HTPPs have significantly improved precision phenotyping and thus our understanding of plant growth, physiology, morphology and development. Remote, non-destructive phenotyping methods can replace time and labour-intensive measurements as well as capture considerably vast unbiased data sets, which have allowed for the development of techniques to phenotype traits previously too costly and laborious to measure.

High-throughput, precision phenotyping still has a number of limitations to overcome, including high data acquisition and processing costs, limited efficient analysis pipelines and complex algorithms needed to identify agronomically relevant traits. Plant phenotyping is based on many interrelated and interdependent traits, which vary depending on crop type, genotypic diversity, developmental stage, environmental and diurnal patterns, affecting the capture and interpretation of relevant data. HTPPs provide additional challenges, such as the efficient storage, collation, mining and retrieval of vast amounts of data, data integration across multiple different platforms and complex image analysis requirements [208]. Recent developments in deep learning and other machine learning algorithm techniques may be the next revolution in image-based phenotyping, helping to reduce processing costs and deepen our understanding of phenotype-genotype-environment interactions [209]. Other developments in computing power and data storage [210] will also assist high-throughput sensor-based phenotyping in becoming an established practice in cereal crop research and breeding programs. Integration of next-generation remote phenotyping technologies with advanced genomics is therefore key to furthering the understanding of crop growth and development under the increasing range of complex agricultural environments and enhance breeding of climate-adapted varieties.

## References


1. Johannsen W (1911) *The genotype concept of heredity.* In: *The American Naturalist*, pp 129–159
2. Tester M, Langridge P (2010) Breeding technologies to increase crop production in a changing world. Science 327:818–822
3. Ray DK, Mueller ND, West PC, Foley JA (2013) Yield trends are insufficient to double global crop production by 2050. PLoS One 8 (6):e66428. https://doi.org/10.1371/journal.pone.0066428
4. Tilman D, Balzer C, Hill J, Befort BL (2011) Global food demand and the sustainable intensification of agriculture. Proc Natl Acad Sci USA 108(50):20260–20264
5. Sultan SE (2000) Phenotypic plasticity for plant development, function and life history. Trends Plant Sci 5(12):537–542
6. Pratap A, Gupta S, Nair R, Gupta S, Schafleitner R, Basu P, Singh C, Prajapati U, Gupta A, Nayyar H, Mishra A, Baek K-H (2019) Using plant phenomics to exploit the gains of genomics. Agronomy 9(3):126. https://doi.org/10.3390/agronomy9030126
7. Houle D, Govindaraju DR, Omholt S (2010) Phenomics: the next challenge. Nat Rev Genet 11(12):855–866. https://doi.org/10.1038/nrg2897
8. Furbank RT, von Caemmerer S, Sheehy J, Edwards G (2009) $C_4$ rice: a challenge for plant phenomics. Funct Plant Biol 36:845–856
9. Reynolds M, Manes Y, Izanloo A, Langridge P (2009) Phenotyping approaches for physiological breeding and gene discovery in wheat. Ann Appl Biol 155(3):309–320. https://doi.org/10.1111/j.1744-7348.2009.00351.x
10. Cabrera-Bosquet L, Crossa J, von Zitzewitz J, Serret MD, Araus JL (2012) High-throughput phenotyping and genomic selection: the frontiers of crop breeding converge. J Integr Plant Biol 54(5):312–320. https://doi.org/10.1111/j.1744-7909.2012.01116.x
11. Reynolds M, Foulkes MJ, Slafer GA, Berry P, Parry MA, Snape JW, Angus WJ (2009) Raising yield potential in wheat. J Exp Bot 60(7):1899–1918. https://doi.org/10.1093/jxb/erp016
12. Fiorani F, Schurr U (2013) Future scenarios for plant phenotyping. Annu Rev Plant Biol 64:267–291. https://doi.org/10.1146/annurev-arplant-050312-120137
13. Bassi FM, Bentley AR, Charmet G, Ortiz R, Crossa J (2016) Breeding schemes for the implementation of genomic selection in wheat (Triticum spp.). Plant Sci 242:23–36. https://doi.org/10.1016/j.plantsci.2015.08.021
14. Gupta PK, Langridge P, Mir RR (2009) Marker-assisted wheat breeding: present status and future possibilities. Mol Breed 26 (2):145–161. https://doi.org/10.1007/s11032-009-9359-7
15. Heffner EL, Jannink J-L, Iwata H, Souza E, Sorrells ME (2011) Genomic selection accuracy for grain quality traits in biparental wheat populations. Crop Sci 51(6):2597–2606. https://doi.org/10.2135/cropsci2011.05.0253
16. Jannink JL, Lorenz AJ, Iwata H (2010) Genomic selection in plant breeding: from theory to practice. Brief Funct Genomics 9 (2):166–177. https://doi.org/10.1093/bfgp/elq001
17. Cobb JN, Declerck G, Greenberg A, Clark R, McCouch S (2013) Next-generation phenotyping: requirements and strategies for enhancing our understanding of genotype-phenotype relationships and its relevance to crop improvement. Theor Appl Genet 126 (4):867–887. https://doi.org/10.1007/s00122-013-2066-0
18. Vargas M, van Eeuwijk FA, Crossa J, Ribaut J-M (2006) Mapping QTLs and QTL x environment interaction for CIMMYT maize drought stress program using factorial regression and partial least squares methods. Theor Appl Genet 112(6):1009–1023. https://doi.org/10.1007/s00122-005-0204-z
19. McMullen MD, Kresovich S, Villeda HS, Bradbury P, Li H, Sun Q, Flint-Garcia S, Thornsberry J, Acharya C, Bottoms C,

Brown P, Browne C, Eller M, Guill K, Harjes C, Kroon D, Lepak N, Mitchell SE, Peterson B, Pressoir G, Romero S, Oropeza Rosas M, Salvo S, Yates H, Hanson M, Jones E, Smith S, Glaubitz JC, Goodman M, Ware D, Holland JB, Buckler ES (2009) *Genetic properties of the maize nested association mapping population.* Science **325** (5941):737–740

20. Araus JL, Cairns JE (2014) Field high-throughput phenotyping: the new crop breeding frontier. Trends Plant Sci 19 (1):52–61. https://doi.org/10.1016/j.tplants.2013.09.008
21. Phillips RL (2010) *Mobilizing science to break yield barriers.* Crop Sci **50**:S-99-S-108. https://doi.org/10.2135/cropsci2009.09.0525
22. Furbank RT, Tester M (2011) Phenomics - technologies to relieve the phenotyping bottleneck. Trends Plant Sci 16(12):635–644. https://doi.org/10.1016/j.tplants.2011.09.005
23. Li L, Zhang Q, Huang D (2014) A review of imaging techniques for plant phenotyping. Sensors 14(11):20078–20111. https://doi.org/10.3390/s141120078
24. Deery D, Jimenez-Berni J, Jones H, Sirault X, Furbank R (2014) Proximal remote sensing buggies and potential applications for field-based phenotyping. Agronomy 5:349–379. https://doi.org/10.3390/agronomy4030349
25. Hanjagi PS, Awaji SM (2018) High-throughput imaging tools to phenotype traits in plants for stress environments. J Pharmacog Phytochem 7(6):97–102
26. Yendrek CR, Tomaz T, Montes CM, Cao Y, Morse AM, Brown PJ, McIntyre LM, Leakey AD, Ainsworth EA (2017) High-throughput phenotyping of maize leaf physiological and biochemical traits using hyperspectral reflectance. Plant Physiol 173(1):614–626. https://doi.org/10.1104/pp.16.01447
27. Blackburn GA (2007) Hyperspectral remote sensing of plant pigments. J Exp Bot 58 (4):855–867. https://doi.org/10.1093/jxb/erl123
28. Fahlgren N, Gehan MA, Baxter I (2015) Lights, camera, action: high-throughput plant phenotyping is ready for a close-up. Curr Opin Plant Biol 24:93–99. https://doi.org/10.1016/j.pbi.2015.02.006
29. Gonzalez-Dugo V, Hernandez P, Solis I, Zarco-Tejada P (2015) Using high-resolution hyperspectral and thermal airborne imagery to assess physiological condition in the context of wheat phenotyping. Remote Sens 7(10):13586–13605. https://doi.org/10.3390/rs71013586
30. Dreccer MF, Barnes LR, Meder R (2014) Quantitative dynamics of stem water soluble carbohydrates in wheat can be monitored in the field using hyperspectral reflectance. Field Crop Res 159:70–80. https://doi.org/10.1016/j.fcr.2014.01.001
31. Zheng Q, Huang W, Cui X, Dong Y, Shi Y, Ma H, Liu L (2018) Identification of wheat yellow rust using optimal three-band spectral indices in different growth stages. Sensors 19 (1):35. https://doi.org/10.3390/s19010035
32. Zarco-Tejada PJ, Guillén-Climent ML, Hernández-Clemente R, Catalina A, González MR, Martín P (2013) Estimating leaf carotenoid content in vineyards using high resolution hyperspectral imagery acquired from an unmanned aerial vehicle (UAV). Agric For Meteorol 171-172:281–294. https://doi.org/10.1016/j.agrformet.2012.12.013
33. Banerjee BP, Joshi S, Thoday-Kennedy E, Pasam RK, Tibbits J, Hayden M, Spangenberg G, Kant S (2020) High-throughput phenotyping using digital and hyperspectral imaging-derived biomarkers for genotypic nitrogen response. J Exp Bot 71(15):4604–4615. https://doi.org/10.1093/jxb/eraa143
34. Das B, Sahoo RN, Pargal S, Krishna G, Verma R, Chinnusamy V, Sehgal VK, Gupta VK, Dash SK, Swain P (2018) Quantitative monitoring of sucrose, reducing sugar and total sugar dynamics for phenotyping of water-deficit stress tolerance in rice through spectroscopy and chemometrics. Spectrochim Acta A Mol Biomol Spectrosc 192:41–51. https://doi.org/10.1016/j.saa.2017.10.076
35. Ansari MS, Young KR, Nicolas ME (2016) *Determining wavelenth for nitrogen and phosphorus nutrients through hyperspectral remote sensing in wheat (Triticum aestivum L.) plant.* Int J Bioresour Stress Manage 7(4):653–662. https://doi.org/10.5958/0976-4038.2016.00125.1
36. Mahajan GR, Sahoo RN, Pandey RN, Gupta VK, Kumar D (2014) Using hyperspectral remote sensing techniques to monitor nitrogen, phosphorus, Sulphur and potassium in wheat (Triticum aestivum L.). Precis Agric 15 (5):499–522. https://doi.org/10.1007/s11119-014-9348-7
37. Uddling J, Gelang-Alfredsson J, Piikki K, Pleijel H (2007) Evaluating the relationship between leaf chlorophyll concentration and

SPAD-502 chlorophyll meter readings. Photosynth Res 91(1):37–46. https://doi.org/10.1007/s11120-006-9077-5
38. Nguyen GN, Panozzo J, Spangenberg G, Kant S (2016) Phenotyping approaches to evaluate nitrogen-use efficiency related traits of diverse wheat varieties under field conditions. Crop Pasture Sci 67(11):1139–1148. https://doi.org/10.1071/cp16173
39. Arregui LM, Lasa B, Lafarga A, Irañeta I, Baroja E, Quemada M (2006) Evaluation of chlorophyll meters as tools for N fertilization in winter wheat under humid Mediterranean conditions. Eur J Agron 24(2):140–148. https://doi.org/10.1016/j.eja.2005.05.005
40. Islam MR, Haque KMS, Akter N, Karim MA (2014) *Leaf chlorophyll dynamics in wheat based on SPAD meter reading and its relationship with grain yield.* Scientia Agriculturae **8** (1):13–18. https://doi.org/10.15192/pscp.sa.2014.4.1.1318
41. Hunt ER, Daughtry CST, Eitel JUH, Long DS (2011) Remote sensing leaf chlorophyll content using a visible band index. Agron J 103(4):1090–1099. https://doi.org/10.2134/agronj2010.0395
42. Kanning M, Kühling I, Trautz D, Jarmer T (2018) High-resolution UAV-based hyperspectral imagery for LAI and chlorophyll estimations from wheat for yield prediction. Remote Sens 10(12):2000. https://doi.org/10.3390/rs10122000
43. Silva-Perez V, Molero G, Serbin SP, Condon AG, Reynolds MP, Furbank RT, Evans JR (2018) Hyperspectral reflectance as a tool to measure biochemical and physiological traits in wheat. J Exp Bot 69(3):483–496. https://doi.org/10.1093/jxb/erx421
44. Scholes JD, Rolfe SA (2009) Chlorophyll fluorescence imaging as tool for understanding the impact of fungal diseases on plant performance: a phenomics perspective. Funct Plant Biol 36:880–892
45. Kuckenberg J, Tartachnyk I, Noga G (2009) Detection and differentiation of nitrogen-deficiency, powdery mildew and leaf rust at wheat leaf and canopy level by laser-induced chlorophyll fluorescence. Biosyst Eng 103 (2):121–128. https://doi.org/10.1016/j.biosystemseng.2008.09.018
46. Kuckenberg J, Tartachnyk I, Noga G (2009) Temporal and spatial changes of chlorophyll fluorescence as a basis for early and precise detection of leaf rust and powdery mildew infections in wheat leaves. Precis Agric 10 (1):34–44. https://doi.org/10.1007/s11119-008-9082-0
47. Bauriegel E, Herppich W (2014) Hyperspectral and chlorophyll fluorescence imaging for early detection of plant diseases, with special reference to Fusarium spec. Infections on wheat. Agriculture 4(1):32–57. https://doi.org/10.3390/agriculture4010032
48. Chen D, Neumann K, Friedel S, Kilian B, Chen M, Altmann T, Klukas C (2014) Dissecting the phenotypic components of crop plant growth and drought responses based on high-throughput image analysis. Plant Cell 26(12):4636–4655. https://doi.org/10.1105/tpc.114.129601
49. Nguyen GN, Kant S (2018) Improving nitrogen use efficiency in plants: effective phenotyping in conjunction with agronomic and genetic approaches. Funct Plant Biol 45 (6):606–619. https://doi.org/10.1071/FP17266
50. Fitzgerald G, Rodriguez D, O'Leary G (2010) Measuring and predicting canopy nitrogen nutrition in wheat using a spectral index—the canopy chlorophyll content index (CCCI). Field Crop Res 116(3):318–324. https://doi.org/10.1016/j.fcr.2010.01.010
51. Eitel JU, Vierling LA, Long DS, Hunt ER (2011) Early season remote sensing of wheat nitrogen status using a green scanning laser. Agric For Meteorol 151(10):1338–1345
52. Erdle K, Mistele B, Schmidhalter U (2011) Comparison of active and passive spectral sensors in discriminating biomass parameters and nitrogen status in wheat cultivars. Field Crop Res 124(1):74–84. https://doi.org/10.1016/j.fcr.2011.06.007
53. Tremblay N, Wang Z, Ma B-L, Belec C, Vigneault P (2009) A comparison of crop data measured by two commercial sensors for variable-rate nitrogen application. Precis Agric 10(2):145–161. https://doi.org/10.1007/s11119-008-9080-2
54. Duan T, Chapman SC, Guo Y, Zheng B (2017) Dynamic monitoring of NDVI in wheat agronomy and breeding trials using an unmanned aerial vehicle. Field Crop Res 210:71–80. https://doi.org/10.1016/j.fcr.2017.05.025
55. Heege HJ, Reusch S, Thiessen E (2008) Prospects and results for optical systems for site-specific on-the-go control of nitrogen-top-dressing in Germany. Precis Agric 9 (3):115–131. https://doi.org/10.1007/s11119-008-9055-3
56. Yang M, Hassan MA, Xu K, Zheng C, Rasheed A, Zhang Y, Jin X, Xia X, Xiao Y, He Z (2020) Assessment of water and nitrogen use efficiencies through UAV-based

multispectral phenotyping in winter wheat. Front Plant Sci 11:927. https://doi.org/10.3389/fpls.2020.00927

57. Yang B, Wang M, Sha Z, Wang B, Chen J, Yao X, Cheng T, Cao W, Zhu Y (2019) Evaluation of aboveground nitrogen content of winter wheat using digital imagery of unmanned aerial vehicles. Sensors 19 (20):4416. https://doi.org/10.3390/s19204416
58. Lee K-J, Lee B-W (2013) Estimation of rice growth and nitrogen nutrition status using color digital camera image analysis. Eur J Agron 48:57–65. https://doi.org/10.1016/j.eja.2013.02.011
59. Li F, Gnyp ML, Jia L, Miao Y, Yu Z, Koppe W, Bareth G, Chen X, Zhang F (2008) Estimating N status of winter wheat using a handheld spectrometer in the North China Plain. Field Crop Res 106(1):77–85. https://doi.org/10.1016/j.fcr.2007.11.001
60. Vigneau N, Ecarnot M, Rabatel G, Roumet P (2011) Potential of field hyperspectral imaging as a non destructive method to assess leaf nitrogen content in wheat. Field Crop Res 122(1):25–31
61. Eitel JUH, Magney TS, Vierling LA, Brown TT, Huggins DR (2014) LiDAR based biomass and crop nitrogen estimates for rapid, non-destructive assessment of wheat nitrogen status. Field Crop Res 159:21–32. https://doi.org/10.1016/j.fcr.2014.01.008
62. Liang L, Di L, Huang T, Wang J, Lin L, Wang L, Yang M (2018) Estimation of leaf nitrogen content in wheat using new hyperspectral indices and a random forest regression algorithm. Remote Sens 10(12):1940. https://doi.org/10.3390/rs10121940
63. Flowers M, Weisz R, Heiniger R (2003) Quantitative approaches for using color infrared photography for assessing in-season nitrogen status in winter wheat. Agron J 95 (5):1189–1200. https://doi.org/10.2134/agronj2003.1189
64. Eitel JUH, Long DS, Gessler PE, Hunt ER, Brown DJ (2009) Sensitivity of ground-based remote sensing estimates of wheat chlorophyll content to variation in soil reflectance. Soil Sci Soc Am J 73(5):1715–1723. https://doi.org/10.2136/sssaj2008.0288
65. Bruning B, Liu H, Brien C, Berger B, Lewis M, Garnett T (2019) The development of hyperspectral distribution maps to predict the content and distribution of nitrogen and water in wheat (Triticum aestivum). Front Plant Sci 10:1380. https://doi.org/10.3389/fpls.2019.01380
66. Tan C, Du Y, Zhou J, Wang D, Luo M, Zhang Y, Guo W (2018) Analysis of different hyperspectral variables for diagnosing leaf nitrogen accumulation in wheat. Front Plant Sci 9:674. https://doi.org/10.3389/fpls.2018.00674
67. Hansen P, Schjoerring J (2003) Reflectance measurement of canopy biomass and nitrogen status in wheat crops using normalized difference vegetation indices and partial least squares regression. Remote Sens Environ 86 (4):542–553
68. Cattivelli L, Rizza F, Badeck F-W, Mazzucotelli E, Mastrangelo AM, Francia E, Marè C, Tondelli A, Stanca AM (2008) Drought tolerance improvement in crop plants: an integrated view from breeding to genomics. Field Crop Res 105(1-2):1–14. https://doi.org/10.1016/j.fcr.2007.07.004
69. Blum A (2017) Osmotic adjustment is a prime drought stress adaptive engine in support of plant production. Plant Cell Environ 40(1):4–10. https://doi.org/10.1111/pce.12800
70. Bartlett MK, Scoffoni C, Ardy R, Zhang Y, Sun S, Cao K, Sack L (2012) Rapid determination of comparative drought tolerance traits: using an osmometer to predict turgor loss point. Methods Ecol Evol 3(5):880–888. https://doi.org/10.1111/j.2041-210X.2012.00230.x
71. Saint Pierre C, González JLB (2012) In: Pask A, Pietragalla J, Mullan D, Reynolds M (eds) *Leaf water potential*, in *Physiological Breeding II: A field guide to wheat phenotyping*. CIMMYT, Mexico, D.F., pp 18–20
72. Govender M, Govender PJ, Weiersbye IM, Witkowski EFT, Ahmed F (2009) *Review of commonly used remote sensing and ground-based technologies to measure plant water stress.* Water SA **35**:5
73. Mullan D, Pietragalla J (2012) In: Pask A, Pietragalla J, Mullan D, Reynolds M (eds) *Leaf relative water content*, in *Physiological Breeding II: A field guide to wheat phenotyping*. CIMMYT, Mexico, D.F., pp 25–27
74. Weber VS, Araus JL, Cairns JE, Sanchez C, Melchinger AE, Orsini E (2012) Prediction of grain yield using reflectance spectra of canopy and leaves in maize plants grown under different water regimes. Field Crop Res 128:82–90. https://doi.org/10.1016/j.fcr.2011.12.016
75. Sun H, Feng M, Xiao L, Yang W, Wang C, Jia X, Zhao Y, Zhao C, Muhammad SK, Li D (2019) *Assessment of plant water status in winter wheat (Triticum aestivum L.) based on*

*canopy spectral indices*. PLoS One **14**(6): e0216890. https://doi.org/10.1371/journal.pone.0216890

76. Gutierrez M, Reynolds MP, Raun WR, Stone ML, Klatt AR (2010) Spectral water indices for assessing yield in elite bread wheat genotypes under well-irrigated, water-stressed, and high-temperature conditions. Crop Sci 50 (1):197–214. https://doi.org/10.2135/cropsci2009.07.0381
77. Zimmermann D, Reuss R, Westhoff M, Gessner P, Bauer W, Bamberg E, Bentrup FW, Zimmermann U (2008) A novel, non-invasive, online-monitoring, versatile and easy plant-based probe for measuring leaf water status. J Exp Bot 59 (11):3157–3167. https://doi.org/10.1093/jxb/ern171
78. Joshi S, Choukimath A, Isenegger D, Panozzo J, Spangenberg G, Kant S (2019) Improved wheat growth and yield by delayed leaf senescence using developmentally regulated expression of a cytokinin biosynthesis gene. Front Plant Sci 10:1285. https://doi.org/10.3389/fpls.2019.01285
79. Bramley H, Bitter R, Zimmermann G, Zimmermann U (2015) Simultaneous recording of diurnal changes in leaf turgor pressure and stem water status of bread wheat reveal variation in hydraulic mechanisms in response to drought. Funct Plant Biol 42 (10):1001–1009. https://doi.org/10.1071/FP15087
80. Kant S, Burch D, Ehrenberger W, Bitter R, Rüger S, Mason J, Zimmermann U, Spangenberg G, Léon J (2014) A novel crop water analysis system: identification of water stress tolerant genotypes of canola (Brassica napus L.) using non-invasive magnetic turgor pressure probes. Plant Breed 133 (5):602–608. https://doi.org/10.1111/pbr.12187
81. Jin X, Xu X, Song X, Li Z, Wang J, Guo W (2013) Estimation of leaf water content in winter wheat using grey relational analysis--partial least squares modeling with hyperspectral data. Agron J 105(5):1385–1392. https://doi.org/10.2134/agronj2013.0088
82. De Bei R, Cozzolino D, Sullivan W, Cynkar W, Fuentes S, Dambergs R, Pech J, Tyerman S (2011) Non-destructive measurement of grapevine water potential using near infrared spectroscopy. Aust J Grape Wine Res 17(1):62–71. https://doi.org/10.1111/j.1755-0238.2010.00117.x
83. Neilson EH, Edwards AM, Blomstedt CK, Berger B, Moller BL, Gleadow RM (2015) Utilization of a high-throughput shoot imaging system to examine the dynamic phenotypic responses of a C4 cereal crop plant to nitrogen and water deficiency over time. J Exp Bot 66(7):1817–1832. https://doi.org/10.1093/jxb/eru526
84. Feng W, Qi S, Heng Y, Zhou Y, Wu Y, Liu W, He L, Li X (2017) Canopy vegetation indices from in situ hyperspectral data to assess plant water status of winter wheat under powdery mildew stress. Front Plant Sci 8:1219. https://doi.org/10.3389/fpls.2017.01219
85. Seelig HD, Hoehn A, Stodieck L, Klaus D, Adams Iii W, Emery W (2008) The assessment of leaf water content using leaf reflectance ratios in the visible, near-, and short-wave-infrared. Int J Remote Sens 29 (13):3701–3713
86. Cohen Y, Alchanatis V, Meron M, Saranga Y, Tsipris J (2005) Estimation of leaf water potential by thermal imagery and spatial analysis. J Exp Bot 56(417):1843–1852. https://doi.org/10.1093/jxb/eri174
87. Gutierrez M, Reynolds MP, Klatt AR (2010) Association of water spectral indices with plant and soil water relations in contrasting wheat genotypes. J Exp Bot 61 (12):3291–3303. https://doi.org/10.1093/jxb/erq156
88. Babar MA, Reynolds M, van Ginkel N, Klatt AR, Ruan WR, Stone ML (2006) Spectral reflectance indices as a potential indirect selection criteria for wheat yield under irrigation. Crop Sci 46(2):578–588
89. Suárez L, Zarco-Tejada PJ, Berni JAJ, González-Dugo V, Fereres E (2009) Modelling PRI for water stress detection using radiative transfer models. Remote Sens Environ 113 (4):730–744. https://doi.org/10.1016/j.rse.2008.12.001
90. Urban J, Ingwers M, McGuire MA, Teskey RO (2017) Stomatal conductance increases with rising temperature. Plant Signal Behaviour 12(8):e1356534. https://doi.org/10.1080/15592324.2017.1356534
91. Zhang X, Wang Y, Huang G, Feng F, Liu X, Guo R, Gu F, Hu X, Yang Z, Zhong X, Mei X (2020) Atmospheric humidity and genotype are key determinants of the diurnal stomatal conductance pattern. J Agron Crop Sci 206 (2):161–168. https://doi.org/10.1111/jac.12375
92. Munns R, James RA, Sirault XR, Furbank RT, Jones HG (2010) New phenotyping methods for screening wheat and barley for beneficial responses to water deficit. J Exp Bot 61 (13):3499–3507. https://doi.org/10.1093/jxb/erq199

93. James RA, von Caemmerer S, Condon AGT, Zwart AB, Munns R (2008) Genetic variation in tolerance to the osmotic stress componentof salinity stress in durum wheat. Funct Plant Biol 35(2):111–123. https://doi.org/10.1071/FP07234
94. Rahnama A, James RA, Poustini K, Munns R (2010) Stomatal conductance as a screen for osmotic stress tolerance in durum wheat growing in saline soil. Funct Plant Biol 37:255–263
95. Jiang Q, Roche D, Monaco TA, Hole D (2006) Stomatal conductance is a key parameter to assess limitations to photosynthesis and growth potential in barley genotypes. Plant Biol 8(4):515–521. https://doi.org/10.1055/s-2006-923964
96. Sirault XRR, James RA, Furbank RT (2009) A new screening method for osmotic component of salinity tolerance in cereals using infrared thermography. Funct Plant Biol 36 (11):970–977. https://doi.org/10.1071/FP09182
97. Zia S, Romano G, Spreer W, Sanchez C, Cairns J, Araus JL, Müller J (2013) Infrared thermal imaging as a rapid tool for identifying water-stress tolerant maize genotypes of different phenology. J Agron Crop Sci 199 (2):75–84. https://doi.org/10.1111/j.1439-037X.2012.00537.x
98. Blum A, Mayer J, Gozlan G (1982) Infrared thermal sensing of plant canopies as a screening technique for dehydration avoidance in wheat. Field Crop Res 5:137–146. https://doi.org/10.1016/0378-4290(82)90014-4
99. Gallé Á, Csiszár J, Benyó D, Laskay G, Leviczky T, Erdei L, Tari I (2013) Isohydric and anisohydric strategies of wheat genotypes under osmotic stress: biosynthesis and function of ABA in stress responses. J Plant Physiol 170(16):1389–1399. https://doi.org/10.1016/j.jplph.2013.04.010
100. Machado S, Paulsen GM (2001) Combined effects of drought and high temperature on water relations of wheat and sorghum. Plant Soil 233(2):179–187. https://doi.org/10.1023/a:1010346601643
101. Baluja J, Diago MP, Balda P, Zorer R, Meggio F, Morales F, Tardaguila J (2012) Assessment of vineyard water status variability by thermal and multispectral imagery using an unmanned aerial vehicle (UAV). Irrig Sci 30 (6):511–522. https://doi.org/10.1007/s00271-012-0382-9
102. Deery DM, Rebetzke GJ, Jimenez-Berni JA, James RA, Condon AG, Bovill WD, Hutchinson P, Scarrow J, Davy R, Furbank RT (2016) Methodology for high-throughput field phenotyping of canopy temperature using airborne thermography. Front Plant Sci 7:1808. https://doi.org/10.3389/fpls.2016.01808
103. Berni J, Zarco-Tejada PJ, Suarez L, Fereres E (2009) Thermal and narrowband multispectral remote sensing for vegetation monitoring from an unmanned aerial vehicle. IEEE Trans Geosci Remote Sens 47(3):722–738. https://doi.org/10.1109/tgrs.2008.2010457
104. Fuentes S, De Bei R, Pech J, Tyerman S (2012) Computational water stress indices obtained from thermal image analysis of grapevine canopies. Irrig Sci 30(6):523–536. https://doi.org/10.1007/s00271-012-0375-8
105. Liu Y, Subhash C, Yan J, Song C, Zhao J, Li J (2011) Maize leaf temperature responses to drought: thermal imaging and quantitative trait loci (QTL) mapping. Environ Exp Bot 71(2):158–165. https://doi.org/10.1016/j.envexpbot.2010.11.010
106. Fehér-Juhász E, Majer P, Sass L, Lantos C, Csiszár J, Turóczy Z, Mihály R, Mai A, Horváth GV, Vass I, Dudits D (2014) Phenotyping shows improved physiological traits and seed yield of transgenic wheat plants expressing the alfalfa aldose reductase under permanent drought stress. Acta Physiol Plant 63:663–673. https://doi.org/10.1007/s11738-013-1445-0
107. Chapman S, Merz T, Chan A, Jackway P, Hrabar S, Dreccer M, Holland E, Zheng B, Ling T, Jimenez-Berni J (2014) Phenocopter: a low-altitude, autonomous remote-sensing robotic helicopter for high-throughput field-based phenotyping. Agronomy 4(2):279–301. https://doi.org/10.3390/agronomy4020279
108. Gracia-Romero A, Kefauver SC, Fernandez-Gallego JA, Vergara-Díaz O, Nieto-Taladriz MT, Araus JL (2019) UAV and ground image-based phenotyping: a proof of concept with durum wheat. Remote Sens 11 (10):1244. https://doi.org/10.3390/rs11101244
109. Jones HG, Serraj R, Loveys BR, Xiong L, Wheaton A, Price AH (2009) Thermal infrared imaging of crop canopies for the remote diagnosis and quantification of plant responses to water stress in the field. Funct Plant Biol 36(11):978–989. https://doi.org/10.1071/FP09123
110. O'Shaughnessy SA, Hebel MA, Evett SR, Colaizzi PD (2011) Evaluation of a wireless infrared thermometer with a narrow field of view. Comput Electron Agric 76(1):59–68.

https://doi.org/10.1016/j.compag.2010.12.017

111. Minervini M, Abdelsamea MM, Tsaftaris SA (2014) Image-based plant phenotyping with incremental learning and active contours. Eco Inform 23:35–48. https://doi.org/10.1016/j.ecoinf.2013.07.004
112. Yu K, Kirchgessner N, Grieder C, Walter A, Hund A (2017) An image analysis pipeline for automated classification of imaging light conditions and for quantification of wheat canopy cover time series in field phenotyping. Plant Methods 13(1):15
113. Tessmer OL, Jiao Y, Cruz JA, Kramer DM, Chen J (2013) Functional approach to high-throughput plant growth analysis. BMC Syst Biol 7(S6):S17
114. Golzarian MR, Frick RA, Rajendran K, Berger B, Roy S, Tester M, Lun DS (2011) Accurate inference of shoot biomass from high-throughput images of cereal plants. Plant Methods 7(1):2
115. Huang C, Yang W, Duan L, Jiang N, Chen G, Xiong L, Liu Q (2013) Rice panicle length measuring system based on dual-camera imaging. Comput Electron Agric 98:158–165. https://doi.org/10.1016/j.compag.2013.08.006
116. Qiu R, Yang C, Moghimi A, Zhang M, Steffenson BJ, Hirsch CD (2019) Detection of fusarium head blight in wheat using a deep neural network and color imaging. Remote Sens 11(22):2658. https://doi.org/10.3390/rs11222658
117. Ma J, Li Y, Chen Y, Du K, Zheng F, Zhang L, Sun Z (2019) Estimating above ground biomass of winter wheat at early growth stages using digital images and deep convolutional neural network. Eur J Agron 103:117–129. https://doi.org/10.1016/j.eja.2018.12.004
118. Banerjee BP, Spangenberg G, Kant S (2020) Fusion of spectral and structural information from aerial images for improved biomass estimation. Remote Sens 12(19):3164. https://doi.org/10.3390/rs12193164
119. Hartmann A, Czauderna T, Hoffmann R, Stein N, Schreiber F (2011) HTPheno: an image analysis pipeline for high-throughput plant phenotyping. BMC Bioinf 12:148
120. Honsdorf N, March TJ, Berger B, Tester M, Pillen K (2014) High-throughput phenotyping to detect drought tolerance QTL in wild barley introgression lines. PLoS One 9(5):e97047. https://doi.org/10.1371/journal.pone.0097047
121. Maphosa L, Thoday-Kennedy E, Vakani J, Phelan A, Badenhorst P, Slater A, Spangenberg G, Kant S (2016) Phenotyping wheat under salt stress conditions using a 3D laser scanner. Israel J Plant Sci:1–8. https://doi.org/10.1080/07929978.2016.1243405
122. Nguyen GN, Maharjan P, Maphosa L, Vakani J, Thoday-Kennedy E, Kant S (2019) A robust automated image-based phenotyping method for rapid vegetative screening of wheat germplasm for nitrogen use efficiency. Front Plant Sci 10:1372
123. Hairmansis A, Berger B, Tester M, Roy SJ (2014) Image-based phenotyping for non-destructive screening of different salinity tolerance traits in rice. Rice 7(1):16
124. Rajendran K, Tester M, Roy SJ (2009) Quantifying the three main components of salinity tolerance in cereals. Plant Cell Environ 32(3):237–249. https://doi.org/10.1111/j.1365-3040.2008.01916.x
125. Tackenberg O (2007) A new method for non-destructive measurement of biomass, growth rates, vertical biomass distribution and dry matter content based on digital image analysis. Ann Bot 99(4):777–783. https://doi.org/10.1093/aob/mcm009
126. Yang W, Guo Z, Huang C, Duan L, Chen G, Jiang N, Fang W, Feng H, Xie W, Lian X, Wang G, Luo Q, Zhang Q, Liu Q, Xiong L (2014) Combining high-throughput phenotyping and genome-wide association studies to reveal natural genetic variation in rice. Nat Commun 5:5087. https://doi.org/10.1038/ncomms6087
127. Holman F, Riche A, Michalski A, Castle M, Wooster M, Hawkesford M (2016) High throughput field phenotyping of wheat plant height and growth rate in field plot trials using UAV based remote sensing. Remote Sens 8(12):1031. https://doi.org/10.3390/rs8121031
128. Meng R, Saade S, Kurtek S, Berger B, Brien C, Pillen K, Tester M, Sun Y (2017) Growth curve registration for evaluating salinity tolerance in barley. Plant Methods 13(1):18
129. Duan T, Chapman SC, Holland E, Rebetzke GJ, Guo Y, Zheng B (2016) Dynamic quantification of canopy structure to characterize early plant vigour in wheat genotypes. J Exp Bot 67(15):4523–4534. https://doi.org/10.1093/jxb/erw227
130. Cabrera-Bosquet L, Molero G, Stellacci A, Bort J, Nogués S, Araus J (2011) NDVI as a potential tool for predicting biomass, plant nitrogen content and growth in wheat genotypes subjected to different water and

nitrogen conditions. Cereal Res Commun 39 (1):147–159. https://doi.org/10.1556/crc.39.2011.1.15

131. Torres-Sánchez J, Peña JM, de Castro AI, López-Granados F (2014) Multi-temporal mapping of the vegetation fraction in early-season wheat fields using images from UAV. Comput Electron Agric 103:104–113. https://doi.org/10.1016/j.compag.2014.02.009
132. Mutanga O, Skidmore AK (2004) Narrow band vegetation indices overcome the saturation problem in biomass estimation. Int J Remote Sens 25(19):3999–4014
133. Aasen H, Burkart A, Bolten A, Bareth G (2015) Generating 3D hyperspectral information with lightweight UAV snapshot cameras for vegetation monitoring: from camera calibration to quality assurance. ISPRS J Photogramm Remote Sens 108:245–259. https://doi.org/10.1016/j.isprsjprs.2015.08.002
134. Bendig J, Bolten A, Bennertz S, Broscheit J, Eichfuss S, Bareth G (2014) Estimating biomass of barley using crop surface models (CSMs) derived from UAV-based RGB imaging. Remote Sens 6(11):10395–10412. https://doi.org/10.3390/rs61110395
135. Bendig J, Yu K, Aasen H, Bolten A, Bennertz S, Broscheit J, Gnyp ML, Bareth G (2015) Combining UAV-based plant height from crop surface models, visible, and near infrared vegetation indices for biomass monitoring in barley. Int J Appl Earth Obs Geoinf 39:79–87. https://doi.org/10.1016/j.jag.2015.02.012
136. Geipel J, Link J, Claupein W (2014) Combined spectral and spatial modeling of corn yield based on aerial images and crop surface models acquired with an unmanned aircraft system. Remote Sens 6(11):10335–10355. https://doi.org/10.3390/rs61110335
137. Jimenez-Berni JA, Deery DM, Rozas-Larraondo P, Condon ATG, Rebetzke GJ, James RA, Bovill WD, Furbank RT, Sirault XRR (2018) High throughput determination of plant height, ground cover, and above-ground biomass in wheat with LiDAR. Front Plant Sci 9:237. https://doi.org/10.3389/fpls.2018.00237
138. Hosoi F, Omasa K (2009) Estimating vertical plant area density profile and growth parameters of a wheat canopy at different growth stages using three-dimensional portable lidar imaging. ISPRS J Photogramm Remote Sens 64(2):151–158. https://doi.org/10.1016/j.isprsjprs.2008.09.003
139. Li W, Niu Z, Huang N, Wang C, Gao S, Wu C (2015) Airborne LiDAR technique for estimating biomass components of maize: a case study in Zhangye City, Northwest China. Ecol Indic 57:486–496. https://doi.org/10.1016/j.ecolind.2015.04.016
140. Vadez V, Kholova J, Hummel G, Zhokhavets U, Gupta SK, Hash CT (2015) LeasyScan: a novel concept combining 3D imaging and lysimetry for high-throughput phenotyping of traits controlling plant water budget. J Exp Bot 66(18):5581–5593. https://doi.org/10.1093/jxb/erv251
141. McCormick RF, Truong SK, Mullet JE (2016) 3D sorghum reconstructions from depth images identify QTL regulating shoot architecture. Plant Physiol 172(2):823–834. https://doi.org/10.1104/pp.16.00948
142. Cabrera-Bosquet L, Fournier C, Brichet N, Welcker C, Suard B, Tardieu F (2016) High-throughput estimation of incident light, light interception and radiation-use efficiency of thousands of plants in a phenotyping platform. New Phytol 212(1):269–281. https://doi.org/10.1111/nph.14027
143. Xie Q, Huang W, Liang D, Chen P, Wu C, Yang G, Zhang J, Huang L, Zhang D (2014) Leaf area index estimation using vegetation indices derived from airborne hyperspectral images in winter wheat. IEEE J Sel Top Appl Earth Obs Remote Sens 7 (8):3586–3594. https://doi.org/10.1109/jstars.2014.2342291
144. Liu J, Pattey E (2010) Retrieval of leaf area index from top-of-canopy digital photography over agricultural crops. Agric For Meteorol 150(11):1485–1490. https://doi.org/10.1016/j.agrformet.2010.08.002
145. Gebbers R, Ehlert D, Adamek R (2011) Rapid mapping of the leaf area index in agricultural crops. Agron J 103(5):1532–1541. https://doi.org/10.2134/agronj2011.0201
146. Anderegg J, Yu K, Aasen H, Walter A, Liebisch F, Hund A (2020) Spectral vegetation indices to track senescence dynamics in diverse wheat germplasm. Front Plant Sci 10:1749. https://doi.org/10.3389/fpls.2019.01749
147. Christopher JT, Veyradier M, Borrell AK, Harvey G, Fletcher S, Chenu K (2014) Phenotyping novel stay-green traits to capture genetic variation in senescence dynamics. Funct Plant Biol 41(11):1035–1048. https://doi.org/10.1071/FP14052
148. Hassan M, Yang M, Rasheed A, Jin X, Xia X, Xiao Y, He Z (2018) Time-series multispectral indices from unmanned aerial vehicle imagery reveal senescence rate in bread wheat. Remote Sens 10(6):809. https://doi.org/10.3390/rs10060809

149. Cai J, Okamoto M, Atieno J, Sutton T, Li Y, Miklavcic SJ (2016) Quantifying the onset and progression of plant senescence by color image analysis for high throughput applications. PLoS One 11(6):e0157102. https://doi.org/10.1371/journal.pone.0157102

150. Camargo AV, Mott R, Gardner KA, Mackay IJ, Corke F, Doonan JH, Kim JT, Bentley AR (2016) Determining phenological patterns associated with the onset of senescence in a wheat MAGIC mapping population. Front Plant Sci 7:1540. https://doi.org/10.3389/fpls.2016.01540

151. Das Choudhury S, Bashyam S, Qiu Y, Samal A, Awada T (2018) Holistic and component plant phenotyping using temporal image sequence. Plant Methods 14(1):35

152. Koh JC, Hayden M, Daetwyler H, Kant S (2019) Estimation of crop plant density at early mixed growth stages using UAV imagery. Plant Methods 15(1):64

153. Xiong X, Yu L, Yang W, Liu M, Jiang N, Wu D, Chen G, Xiong L, Liu K, Liu Q (2017) A high-throughput stereo-imaging system for quantifying rape leaf traits during the seedling stage. Plant Methods 13(1):7

154. Jin X, Liu S, Baret F, Hemerlé M, Comar A (2017) Estimates of plant density of wheat crops at emergence from very low altitude UAV imagery. Remote Sens Environ 198:105–114. https://doi.org/10.1016/j.rse.2017.06.007

155. Liu S, Baret F, Allard D, Jin X, Andrieu B, Burger P, Hemmerlé M, Comar A (2017) A method to estimate plant density and plant spacing heterogeneity: application to wheat crops. Plant Methods 13(1):38

156. Liu S, Baret F, Andrieu B, Burger P, Hemmerle M (2017) Estimation of wheat plant density at early stages using high resolution imagery. Front Plant Sci 8:739. https://doi.org/10.3389/fpls.2017.00739

157. Sankaran S, Khot LR, Carter AH (2015) Field-based crop phenotyping: multispectral aerial imaging for evaluation of winter wheat emergence and spring stand. Comput Electron Agric 118:372–379

158. Bendig J, Bolten A, Bareth G (2013) UAV-based imaging for multi-temporal, very high resolution crop surface models to monitor crop growth variability. Photogrammetrie-Fernerkundung-Geoinformation 2013(6):551–562

159. Bareth G, Bendig J, Tilly N, Hoffmeister D, Aasen H, Bolten A (2016) A comparison of UAV- and TLS-derived plant height for crop monitoring: using polygon grids for the analysis of crop surface models (CSMs). Photogrammetrie-Fernerkundung-Geoinformation 2016(2):85–94. https://doi.org/10.1127/pfg/2016/0289

160. Singh D, Wang X, Kumar U, Gao L, Noor M, Imtiaz M, Singh RP, Poland J (2019) High-throughput phenotyping enabled genetic dissection of crop lodging in wheat. Front Plant Sci 10:394. https://doi.org/10.3389/fpls.2019.00394

161. Bi K, Jiang P, Li L, Shi B, Wang C (2010) Non-destructive measurement of wheat spike characteristics based on morphological image processing. Trans Chin Soc Agric Eng 26(12):212–216

162. Korohou T, Okinda C, Li H, Cao Y, Nyalala I, Huo L, Potcho M, Li X, Ding Q (2020) Wheat grain yield estimation based on image morphological properties and wheat biomass. J Sens 2020:1–11. https://doi.org/10.1155/2020/1571936

163. Ikeda M, Hirose Y, Takashi T, Shibata Y, Yamamura T, Komura T, Doi K, Ashikari M, Mastsuoka M, Kitano H (2010) Analysis of rice panicle traits and detection of QTLs using an image analyzing method. Breed Sci 60:55–64

164. Qiongyan L, Cai J, Berger B, Okamoto M, Miklavcic SJ (2017) Detecting spikes of wheat plants using neural networks with Laws texture energy. Plant Methods 13(1):83

165. Alkhudaydi T, Reynolds D, Griffiths S, Zhou J, de la Iglesia B (2019, **2019**) *An exploration of deep-learning based phenotypic analysis to detect spike regions in field conditions for UK bread wheat.* Plant Phenomics:1–17. https://doi.org/10.34133/2019/7368761

166. Hasan MM, Chopin JP, Laga H, Miklavcic SJ (2018) Detection and analysis of wheat spikes using convolutional neural networks. Plant Methods 14(1):100

167. Madec S, Jin X, Lu H, De Solan B, Liu S, Duyme F, Heritier E, Baret F (2019) Ear density estimation from high resolution RGB imagery using deep learning technique. Agric For Meteorol 264:225–234. https://doi.org/10.1016/j.agrformet.2018.10.013

168. Sadeghi-Tehran P, Sabermanesh K, Virlet N, Hawkesford MJ (2017) Automated method to determine two critical growth stages of wheat: heading and flowering. Front Plant Sci 8:252. https://doi.org/10.3389/fpls.2017.00252

169. Sadeghi-Tehran P, Virlet N, Ampe EM, Reyns P, Hawkesford MJ (2019) DeepCount: in-field automatic quantification of wheat

spikes using simple linear iterative clustering and deep convolutional neural networks. Front Plant Sci 10:1176. https://doi.org/10.3389/fpls.2019.01176

170. Velumani K, Madec S, de Solan B, Lopez-Lozano R, Gillet J, Labrosse J, Jezequel S, Comar A, Baret F (2020) An automatic method based on daily in situ images and deep learning to date wheat heading stage. Field Crop Res 252:107793. https://doi.org/10.1016/j.fcr.2020.107793

171. Virlet N, Sabermanesh K, Sadeghi-Tehran P, Hawkesford MJ (2017) Field Scanalyzer: an automated robotic field phenotyping platform for detailed crop monitoring. Funct Plant Biol 44(1):143–153. https://doi.org/10.1071/FP16163

172. Xiong H, Cao Z, Lu H, Madec S, Liu L, Shen C (2019) TasselNetv2: in-field counting of wheat spikes with context-augmented local regression networks. Plant Methods 15 (1):150

173. Zhou C, Liang D, Yang X, Xu B, Yang G (2018) Recognition of wheat spike from field based phenotype platform using multi-sensor fusion and imaximum entropy segmentation algorithms. Remote Sens 10 (2):246. https://doi.org/10.3390/rs10020246

174. Wang X, Xuan H, Evers B, Shrestha S, Pless R, Poland J (2019) High-throughput phenotyping with deep learning gives insight into the genetic architecture of flowering time in wheat. Gigascience 8:1–11. https://doi.org/10.1093/gigascience/giz120

175. Marti J, Bort J, Slafer GA, Araus JL (2007) Can wheat yield be assessed by early measurements of normalized difference vegetation index? Ann Appl Biol 150(2):253–257. https://doi.org/10.1111/j.1744-7348.2007.00126.x

176. Du M, Noguchi N (2017) Monitoring of wheat growth status and mapping of wheat yield's within-field spatial variations using color images acquired from UAV-camera system. Remote Sens 9(3):289. https://doi.org/10.3390/rs9030289

177. Hayes BJ, Panozzo J, Walker CK, Choy AL, Kant S, Wong D, Tibbits J, Daetwyler HD, Rochfort S, Hayden MJ, Spangenberg GC (2017) Accelerating wheat breeding for end-use quality with multi-trait genomic predictions incorporating near infrared and nuclear magnetic resonance-derived phenotypes. Theor Appl Genet 130 (12):2505–2519. https://doi.org/10.1007/s00122-017-2972-7

178. Fox GP, Kelly A, Sweeney N, Hocroft DM (2011) *Development of a single kernel NIR barley protein calibration and assessment of variation in protein on grain quality.* J Inst Brew 117(4):582–586

179. Mahesh S, Manickavasagan A, Jayas DS, Paliwal J, White NDG (2008) Feasibility of near-infrared hyperspectral imaging to differentiate Canadian wheat classes. Biosyst Eng 101(1):50–57. https://doi.org/10.1016/j.biosystemseng.2008.05.017

180. Clohessy JW, Pauli D, Kreher KM, Buckler ES, Armstrong PR, Wu T, Hoekenga OA, Jannink JL, Sorrells ME, Gore MA (2018) A low-cost automated system for high-throughput phenotyping of single oat seeds. Plant Phenom J 1(1):1–13. https://doi.org/10.2135/tppj2018.07.0005

181. ElMasry G, Mandour N, Al-Rejaie S, Belin E, Rousseau D (2019) Recent applications of multispectral imaging in seed phenotyping and quality monitoring-an overview. Sensors 19(5):1090. https://doi.org/10.3390/s19051090

182. Singh CB, Jayas DS, Paliwal J, White ND (2010) Identification of insect-damaged wheat kernels using short-wave near-infrared hyperspectral and digital colour imaging. Comput Electron Agric 73(2):118–125

183. Kun L, Jiani H, Ruiyin H, Qiujin W, Yinyin C, Mingxia S (2020) Comparison of Vis-NIR and SWIR hyperspectral imaging for the non-destructive detection of DON levels in fusarium head blight wheat kernels and wheat flour. Infrared Phys Technol 106:103281

184. Manschadi AM, Christopher J, de Voil P, Hammer GL (2006) *The role of root architectural traits in adaptation of wheat to water-limited environments.* Funct Plant Biol 33 (9):823–837. https://doi.org/10.1071/FP06055

185. Manschadi AM, Hammer GL, Christopher JT (2008) And deVoil, P., *Genotypic variation in seedling root architectural traits and implications for drought adaptation in wheat (Triticum aestivum L.).* Plant Soil 303 (1-2):115–129. https://doi.org/10.1007/s11104-007-9492-1

186. Richard CA, Hickey LT, Fletcher S, Jennings R, Chenu K, Christopher JT (2015) High-throughput phenotyping of seminal root traits in wheat. Plant Methods 11(1):1–11

187. Atkinson JA, Wingen LU, Griffiths M, Pound MP, Gaju O, Foulkes MJ, Le Gouis J, Griffiths S, Bennett MJ, King J, Wells DM (2015) Phenotyping pipeline reveals major

seedling root growth QTL in hexaploid wheat. J Exp Bot 66(8):2283–2292. https://doi.org/10.1093/jxb/erv006

188. Adeleke E, Millas R, McNeal W, Faris J, Taheri A (2020) Variation analysis of root system development in wheat seedlings using root phenotyping system. Agronomy 10 (2):206. https://doi.org/10.3390/agronomy10020206

189. Rich SM, Christopher J, Richards R, Watt M (2020) Root phenotypes of young wheat plants grown in controlled environments show inconsistent correlation with mature root traits in the field. J Exp Bot 71 (16):4751–4762. https://doi.org/10.1093/jxb/eraa201

190. Hargreaves CE, Gregory PJ, Bengough AG (2009) Measuring root traits in barley (Hordeum vulgare ssp. vulgare and ssp. spontaneum) seedlings using gel chambers, soil sacs and X-ray microtomography. Plant Soil 316 (1-2):285–297. https://doi.org/10.1007/s11104-008-9780-4

191. Chen YL, Dunbabin VM, Diggle AJ, Siddique KH, Rengel Z (2011) Development of a novel semi-hydroponic phenotyping system for studying root architecture. Funct Plant Biol 38(5):355–363

192. Topp CN, Iyer-Pascuzzi AS, Anderson JT, Lee CR, Zurek PR, Symonova O, Zheng Y, Bucksch A, Mileyko Y, Galkovskyi T, Moore BT, Harer J, Edelsbrunner H, Mitchell-Olds T, Weitz JS, Benfey PN (2013) 3D phenotyping and quantitative trait locus mapping identify core regions of the rice genome controlling root architecture. Proc Natl Acad Sci USA 110(18):E1695–E1704. https://doi.org/10.1073/pnas.1304354110

193. Hassouni K, Alahmad S, Belkadi B, Filali-Maltouf A, Hickey LT, Bassi FM (2018) Root system architecture and its association with yield under different water regimes in durum wheat. Crop Sci 58(6):2331–2346. https://doi.org/10.2135/cropsci2018.01.0076

194. Nagel KA, Bonnett D, Furbank R, Walter A, Schurr U, Watt M (2015) Simultaneous effects of leaf irradiance and soil moisture on growth and root system architecture of novel wheat genotypes: implications for phenotyping. J Exp Bot 66(18):5441–5452. https://doi.org/10.1093/jxb/erv290

195. Nagel KA, Putz A, Gilmer F, Heinz K, Fischbach A, Pfeifer J, Faget M, Blossfeld S, Ernst M, Dimaki C, Kastenholz B, Kleinert AK, Galinski A, Scharr H, Fiorani F, Schurr U (2012) GROWSCREEN-Rhizo is a novel phenotyping robot enabling simultaneous measurements of root and shoot growth for plants grown in soil-filled rhizotrons. Funct Plant Biol 39(11):891–904. https://doi.org/10.1071/FP12023

196. Jeudy C, Adrian M, Baussard C, Bernard C, Bernaud E, Bourion V, Busset H, Cabrera-Bosquet L, Cointault F, Han S (2016) RhizoTubes as a new tool for high throughput imaging of plant root development and architecture: test, comparison with pot grown plants and validation. Plant Methods 12 (1):31

197. Bodner G, Nakhforoosh A, Arnold T, Leitner D (2018) Hyperspectral imaging: a novel approach for plant root phenotyping. Plant Methods 14(1):84

198. Trachsel S, Kaeppler SM, Brown KM, Lynch JP (2010) Shovelomics: high throughput phenotyping of maize (Zea mays L.) root architecture in the field. Plant Soil 341 (1-2):75–87. https://doi.org/10.1007/s11104-010-0623-8

199. York LM, Slack S, Bennett MJ, Foulkes MJ (2018) *Wheat shovelomics I: A field phenotyping approach for characterising the structure and function of root systems in tillering species.* BioRxiv:280875

200. Wasson A, Bischof L, Zwart A, Watt M (2016) A portable fluorescence spectroscopy imaging system for automated root phenotyping in soil cores in the field. J Exp Bot 67 (4):1033–1043. https://doi.org/10.1093/jxb/erv570

201. Chen X, Ding Q, Blaszkiewicz Z, Sun J, Sun Q, He R, Li Y (2017) Phenotyping for the dynamics of field wheat root system architecture. Sci Rep 7:37649. https://doi.org/10.1038/srep37649

202. Chen X, Li Y, He R, Ding Q (2018) Phenotyping field-state wheat root system architecture for root foraging traits in response to environment× management interactions. Sci Rep 8(1):2642

203. Whalley WR, Binley A, Watts CW, Shanahan P, Dodd IC, Ober ES, Ashton RW, Webster CP, White RP, Hawkesford MJ (2017) Methods to estimate changes in soil water for phenotyping root activity in the field. Plant Soil 415(1):407–422. https://doi.org/10.1007/s11104-016-3161-1

204. Lontoc-Roy M, Dutilleul P, Prasher SO, Han L, Brouillet T, Smith DL (2006) Advances in the acquisition and analysis of CT scan data to isolate a crop root system from the soil medium and quantify root system complexity in 3-D space. Geoderma 137 (1-2):231–241. https://doi.org/10.1016/j.geoderma.2006.08.025

205. Mairhofer S, Zappala S, Tracy S, Sturrock C, Bennett MJ, Mooney SJ, Pridmore TP (2013) Recovering complete plant root system architectures from soil via X-ray μ-computed tomography. Plant Methods 9 (1):8
206. Metzner R, Eggert A, Van Dusschoten D, Pflugfelder D, Gerth S, Schurr U, Uhlmann N, Jahnke S (2015) Direct comparison of MRI and X-ray CT technologies for 3D imaging of root systems in soil: potential and challenges for root trait quantification. Plant Methods 11(1):17
207. van Dusschoten D, Metzner R, Kochs J, Postma JA, Pflugfelder D, Buhler J, Schurr U, Jahnke S (2016) Quantitative 3D analysis of plant roots growing in soil using magnetic resonance imaging. Plant Physiol 170(3):1176–1188. https://doi.org/10.1104/pp.15.01388
208. Minervini M, Scharr H, Tsaftaris SA (2015) *Image analysis: The new bottleneck in plant phenotyping*. IEEE Signal Process Mag 32 (4):126–131. https://doi.org/10.1109/msp.2015.2405111
209. Singh AK, Ganapathysubramanian B, Sarkar S, Singh A (2018) Deep learning for plant stress phenotyping: trends and future perspectives. Trends Plant Sci 23 (10):883–898. https://doi.org/10.1016/j.tplants.2018.07.004
210. Debauche O, Mahmoudi SA, De Cock N, Mahmoudi S, Manneback P, Lebeau F (2020) Cloud architecture for plant phenotyping research. Concurrency Comput Pract Exp 32(17):e5661. https://doi.org/10.1002/cpe.5661

# Index

## A



## B



## C



Andriy Bilichak and John D. Laurie (eds.), *Accelerated Breeding of Cereal Crops*, Springer Protocols Handbooks, https://doi.org/10.1007/978-1-0716-1526-3, © Springer Science+Business Media, LLC, part of Springer Nature 2022

## D



## E



## F



## G

## N



## O



## P

## Q



## R



## S

MIX
Papier aus verantwortungsvollen Quellen
Paper from responsible sources
FSC® C105338

If you have any concerns about our products,
you can contact us on
**ProductSafety@springernature.com**

In case Publisher is established outside the EU,
the EU authorized representative is:
**Springer Nature Customer Service Center GmbH**
**Europaplatz 3, 69115 Heidelberg, Germany**

Printed by Libri Plureos GmbH
in Hamburg, Germany